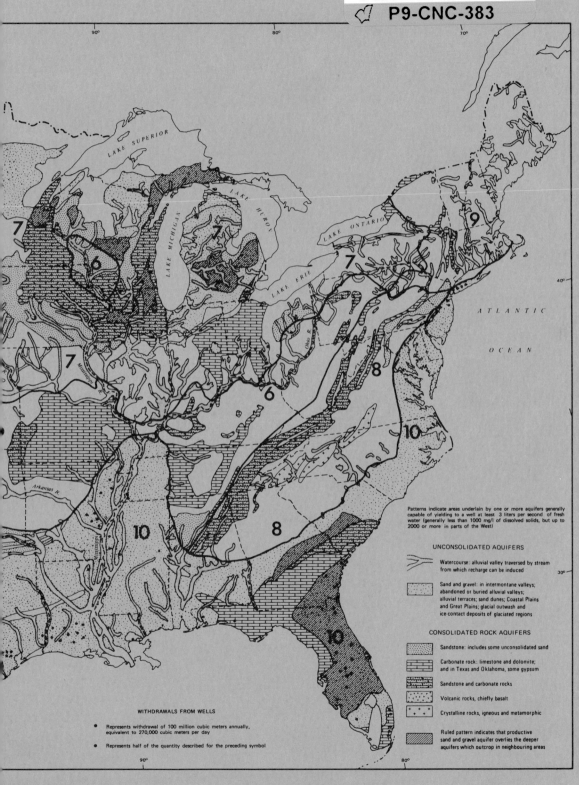

Patterns indicate areas underlain by one or more aquifers generally capable of yielding to a well at least 3 liters per second of fresh water (generally less than 1000 mg/l of dissolved solids, but up to 2000 or more in parts of the West)

UNCONSOLIDATED AQUIFERS

Watercourse: alluvial valley traversed by stream from which recharge can be induced

Sand and gravel: in intermontane valleys; abandoned or buried alluvial valleys; alluvial terraces; sand dunes; Coastal Plains and Great Plains; glacial outwash and ice-contact deposits of glaciated regions

CONSOLIDATED ROCK AQUIFERS

Sandstone: includes some unconsolidated sand

Carbonate rock: limestone and dolomite; and in Texas and Oklahoma, some gypsum

Sandstone and carbonate rocks

Volcanic rocks, chiefly basalt

Crystalline rocks, igneous and metamorphic

Ruled pattern indicates that productive sand and gravel aquifer overlies the deeper aquifers which outcrop in neighbouring areas

WITHDRAWALS FROM WELLS

Represents withdrawal of 100 million cubic meters annually, equivalent to 270,000 cubic meters per day

Represents half of the quantity described for the preceding symbol

Groundwater
Hydrology

Groundwater Hydrology ....

SECOND EDITION

David Keith Todd

UNIVERSITY OF CALIFORNIA, BERKELEY
and
DAVID KEITH TODD,
CONSULTING ENGINEERS, INC.

John Wiley & Sons
New York Chichester Brisbane Toronto

Endpaper: Productive aquifers and withdrawals from wells in
the United States after Dept. of Economic and Social Affairs,
Ground water in the Western Hemisphere, United Nations,
New York, 1976.

Library of Congress Cataloging in the Publication Data:

Todd, David Keith, 1923-
    Groundwater hydrology.

    Includes bibliographical references and index.
    1. Water, Underground. I. Title.
GB1003.2.T6 1980   551.49   80–11831
ISBN 0–471–87616–X

Printed in the United States of America

10 9 8 7 6 5 4 3 2 1

Preface

Water is essential to people, and the largest available source of fresh water lies underground. Increased demands for water have stimulated development of underground water resources. As a result, techniques for investigating the occurrence and movement of groundwater have been improved, better equipment for extracting groundwater has been developed, concepts of resource management have been established, and research has contributed to a better understanding of the subject. Thus, knowledge of groundwater hydrology, once veiled in mystery, has grown rapidly in recent decades.

This is the second edition of *Groundwater Hydrology*. My purpose, as with the first edition, has been to present the fundamentals of groundwater hydrology in a manner understandable to those most concerned with such knowledge. Few persons specialize in groundwater hydrology, yet, because groundwater is a major natural resource, the subject is important to students and professionals in several fields. Chief among these are civil engineers (including specialists in hydraulics, hydrology, sanitary engineering, soil mechanics, and water resources), geologists, agricultural and irrigation engineers, and water-well drillers. Personnel in charge of municipal and industrial water supplies, environmentalists, and planners often have a vital interest in groundwater. Persons indirectly concerned can be found in the fields of economics, mining, petroleum engineering, forestry, public health, and law, among others. Although it is impossible to present a subject fitted to such a diversity of interests, the common need of all is an understanding of the fundamental principles, methods, and problems encountered in the field as a whole. Thus, this book represents an effort to make available a unified presentation of groundwater hydrology.

This book presupposes only a background of mathematics through calculus and an elementary knowledge of geology. I believe previous instruction in fluid mechanics and hydrology to be desirable but not essential.

It has been stated that information available to people is now doubling every 10 years. Certainly since the first edition of this book appeared, developments of great significance to groundwater have occurred. In the United States, for example, legislation on research in water resources and control of water pollution, university programs stressing the environment and natural resources, the energy

v

shortage, droughts, and recurring problems of inadequate water supplies have all had an impact on groundwater. The National Water Well Association has become a leading force in upgrading the water-well drilling industry and in creating an awareness of groundwater within the general public. New journals such as *Ground Water, Journal of Hydrology,* and *Water Resources Research* have emerged; they serve as focal points for contributions to the groundwater literature.

Persons familiar with the first edition of this book will note many changes in the second edition. In fact, the changes necessary to keep up with the growth of the subject matter have been so numerous that I prefer to think of this as a new book with the same title rather than as a revised edition. All chapters have been extensively rewritten and expanded. A chapter on groundwater pollution has been added, while the previous one on water rights has been omitted. Nearly all references have been replaced; selection of those to include has been on the basis of significance and general availability. Metric units have been employed exclusively because of their international importance. A conversion table for English units is included as an appendix.

Experience has shown that this book serves both as a reference book and as a textbook; therefore, from considerations of clarity and space, illustrative examples describing field situations have been minimized. The groundwater literature today is voluminous so that many project and resource studies are readily available. Publications of the United States Geological Survey are extensive and contain a wealth of material from field investigations. Also, many state water resource agencies have published comprehensive reports on local groundwater situations. For instructors using this book as a text, I recommend that data from these sources be presented in the form of illustrations and problems to supplement the text material.

During the interval between editions of this book, I have participated intermittently as a consultant on groundwater investigations on a worldwide basis. These opportunities have broadened my background in groundwater hydrology, and for this experience I wish to recognize at least collectively the contribution of my many professional associations. Comments and questions of students have guided me in the presentation of many topics. Colleagues at the University of California, Berkeley, and members of the U.S. Geological Survey have assisted me, directly and indirectly, in the preparation of this book. Specific chapters were reviewed by J. F. Mann, Jr., and K. D. Schmidt; Brian Todd assisted with technical computations. To all of these, and last but not least to my patient wife, I should like to acknowledge my indebtedness.

David Keith Todd

Berkeley, California
June 1980

Contents

Groundwater
Hydrology

Introduction

Groundwater hydrology may be defined as the science of the occurrence, distribution, and movement of water below the surface of the earth. Geohydrology has an identical connotation, and hydrogeology differs only by its greater emphasis on geology. Utilization of groundwater dates from ancient times, although an understanding of the occurrence and movement of subsurface water as part of the hydrologic cycle has come only relatively recently.

Scope

Groundwater referred to without further specification is commonly understood to mean water occupying all the voids within a geologic stratum. This saturated zone is to be distinguished from an unsaturated, or aeration, zone where voids are filled with water and air. Water contained in saturated zones is important for engineering works, geologic studies, and water supply developments; consequently, the occurrence of water in these zones will be emphasized here. Unsaturated zones are usually found above saturated zones and extend upward to the ground surface. Because this water includes soil moisture within the root zone, it is a major concern of agriculture, botany, and soil science. No rigid demarcation of waters

1

between the two zones is possible, for they possess an interdependent boundary, and water can move from zone to zone in either direction. The interrelationships are described more fully in Chapter 2.

Groundwater plays an important part in petroleum engineering. Two-fluid systems, involving oil and water, and three-fluid systems, involving gas, oil, and water, occur frequently in development of petroleum. Although the same hydrodynamic laws govern flows of these systems and groundwater, the distinctive nature of water in petroleum reservoirs sets it apart from other groundwater. Major differences exist in water quality, depth of occurrence, and methods of development and utilization, all of which contribute to a separation of interests and applications. Therefore, groundwater in petroleum reservoirs will not be treated specifically here. It should be noted, however, that groundwater hydrology has gained immeasurably from research conducted by the petroleum industry.

Historical Background

Qanats.   Groundwater development dates from ancient times.[11]* The Old Testament contains numerous references to groundwater, springs, and wells. Other than dug wells, groundwater in ancient times was supplied from horizontal wells known as *qanats.*† These persist to the present day and can be found in a band across the arid regions of Southwestern Asia and North Africa extending from Afghanistan to Morocco. A cross section along a qanat is shown in Fig. 1.1. Typically, a gently sloping tunnel dug through alluvial material

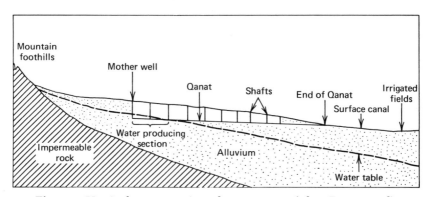

Fig. 1.1   Vertical cross section along a qanat (after Beaumont[9]).

*Superscript numbers refer to references at the end of the chapter.

†*Qanat* comes from a Semitic word meaning "to dig." There are several variants of the name, including *karez, foggara,* and *falaj,* depending on location; in addition, there are numerous differences in spelling.[23]

leads water by gravity flow from beneath the water table at its upper end to a ground surface outlet and irrigation canal at its lower end.[10] Vertical shafts dug at closely spaced intervals provide access to the tunnel.[67] Qanats are laboriously hand constructed by skilled workers employing techniques that date back 3000 years.*

Iran possesses the greatest concentration of qanats; here some 22,000 qanats supply 75 percent of all water used in the country. Lengths of qanats extend up to 30 km, but most are less than 5 km.[10] The depth of the qanat mother well (see Fig. 1.1) is normally less than 50 m, but instances of depths exceeding 250 m have been reported. Discharges of qanats vary seasonally with water table fluctuations and seldom exceed 100 m^3/hr. Indicative of the density of qanats is the map in Fig. 1.2. Based on aerial photographs of the Varamin Plain, located 40 km southeast of Tehran, this identifies 266 qanats within an area of 1300 km^2.

Groundwater Theories. Utilization of groundwater greatly preceded understanding of its origin, occurrence, and movement. The writings of Greek and Roman philosophers to explain origins of springs and groundwater contain theories ranging from fantasy to nearly correct accounts.[1,5] As late as the seventeenth century it was generally assumed that water emerging from springs could not be derived from rainfall, for it was believed that the quantity was inadequate and the earth too impervious to permit penetration of rainwater far below the surface. Thus, early Greek philosophers such as Homer, Thales, and Plato hypothesized that springs were formed by seawater conducted through subterranean channels below the mountains, then purified and raised to the surface. Aristotle suggested that air enters cold dark caverns under the mountains where it condenses into water and contributes to springs.

The Roman philosophers, including Seneca and Pliny, followed the Greek ideas and contributed little to the subject. An important step forward, however, was made by the Roman architect Vitruvius. He explained the now-accepted infiltration theory that the mountains receive large amounts of rain that percolate through the rock strata and emerge at their base to form streams.

The Greek theories persisted through the Middle Ages with no advances until the end of the Renaissance. The French potter and philosopher Bernard Palissy (c. 1510–1589) reiterated the infiltration

---

*Illustrative of the tremendous human effort expended to construct a qanat is a calculation by Beaumont.[10] The longest qanat near Zarand, Iran, is 29 km long with a mother well depth of 96 m and with 966 shafts along its length; the total volume of material excavated is estimated at 75,400 m^3.

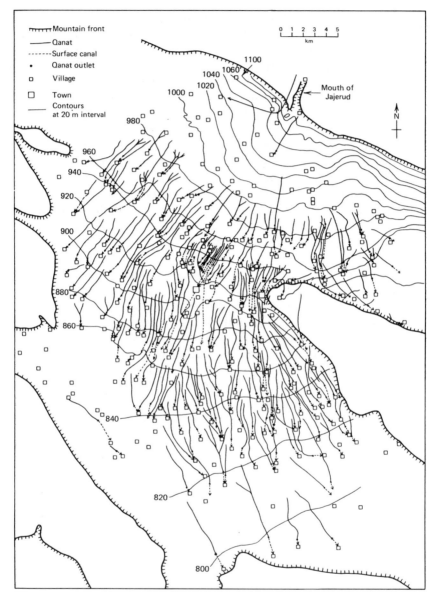

Fig. 1.2 Map of qanats on the Varamin Plain, Iran (after Beaumont[9]).

theory in 1580, but his teachings were generally ignored. The German astronomer Johannes Kepler (1571–1630) was a man of strong imagination who likened the earth to a huge animal that takes in water of the ocean, digests and assimilates it, and discharges the end products of these physiological processes as groundwater and springs. The

seawater theory of the Greeks, supplemented by the ideas of vapor-
ization and condensation processes within the earth, was restated by
the French philosopher René Descartes (1596–1650).

A clear understanding of the hydrologic cycle was achieved by
the latter part of the seventeenth century. For the first time theories
were based on observations and quantitative data. Three Europeans
made notable contributions, although others contributed to and supported these advances. Pierre Perrault* (1611–1680) measured rainfall during three years and estimated runoff of the upper Seine River
drainage basin. He reported in 1674 that precipitation on the basin
was about six times the river discharge, thereby demonstrating as
false the early assumption of inadequate rainfall.[47] The French physicist Edme Mariotte (c. 1620–1684) made measurements of the Seine
at Paris and confirmed Perrault's work. His publications appeared in
1686, after his death, and contained factual data strongly supporting
the infiltration theory. Meinzer[42] once stated, "Mariotte ... probably
deserves more than any other man the distinction of being regarded
as the founder of ground-water hydrology, perhaps I should say of
the entire science of hydrology." The third contribution came from
the English astronomer Edmund Halley (1656–1742), who reported in
1693 on measurements of evaporation, demonstrating that sea evaporation was sufficient to account for all springs and stream flow.

Recent Centuries. During the eighteenth century, fundamentals
in geology were established that provided a basis for understanding
the occurrence and movement of groundwater. During the first half
of the nineteenth century many artesian wells were drilled in France,
stimulating interest in groundwater. The french hydraulic engineer
Henry Darcy (1803–1858) studied the movement of water through
sand. His treatise of 1856 defined the relation, now known as Darcy's
law, governing groundwater flow in most alluvial and sedimentary
formations. Later European contributions of the nineteenth century
emphasized the hydraulics of groundwater development. Significant
contributions were made by J. Boussinesq, G. A. Daubrée, J. Dupuit,

*Pierre Perrault was a lawyer by profession and held administrative and financial
positions in the French government; hence he is not well known in scientific circles.
His interest in groundwater, leading to publication of *De l'Origine des Fontaines* in
1674, can be traced to the stimulus of the Dutch mathematician, astronomer, and
physicist, Christiaan Huygens, who was then living in Paris and to whom the book
is dedicated. Also, Pierre Perrault is often overshadowed by his four distinguished
brothers: Jean (c. 1610–1669), a lawyer; Nicolas (1624–1662), a noted theologian; Claude
(1613–1688), a physician, architect, and scientist, who is regarded as one of the most
eminent French scholars of his time; and Charles (1628–1703), author and critic, who
is best known for his fairy tales of Mother Goose.

Fig. 1.3 Villagers laboriously lifting and carrying water from a
deep dug well in northern India (photo by David K. Todd).

P. Forchheimer, and A. Thiem. In the twentieth century, increased
activity in all phases of groundwater hydrology has occurred. Many
Europeans have participated with publications of either specialized
or comprehensive works. There are too many people to mention them
all, but R. Dachler, E. Imbeaux, K. Keilhack, W. Koehne, J. Kozeny,
E. Prinz, H. Schoeller, and G. Thiem are best known in the United
States.

American contributions to groundwater hydrology date from near
the end of the nineteenth century. In the past 90 years, tremendous
advances have been made. Important early theoretical contributions
were made by A. Hazen, F. H. King, and C. S. Slichter, while detailed
field investigations were begun by men such as T. C. Chamberlin,
N. H. Darton, W. T. Lee, and W. C. Mendenhall. O. E. Meinzer, through
his consuming interest in groundwater and his dynamic leadership
of groundwater activities of the U.S. Geological Survey, stimulated
many individuals in the quest for groundwater knowledge. In recent
decades the publications of M. S. Hantush, C. E. Jacob, G. B. Maxey,
C. L. McGuinness, and R. W. Stallman are noteworthy. Within the last
20 years the surge in university research on groundwater problems,
the establishment of professional consulting firms specializing in
water resources, and the advent of the digital computer have jointly
produced a competence for development and management of groundwater resources that was nonexistent heretofore.

Utilization of Groundwater

Groundwater is an important source of water supply throughout the world. Its use in irrigation, industries, municipalities, and rural homes continues to increase.[44] Fig. 1.3 strikingly illustrates the dependence of an Indian village on its only water source—groundwater from a single dug well. Cooling and air-conditioning have made heavy demands on groundwater because of its characteristic uniformity in temperature. Shortages of groundwater in areas where excessive withdrawals have occurred emphasize the need for accurate estimates of the available subsurface resources and the importance of proper planning to ensure the continued availability of water supplies.

There is a tendency to think of groundwater as being the primary water source in arid regions and of surface water in humid regions. But a study of groundwater use in the United States, for example, reveals that groundwater serves as an important resource in all climatic zones.[45] Reasons for this include its convenient availability near the point of use, its excellent quality (which typically requires little treatment), and its relatively low cost of development. Furthermore, in humid locales such as Barbados, Jamaica, and Hawaii, groundwater predominates as the water source because the high infiltration capacity of the soils sharply reduces surface runoff.

Fig. 1.4 Signs in Berkeley, California, during the 1977 drought announcing use of self-supplied groundwater for landscape irrigation. During this period municipal water supplies from surface water sources were severely rationed (photos by David K. Todd).

During the dramatic drought of 1976–1977 in California, surface
water resources all but disappeared in many areas. Emergency measures of many types were instituted to sustain public water supplies,
but the drilling of thousands of new wells* became a key factor in
meeting restricted water demands during the critical period (see
Fig. 1.4)

Estimates of water use in the United States as of 1975 were prepared by the U.S. Geological Survey. The largest single demand on
groundwater is irrigation, amounting to 71 percent of all groundwater used. More than 90 percent of this water is pumped in the

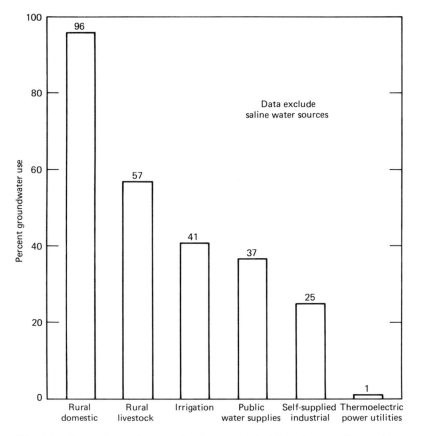

Fig. 1.5 Percentage of groundwater use to total water use for various
types of use in the United States, 1975 (after Murray and Reeves[44]).

*For many years the number of new wells drilled annually in California has been in
the range of 8,000 to 9,000. But an unprecedented number were drilled or deepened
during the drought: 11,300 wells during 1976 and 19,950 wells during 1977.

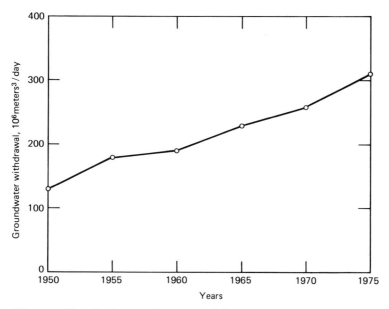

Fig. 1.6 Trend of groundwater withdrawal in the United States (after Murray and Reeves[44]).

western states, where arid and semiarid conditions have fostered extensive irrigation development. The relative importance of groundwater in relation to various types of water use is indicated in Fig. 1.5. It can be seen that 96 percent of rural homes are supplied by groundwater. The largest industrial users of groundwater include, in order of water requirements, oil refineries, paper manufacturers, metalworking plants, chemical manufacturers, air-conditioning and refrigerating units, and distilleries.[37]

There has been a steady increase in the production of groundwater in recent years, as demonstrated in Fig. 1.6. Furthermore, the proportion of groundwater use to total water use has been rising. This may be attributed to a reduction in physical and environmental opportunities to develop additional fresh surface-water supplies. In addition, a growing awareness of the ready availability, the low cost, and the high quality of groundwater is a contributing factor.

Table 1.1 presents a breakdown of groundwater use by states and by type of demand as of 1975.* States are listed in order of groundwater use.

*Power generation uses have been omitted from these data because essentially all hydroelectric and thermoelectric production plants employ large quantities of surface water; including these uses would have distorted the role of groundwater for all other purposes.

TABLE 1.1 Use of Groundwater in the United States, 1975 (after Murray and Reeves[45])[a] (quantities in 1000 m³/day)

State	Public Supplies	Rural	Irrigation	Self-supplied Industrial	Total Groundwater Use	Ratio of Groundwater Use to Total Water Use, percent
California	6,400	610	64,000	1,500	73,000	49
Texas	3,200	870	36,000	1,600	42,000	73
Nebraska	830	420	20,000	320	22,000	72
Idaho	420	180	13,000	7,200	21,000	32
Kansas	530	330	17,000	490	18,000	87
Arizona	1,000	230	16,000	720	18,000	61
Florida	3,700	950	4,500	3,000	12,000	62
Colorado	200	130	9,500	220	10,050	27
Arkansas	340	280	8,000	1,300	9,900	79
New Mexico	640	140	4,900	250	5,900	49
Louisiana	760	190	3,100	1,700	5,800	25
Oklahoma	530	130	3,800	220	4,700	63
Oregon	250	610	3,500	300	4,700	18
Mississippi	680	130	2,100	1,300	4,200	76
Ohio	1,500	490	21	1,900	3,900	26
Illinois	2,700	170	110	910	3,900	27
Pennsylvania	1,300	640	23	1,300	3,300	13
New York	2,100	570	72	490	3,200	19
Georgia	570	260	91	2,100	3,000	44
New Jersey	1,400	420	420	770	3,000	42

Utah	680	220	1,800	250	3,000	19
Hawaii	640	22	1,600	370	2,600	53
Washington	1,000	170	870	490	2,500	9
Nevada	270	45	2,000	210	2,500	20
Minnesota	680	570	91	830	2,200	49
Iowa	830	530	68	680	2,100	70
North Carolina	220	720	200	950	2,100	36
Indiana	870	490	91	530	2,000	13
Michigan	910	680	91	230	1,900	17
Wisconsin	720	490	190	350	1,800	47
Missouri	450	310	340	640	1,700	40
Wyoming	83	45	1,000	450	1,600	6
Tennessee	640	180	13	570	1,400	20
Alabama	450	280	25	610	1,400	18
Massachusetts	640	99	42	530	1,300	26
South Dakota	150	420	190	68	830	41
Montana	160	140	420	99	820	2
Virginia	290	340	14	160	800	15
South Carolina	220	280	34	200	730	24
Kentucky	140	140	0.4	280	560	24
Maryland	180	230	16	120	550	14
Connecticut	130	190	1.5	120	440	17
North Dakota	91	110	180	12	390	40
West Virginia	130	83	0	99	310	10
Delaware	110	49	45	91	300	34

TABLE 1.1 (Continued)

State	Public Supplies	Rural	Irrigation	Self-supplied Industrial	Total Groundwater Use	Ratio of Groundwater Use to Total Water Use, percent
New Hampshire	140	32	0	49	220	20
Maine	72	53	0	45	170	8
Vermont	61	83	1.5	20	170	49
Alaska	130	23	0	0	150	23
Rhode Island	53	17	1.5	30	100	18
District of Columbia	0	0	0	3.0	3.0	1
United States	42,000	15,000	220,000	36,000	310,000	39

[a]Excluding hydroelectric and thermoelectric power generation and saline water uses.

TABLE 1.2 Leading States as to Intensity of Groundwater Use
(based on data from Table 1.1)

Rank	State	Intensity of Groundwater Use, $m^3/day/km^2$
1	California	180
2	Hawaii	160
3	New Jersey	150
4	Nebraska	110
5	Idaho	98
6	Kansas	89
7	Florida	86
8	Arkansas	73
9	Massachusetts	64
10	Arizona	61

The proportion of groundwater use relative to total water use for each state is shown on the map in Fig. 1.7. This indicates the significant contribution that groundwater makes in the semiarid central states and the arid southwestern states. A further interesting group of statistics appears in Table 1.2 listing states where groundwater has been most intensely developed. Here states are ranked by groundwater production rate per area. It can be noted that humid, populous eastern states such as New Jersey and Massachusetts, which use substantial quantities of groundwater for urban water supply purposes, outrank large, arid western states such as Arizona and Texas where irrigation is the predominant water demand.

Groundwater in the Hydrologic Cycle

Groundwater constitutes one portion of the earth's water circulatory system known as the hydrologic cycle. Figure 1.8 illustrates some of the many facets involved in this cycle. Water-bearing formations of the earth's crust act as conduits for transmission and as reservoirs for storage of water. Water enters these formations from the ground surface or from bodies of surface water, after which it travels slowly for varying distances until it returns to the surface by action of natural flow, plants, or humans. The storage capacity of groundwater reservoirs combined with small flow rates provide large, extensively distributed sources of water supply. Groundwater emerging into surface streams channels aids in sustaining streamflow when surface runoff is low or nonexistent. Similarly, water pumped from wells represents the sole water source in many regions during much of every year.

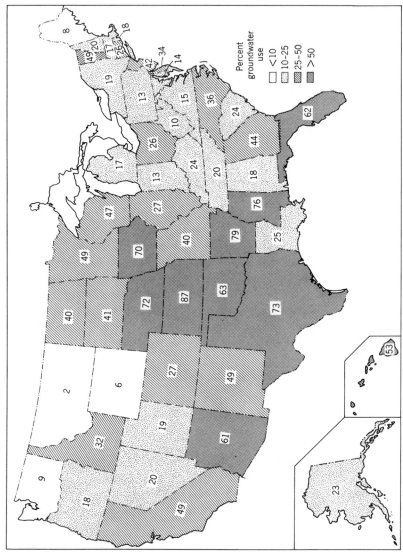

Fig. 1.7 Groundwater use relative to total water use in the United States, excluding power generation and saline water uses (after Murray and Reeves[44]).

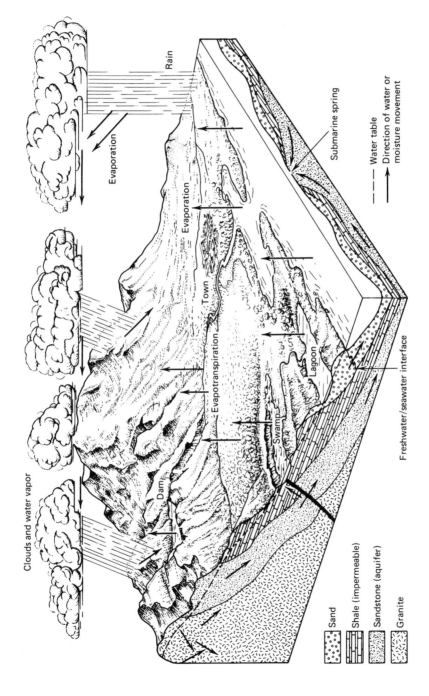

Fig. 1.8 Schematic diagram of the hydrologic cycle (courtesy Australian Water Resources Council).

**TABLE 1.3   Approximate Distribution of Water in the United States
(after Ad Hoc Panel on Hydrology[2])[a]**

	Volume, $10^9 m^3$	Annual Circulation, $10^9 m^3/yr$	Detention Period, years
Frozen water			
Glaciers	67	1.6	40
Liquid water			
Freshwater lakes	19,000	190	100
Salt lakes	58	5.7	10
Average in-stream channels	50	1900	0.03
Groundwater			
Shallow (<800 m deep)	63,000	310	200
Deep (>800 m deep)	63,000	6.2	10,000
Soil moisture			
1-m root zone	630	3100	0.2
Gaseous water			
Atmosphere	190	6,200	0.03

[a]Excluding Alaska and Hawaii.

Practically all groundwater originates as surface water. Principal sources of natural recharge include precipitation, streamflow, lakes, and reservoirs. Other contributions, known as artificial recharge, occur from excess irrigation, seepage from canals, and water purposely applied to augment groundwater supplies. Even seawater can enter underground along coasts where hydraulic gradients slope downward in an inland direction. Water within the ground moves downward through the unsaturated zone under the action of gravity, whereas in the saturated zone it moves in a direction determined by the surrounding hydraulic situation.

Discharge of groundwater occurs when water emerges from underground. Most natural discharge occurs as flow into surface water bodies, such as streams, lakes, and oceans; flow to the surface appears as a spring. Groundwater near the surface may return directly to the atmosphere by evaporation from within the soil and by transpiration from vegetation. Pumpage from wells constitutes the major artificial discharge of groundwater.

Data in Table 1.3 show the approximate distribution of all water in the United States. The first column lists the volume of water in storage within each domain, while the second column lists the annual volume of water circulating within each domain. Dividing the first column by the second gives an average detention period for water within each domain. It can be noted that groundwater constitutes about 86 percent of the total water stored within the United States, but at the same time its annual replacement rate amounts to only 3 percent of the total water circulation. Thus, groundwater represents a large reserve water source that is little affected by the eccentricities of rainfall from year to year. Furthermore, Table 1.3 indicates that shallow groundwater, into which most wells penetrate, has an average residence time of some 200 years; this contrasts with a period of several days for surface water.

Literature and Data Sources

U.S. Geological Survey Publications.   In the United States a majority of the field measurements and investigations of groundwater have been conducted by the U.S. Geological Survey. Most work has been on a cooperative basis with individual states. Results are published by the Survey as *Circulars, Professional Papers,* and *Water-Supply Papers.*[20] Since 1935 records of groundwater measurements in key observation wells have been published in *Water-Supply Papers* under the title *Ground-Water Levels in the United States.* Prior to 1940 records for each year were published in a single volume, but since then records have been published in six volumes covering sections of the United States shown in Fig. 1.9. Beginning in 1956 only

Fig. 1.9   Outline map of the United States showing areas included in *U.S. Geological Survey Water-Supply Papers* on water levels in observation wells.

water-level data from a basic network of observation wells have been published. This network contains observation wells so located that the most significant data are obtained from the fewest wells in the most important groundwater zones. In some states the basic network forms the skeletal index of a more comprehensive observation-well program. The U.S. Geological Survey publishes at irregular intervals other papers on the geology and groundwater resources of local areas. Invariably these intensive investigations concern areas containing important groundwater problems and are carried out in cooperation with local agencies. Published information on a particular problem or area can be found in issues of *Publications of the Geological Survey.*

WATSTORE. The National Water Data Storage and Retrieval System (WATSTORE) was established by the U.S. Geological Survey in 1971 to provide a large-scale computerized system for the storage and retrieval of water data. These data are available to the public through any of the district offices of the Survey.[20] The portions of WATSTORE pertaining to groundwater include:

1. *Water Quality File*—Analyses of biological, chemical, physical, and radiochemical characteristics of 200 different constituents in groundwaters.

2. *Groundwater Site Inventory File*—Data on geohydrologic characteristics, one-time field measurements such as water temperature, site location and identification, and well-construction history for wells, springs, and other sources of groundwater.

Data products provided by WATSTORE encompass computer-printed tables, computer-printed graphs, statistical analyses, digital plotting, and data in machine-readable form.

State Publications. A second source of basic data on groundwater is state geological and water resources agencies. Various states differ widely in their degree of activity, but California and Illinois, for example, maintain large water resources agencies and have made extensive groundwater investigations. Bibliographies of all water publications by state agencies through 1974 are available.[32,33]

Professional Literature. The professional literature on groundwater hydrology embraces many fields of interest. Important contributions can be found in journals of civil engineering, water resources, water supply, geology, geophysics, agriculture, and soil science.[31] Pertinent papers on flow in porous media also appear in chemical engineering, mechanics, and physics journals. The only organization in the United States recognizing hydrology as a distinct science is the American Geophysical Union; its *Transactions, Journal of Geophysical Research,* and *Water Resources Research* contain a wealth of groundwater papers. Similarly, publications of the International Association of Hydrological Sciences, an organization within the International Union of Geodesy and Geophysics, and *Journal of Hydrology* serve as the major media for exchange of groundwater information on a worldwide basis.

Finally, and most importantly, the serials *Ground Water* and *Water Well Journal,* published by the National Water Well Association, have become indispensable reading for professionals concerned with the development and management of groundwater resources.

Selected references are listed at the end of each chapter in this book. These provide sources for additional material on topics treated within each chapter. Several general texts pertaining to groundwater are included in the references for this chapter.

References

1. Adams, F. D., Origin of springs and rivers—an historical review, *Fennia,* v. 50, no. 1, 16 pp., 1928.
2. Ad Hoc Panel on Hydrology, *Scientific hydrology,* Federal Council for Science and Technology, Washington, D. C., 37 pp., 1962.
3. Amer. Soc. Civil Engrs., *Ground water management,* Manual Engrng. Practice 40, New York, 216 pp., 1972.
4. American Water Works Assoc., *Ground water,* AWWA Manual M21, New York, 130 pp., 1973.
5. Baker, M. N., and R. E. Horton, Historical development of ideas regarding the origin of springs and ground-water, *Trans. Amer. Geophysical Union,* v. 17, pp. 395–400, 1936.
6. Baldwin, G. V., and C. L. McGuinness, *A primer on ground water,* U.S. Geological Survey, 26 pp., 1963.
7. Bear, J., *Dynamics of fluids in porous media,* Amer. Elsevier, New York, 764 pp., 1972.
8. Bear, J., et al., *Physical principles of water percolation and seepage,* UNESCO, Paris, 465 pp., 1968.
9. Beaumont, P., Qanats on the Varamin Plain, Iran, *Trans. Inst. British Geographers,* Publ. no. 45, pp. 169–179, 1968.
10. Beaumont, P., Qanat systems in Iran, *Bull. Intl. Assoc. Sci. Hydrology,* v. 16, pp. 39–50, 1971.
11. Biswas, A. K., *History of hydrology,* Amer. Elsevier, New York, 348 pp., 1970.
12. Bouwer, H., *Groundwater hydrology,* McGraw-Hill, New York, 480 pp., 1978.

13. Brown, R. H., et al. (eds.), *Ground-water studies,* UNESCO, Paris, vars. pp. and suppls., 1972–.
14. Bureau de Recherches Géologiques et Minières, *Méthodes d'études et de recherches des nappes aquifères,* Paris, 158 pp., 1962.
15. Bureau of Reclamation, *Ground water manual,* U.S. Dept. Interior, 480 pp., 1977.
16. Carlston, C. W., Notes on the early history of water-well drilling in the United States, *Econ. Geol.,* v. 38, pp. 119–136, 1943.
17. Castany, G., *Traité pratique des eaux souterraines,* Dunod, Paris, 657 pp., 1963.
18. Cedergren, H. R., *Seepage, drainage, and flow nets,* 2nd ed., John Wiley & Sons, New York, 534 pp., 1977.
19. Chow, V. T. (ed.), *Handbook of applied hydrology,* McGraw-Hill, New York, 1453 pp., 1964.
20. Clarke, P. F., et al., A guide to obtaining information from the USGS 1978, *U.S. Geological Survey Circ. 777,* 36 pp., 1978.
21. Collins, R. E., *Flow of fluids through porous materials,* Reinhold, New York, 270 pp., 1961.
22. Cooley, R. L., et al., Principles of ground-water hydrology, *Hydrologic Engineering Methods for Water Resources Development,* v. 10, Corps of Engrs., U.S. Army, Davis, Calif., 337 pp., 1972.
23. Cressey, G. B., Qanats, karez and foggaras, *Geogr. Review,* v. 48, pp. 27–44, 1958.
24. Davis, S. N., and R. J. M. DeWiest, *Hydrogeology,* John Wiley & Sons, New York, 463 pp., 1966.
25. Dept. Economic and Social Affairs, *Ground water in the Western Hemisphere,* Natural Resources, Water Ser. no. 4, United Nations, New York, 337 pp., 1976.
26. DeWiest, R. J. M., *Geohydrology,* John Wiley & Sons, New York, 366 pp., 1965.
27. DeWiest, R. J. M. (ed.), *Flow through porous media,* Academic, New York, 530 pp., 1969.
28. Domenico, P. A., *Concepts and models in groundwater hydrology,* McGraw-Hill, New York, 405 pp., 1972.
29. Freeze, R. A., and J. A. Cherry, *Groundwater,* Prentice-Hall, Englewood Cliffs, N.J., 604 pp., 1979.
30. Geraghty, J. J., et al., *Water atlas of the United States,* 3d ed., Water Information Center, Port Washington, N.Y., 122 pl., 1973.
31. Giefer, G. J., *Sources of information in water resources,* Water Information Center, Port Washington, N.Y., 290 pp., 1976.
32. Giefer, G. J., and D. K. Todd (eds.), *Water publications of state agencies,* Water Information Center, Port Washington, N.Y., 319 pp., 1972.
33. Giefer, G. J., and D. K. Todd, *Water publications of state agencies, First supplement, 1971–1974,* Water Information Center, Huntington, N.Y., 189 pp., 1976.
34. Harr, M. E., *Groundwater and seepage,* McGraw-Hill, New York, 315 pp., 1962.

35. Heath, R. C., and F. W. Trainer, *Introduction to ground-water hydrology*, John Wiley & Sons, New York, 284 pp., 1968.
36. Huisman, L., *Groundwater recovery*, Winchester Press, New York, 336 pp., 1972.
37. MacKichan, K. A., Estimated use of water in the United States, 1955, *Jour. Amer. Water Works Assoc.*, v. 49, pp. 369–391, 1957.
38. Matthess, G., Die Beschaffenheit des Grundwassers, *in* Richter, W. (ed.), *Lehrbuch der Hydrogeologie*, v. 2, Gebrüder Borntraeger, Berlin, 324 pp., 1973.
39. McGuinness, C. L., The role of ground water in the national water situation, *U.S. Geological Survey Water-Supply Paper* 1800, 1121 pp., 1963.
40. McWhorter, D. G., and D. K. Sunada, *Ground-water hydrology and hydraulics*, Water Resources Publs., Fort Collins, Colo., 290 pp., 1977.
41. Meinzer, O. E., Outline of ground-water hydrology with definitions, *U.S. Geological Survey Water-Supply Paper* 494, 71 pp., 1923.
42. Meinzer, O. E., The history and development of ground-water hydrology, *Jour. Washington Acad. Sci.*, v. 24, pp. 6–32, 1934.
43. Meyboom, P., Current trends in hydrogeology, *Earth-Science Reviews*, v. 2, pp. 345–364, 1966.
44. Meyer, G., and G. G. Wyrick, Regional trends in water-well drilling in the United States, *U.S. Geological Survey Circ.* 533, 8 pp., 1966.
45. Murray, C. R., and E. B. Reeves, Estimated use of water in the United States in 1975, *U.S. Geological Survey Circ.* 765, 39 pp., 1977.
46. Muskat, M., *The flow of homogeneous fluids through porous media*, McGraw-Hill, New York, 763 pp., 1937.
47. Perrault, P., *On the origin of springs*, Trans. by A. LaRocque, Hafner, New York, 209 pp., 1957.
48. Polubarinova-Kochina, P. Y., *Theory of groundwater movement*, Princeton Univ. Press, Princeton, N.J., 613 pp., 1962.
49. Rau, J. L., *Ground water hydrology for water well drilling contractors*, National Water Well Assoc., Columbus, Ohio, 259 pp., 1969.
50. Randkivi, A. J., and R. A. Callander, *Analysis of groundwater flow*, John Wiley & Sons, New York, 214 pp., 1976.
51. Richter, W., and R. Wager, Hydrogeologie, *in* Bentz, A. and H. J. Martini (eds.), *Lehrbuch der Angewandten Geologie*, v. II, pt. 2, pp. 1357–1546, Ferdinand Enke, Stuttgart, 1969.
52. Saleem, Z. A. (ed.), *Advances in groundwater hydrology*, Amer. Water Resources Assoc., Minneapolis, 333 pp., 1976.
53. Scheidegger, A. E., *The physics of flow through porous media*, 3rd ed., Univ. of Toronto, Toronto, 353 pp., 1974.
54. Schoeller, H., *Les eaux souterraines*, Masson & Cie, Paris, 642 pp., 1962.
55. Silinbektchurin, A. I., *Dynamics of underground water (with elements of hydraulics)*, 2nd ed., Moscow University, Moscow, 380 pp., 1965.
56. Stow, D. A. V., et al., *Preliminary bibliography on groundwater in developing countries 1970 to 1976*, The Geosciences in Intl. Development Rept. 4, Assoc. of Geoscientists for Intl. Development, St. John's, Newfoundland, 305 pp., 1976.

57. Subcommittee on Hydrology, Inter-Agency Committee on Water Resources, *Inventory of federal sources of ground-water data,* Bull. no. 12, Washington, D.C., 294 pp., 1966.

58. Thomas, H. E., *The conservation of ground water,* McGraw-Hill, New York, 327 pp., 1951.

59. Thomas, H. E., and L. B. Leopold, Groundwater in North America, *Science,* v. 143, pp. 1001–1006, 1964.

60. Thurner, A., *Hydrogeologie,* Springer, Vienna, 350 pp., 1967.

61. Todd, D. K., *The water encyclopedia,* Water Information Center, Port Washington, N.Y., 559 pp., 1970.

62. Tolman, C. F., *Ground water,* McGraw-Hill, New York, 593 pp., 1937.

63. U.S. Geological Survey, Ground-water levels in the United States, *Water-Supply Papers,* published intermittently.

64. van der Leeden, F., *Ground water, a selected bibliography,* 2nd ed., Water Information Center, Port Washington, N.Y., 146 pp., 1974.

65. Verruijt, A., *Theory of groundwater flow,* Gordon and Breach, New York, 190 pp., 1970.

66. Walton, W. C., *Groundwater resource evaluation,* McGraw-Hill, New York, 664 pp., 1970.

67. Wulff, H. E., The qanats of Iran, *Sci. Amer.,* v. 218, pp. 94–100, 105, 1968.

Occurrence
of Groundwater

To describe the occurrence of groundwater necessitates a review of where and how groundwater exists; subsurface distribution, in both vertical and areal extents, needs to be considered. The geologic zones important to groundwater must be identified as well as their structure in terms of water-holding and water-yielding capabilities. Assuming hydrologic conditions furnish water to the underground zone, the subsurface strata govern its distribution and movement; hence the important role of geology in groundwater hydrology cannot be overemphasized. Springs, hydrothermal phenomena, and water in permanently frozen ground constitute special groundwater occurrences.

Origin and Age of Groundwater

Almost all groundwater can be thought of as a part of the hydrologic cycle, including surface and atmospheric (meteoric) waters. Relatively minor amounts of groundwater may enter this cycle from other origins.

Water that has been out of contact with the atmosphere for at least an appreciable part of a geologic period is termed *connate water;* essentially, it consists of fossil interstitial water that has

23

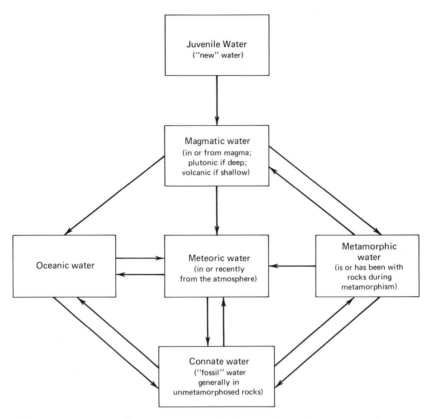

Fig. 2.1 Diagram illustrating relationships of genetic types of water (after White[49]; courtesy The Geological Society of America, 1957).

migrated from its original burial location.[49] This water may have been derived from oceanic or freshwater sources and, typically, is highly mineralized.[15] *Magmatic water* is water derived from magma; where the separation is deep, the term *plutonic water* is applied, while *volcanic water* designates water from relatively shallow depth (perhaps 3 to 5 km).[50] New water of magmatic or cosmic origin that has not previously been a part of the hydrosphere is referred to as *juvenile water*. And finally, *metamorphic water* is water that is or has been associated with rocks during their metamorphism. The diagram in Fig. 2.1 illustrates the interrelations of these genetic types of groundwater.

The residence time of water underground has always been a topic of considerable speculation. But with the advent of radioisotopes, determinations of the age of groundwater have become possible.

Hydrogen-3 (tritium) and carbon-14 are the two isotopes that have proved most useful. Tritium with a half-life of 12.3 years is produced in the upper atmosphere by cosmic radiation; carried to earth by rainfall and hence underground, this natural level of tritium begins to decay as a function of time, such that

$$A = A_o\,e^{-\lambda t} \qquad\qquad (2.1)$$

where A is the observed radioactivity, A_o is the activity at the time the water entered the aquifer, λ is the decay constant, and t is the age of the water. Carbon-14 has a half-life of 5730 years and is also produced at an established constant level in the atmosphere. This isotope is present in groundwater as dissolved bicarbonate originating from the biologically active layers of the soil where CO_2 is generated by root respiration and the decay of humus.[46,53] Tritium is applicable for estimating groundwater residence times of up to 50 years, while carbon-14 spans the age bracket of a several hundred to about 50,000 years.

Applications of the age-dating techniques have revealed groundwaters ranging in age from a few years or less to many thousand years.* Measurements of water samples taken from deep wells in deserts of the United Arab Republic and Saudi Arabia indicate ages of 20,000 to 30,000 years.[43] This period is compatible with the Wisconsin ice age, when these desert areas last possessed a high rainfall capable of recharging the underlying major aquifers.

Rock Properties Affecting Groundwater

Aquifers. Groundwater occurs in many types of geologic formations; those known as aquifers are of most importance. An *aquifer* may be defined as a formation that contains sufficient saturated permeable material to yield significant quantities of water to wells and springs.[25] This implies an ability to store and to transmit water; unconsolidated sands and gravels are a typical example. Furthermore, it is generally understood that an aquifer includes the unsaturated portion of the permeable unit. Synonyms frequently employed include *groundwater reservoir* and *water-bearing formation.* Aquifers are generally areally extensive and may be overlain or underlain by a *confining bed,* which may be defined as a relatively impermeable material stratigraphically adjacent to one or more aquifers.

---

*The fallout of bomb tritium and C-14 in precipitation since the advent of nuclear weapon testing in 1952 has greatly complicated much of the dating of groundwater because recent levels greatly exceed the prebomb level.

Clearly, there are various types of confining beds; the following types are well established in the literature:

1. *Aquiclude*—A saturated but relatively impermeable material that does not yield appreciable quantities of water to wells; clay is an example.

2. *Aquifuge*—A relatively impermeable formation neither containing nor transmitting water; solid granite belongs in this category.

3. *Aquitard*—A saturated but poorly permeable stratum that impedes groundwater movement and does not yield water freely to wells, but that may transmit appreciable water to or from adjacent aquifers and, where sufficiently thick, may constitute an important groundwater storage zone; sandy clay is an example.[35]*

Porosity. Those portions of a rock or soil not occupied by solid mineral matter can be occupied by groundwater. These spaces are known as voids, interstices, pores, or pore space. Because interstices serve as water conduits, they are of fundamental importance to the study of groundwater. Typically, they are characterized by their size, shape, irregularity, and distribution. Original interstices were created by geologic processes governing the origin of the geologic formation and are found in sedimentary and igneous rocks. Secondary interstices developed after the rock was formed; examples include joints, fractures, solution openings, and openings formed by plants and animals. With respect to size, interstices may be classed as capillary, supercapillary, and subcapillary. Capillary interstices are sufficiently small that surface tension forces will hold water within them; supercapillary interstices are those larger than capillary ones; and subcapillary interstices are so small that water is held primarily by adhesive forces. Depending on the connection of interstices with others, they may be classed as communicating or isolated.

The *porosity* of a rock or soil is a measure of the contained interstices or voids expressed as the ratio of the volume of interstices to the total volume. If α is the porosity, then

$$\alpha = v_i/V \tag{2.2}$$

where v_i is the volume of interstices and V is the total volume. Porosity may also be expressed by

$$\alpha = \frac{\rho_m - \rho_d}{\rho_m} = 1 - \frac{\rho_d}{\rho_m} \tag{2.3}$$

---

*The word *aquifer* can be traced to its Latin origin. *Aqui-* is a combining form of *aqua* ("water") and *-fer* comes from *ferre* ("to bear"). Hence, an aquifer is a water-bearer. The suffix *-clude* of aquiclude is derived from the Latin *claudere* ("to shut or close"). Similarly, the suffix *-fuge* of aquifuge comes from *fugere* ("to drive away"), while the suffix *-tard* of aquitard follows from the Latin *tardus* ("slow").

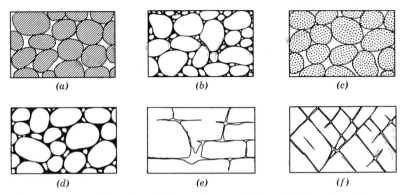

Fig. 2.2 Examples of rock interstices and the relation of rock texture to porosity. (a) Well-sorted sedimentary deposit having high porosity. (B) Poorly sorted sedimentary deposit having low porosity. (c) Well-sorted sedimentary deposit consisting of pebbles that are themselves porous, so that the deposit as a whole has a very high porosity. (d) Well-sorted sedimentary deposit whose porosity has been diminished by the deposition of mineral matter in the interstices. (e) Rock rendered porous by solution. (f) Rock rendered porous by fracturing (after Meinzer[31]).

where ρ_m is the density of mineral particles (grain density)* and ρ_d is the bulk density.

The term *effective porosity* refers to the amount of interconnected pore space available for fluid flow and is also expressed as a ratio of interstices to total volume. For unconsolidated porous media and for many consolidated rocks, the two porosities are identical. Porosity may also be expressed as a percentage by multiplying the right-hand side of Eq. 2.2 or 2.3 by 100. The terms *primary* and *secondary porosity* are associated with original and secondary interstices, respectively.

Figure 2.2 shows several types of interstices and their relation to porosity. In terms of groundwater supply, granular sedimentary deposits are of major importance. Porosities in these deposits depend on the shape and arrangement of individual particles, distribution by size, and degree of cementation and compaction. In consolidated formations, removal of mineral matter by solution and degree of fracture are also important. Porosities range from near zero to more than 50 percent, depending on the above factors and the type of material. Representative porosity values for various geologic mate-

---

*The density of solid rock varies with the type of mineral. For alluvium where quartz is the predominant mineral, a value of 2.65 g/cm³ is typical; limestone and granite fall in the range 2.7–2.8 g/cm³, and basalt can approach 3.0 g/cm³.

rials are listed in Table 2.1. It should be recognized that porosities for a particular soil or rock can vary considerably from these values.

In sedimentary rocks subject to compaction, measurements show that porosity decreases with depth of burial.[26] Thus, a typical relation has the form

$$\alpha_z = \alpha_o \, e^{-az} \tag{2.4}$$

where α_z is the porosity at depth z, α_o is the porosity at the surface, a is a constant, and e is the base of Naperian logarithms.

Soil Classification.   Unconsolidated geologic materials are normally classified according to their size and distribution. A commonly employed system based on particle, or grain, size is listed in Table 2.2. Evaluation of the distribution of sizes is accomplished by mechanical analysis. This involves sieving particles coarser than 0.05 mm and measuring rates of settlement for smaller particles in suspension. Results are plotted on a particle-size distribution graph such as that shown in Fig. 2.3. The percentage finer scale on the ordinate shows the percentage of material smaller than that of a given size particle on a dry-weight basis.

The effective particle size is the 10 percent finer than value (d_{10}). The distribution of particles is characterized by the *uniformity coefficient U_c* as

$$U_c = d_{60}/d_{10} \tag{2.5}$$

**TABLE 2.1   Representative Values
of Porosity (after Morris and Johnson[35])**

Material	Porosity, Percent	Material	Porosity, Percent
Gravel, coarse	28[a]	Loess	49
Gravel, medium	32[a]	Peat	92
Gravel, fine	34[a]	Schist	38
Sand, coarse	39	Siltstone	35
Sand, medium	39	Claystone	43
Sand, fine	43	Shale	6
Silt	46	Till,	
Clay	42	predominantly silt	34
Sandstone,		Till,	
fine-grained	33	predominantly sand	31
Sandstone,		Tuff	41
medium-grained	37	Basalt	17
Limestone	30	Gabbro, weathered	43
Dolomite	26	Granite, weathered	45
Dune sand	45		

[a]These values are for repacked samples; all others are undisturbed.

**TABLE 2.2 Soil Classification Based
on Particle Size
(after Morris and Johnson[35])**

Material	Particle Size, mm
Clay	<0.004
Silt	0.004–0.062
Very fine sand	0.062–0.125
Fine sand	0.125–0.25
Medium sand	0.25–0.5
Coarse sand	0.5–1.0
Very coarse sand	1.0–2.0
Very fine gravel	2.0–4.0
Fine gravel	4.0–8.0
Medium gravel	8.0–16.0
Coarse gravel	16.0–32.0
Very coarse gravel	32.0–64.0

where d_{60} is the 60 percent finer than value. A *uniform material* has a low uniformity coefficient (the dune sand in Fig. 2.3), while a *well-graded material* has a high uniformity coefficient (the alluvium in Fig. 2.3).

The texture of a soil is defined by the relative proportions of sand, silt, and clay present in the particle-size analysis. This can be expressed by the soil-textural triangle in Fig. 2.4. Note, for example, that a soil composed of 30 percent clay, 60 percent silt, and 10 percent sand constitutes a silty clay loam.

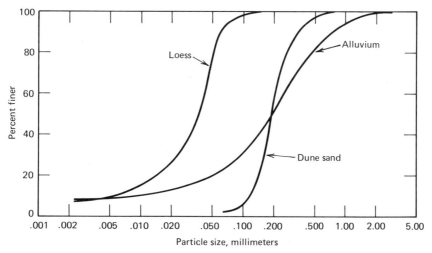

Fig. 2.3 Particle-size distribution graph for three geologic samples (data from U.S. Geological Survey).

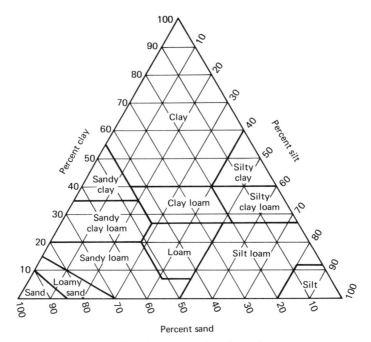

Fig. 2.4 Triangle of soil textures for describing various combinations of sand, silt, and clay (after Soil Survey Staff[39]).

Specific Surface. The water retentive property of a soil or rock is markedly influenced by its surface area. This area depends on particle size and shape and on type of clay minerals present. The term *specific surface* refers to the area per unit weight of the material, usually expressed as m^2/g. Relative methods for measuring specific surface are based on retention of a polar organic molecule such as ethylene glycol; these have been related to absolute values derived from statistical calculations of surface area.[3] Clay particles contribute the greatest amount of surface area in unconsolidated formations. Nonswelling clays such as kaolinite have only an external surface and exhibit specific surfaces in the range of 10–30 m^2/g; however, swelling clays such as montmorillonite and vermiculite have internal and external surfaces that yield specific surface values near 800 m^2/g.

An illustration of the importance of particle size to specific surface is presented in Table 2.3. Here, considering only uniform spheres, it can be seen that when a given volume is transformed into 100 small spheres totaling the same volume, the specific surface increases by a factor of 100. Furthermore, it can be shown that when the volume is deformed into rod, disk, or plate shapes, specific surface increases even more.[2]

**TABLE 2.3 Relation of Surface Area
to Particle Size for Uniform Spheres**

Diameter of Particle, mm	Soil Classification	Number of Particles per cm^3	Total Surface Area, cm^2
10	Medium gravel	1	3.14
1	Coarse sand	1×10^3	31.4
0.1	Very fine sand	1×10^6	314
0.02	Silt	1.25×10^8	1,570
0.002	Clay	1.25×10^{11}	15,700

NOTE: Rectangular packing is assumed in a cubic container 1 cm on a side so that the total volume, and weight, of spheres remains constant at $\pi/6$ cm^3.

Vertical Distribution of Groundwater

The subsurface occurrence of groundwater may be divided into zones of aeration and saturation. The zone of aeration consists of interstices occupied partially by water and partially by air. In the zone of saturation all interstices are filled with water under hydrostatic pressure. On most of the land masses of the earth, a single zone of aeration overlies a single zone of saturation and extends upward to the ground surface, as shown in Fig. 2.5.

In the zone of aeration, *vadose water** occurs. This general zone may be further subdivided into the soil water zone, the intermediate vadose zone, and the capillary zone (Fig. 2.5).[10]

The saturated zone extends from the upper surface of saturation down to underlying impermeable rock. In the absence of overlying impermeable strata, the *water table,* or *phreatic surface,*† forms the upper surface of the zone of saturation. This is defined as the surface of atmospheric pressure and appears as the level at which water stands in a well penetrating the aquifer. Actually, saturation extends slightly above the water table due to capillary attraction; however, water is held here at less than atmospheric pressure. Water occurring in the zone of saturation is commonly referred to simply as *groundwater,* but the term *phreatic water* is also employed.

Zone of Aeration

Soil-Water Zone. Water in the soil-water zone exists at less than saturation except temporarily when excessive water reaches the ground surface as from rainfall or irrigation. The zone extends from the ground surface down through the major root zone. Its thickness varies with soil type and vegetation. Because of the agricultural

*Vadose is derived from the Latin *vadosus* ("shallow").

†*Phreatic* is derived from the Greek *phrear, -atos* ("a well").

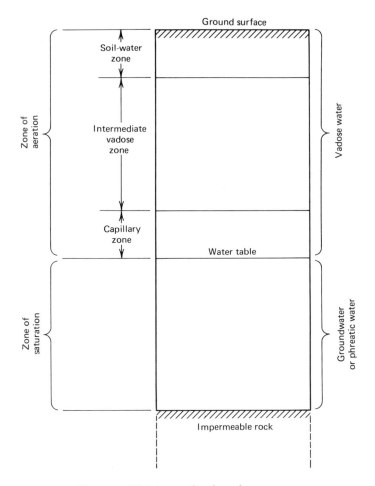

Fig. 2.5 Divisions of subsurface water.

importance of soil water in supplying moisture to roots, agricultur-
ists and soil scientists have studied soil moisture distribution and
movement extensively.

The amount of water present in the soil-water zone depends primarily on the recent exposure of the soil to moisture. Under hot,
arid conditions a water-vapor equilibrium tends to become established between the ambient air and the surfaces of fine-grained soil
particles. As a result, only thin films of moisture—known as *hygroscopic water*—remain adsorbed on the surfaces. For coarse-grained
materials and where additional moisture is available, water also
forms liquid rings surrounding contacts between grains, as sketched
in Fig. 2.6. This water is held by surface tension forces and is some-

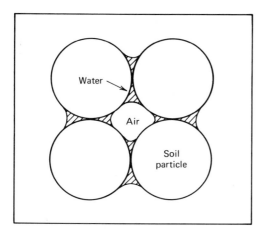

Fig. 2.6 Illustration of vadose water held at contact points of particles in the unsaturated zone.

times referred to as *capillary water*. Temporarily, the soil-water zone may contain water in excess of capillary water from rainfall or irrigation; this *gravitational water* drains through the soil under the influence of gravity.

Intermediate Vadose Zone. The intermediate vadose zone extends from the lower edge of the soil-water zone to the upper limit of the capillary zone (Fig. 2.5). The thickness may vary from zero, where the bounding zones merge with a high water table approaching ground surface, to more than 100 m under deep water table conditions. The zone serves primarily as a region connecting the zone near ground surface with that near the water table through which water moving vertically downward must pass. Nonmoving vadose water is held in place by hygroscopic and capillary forces. Temporary excesses of water migrate downward as gravitational water.

Capillary Zone. The capillary zone (or capillary fringe) extends from the water table up to the limit of capillary rise of water. If a pore space could be idealized to represent a capillary tube, the capillary rise h_c (Fig. 2.7) can be derived from an equilibrium between surface tension of water and the weight of water raised. Thus,

$$h_c = \frac{2\tau}{r\gamma} \cos \lambda \qquad (2.6)$$

where τ is surface tension, γ is the specific weight of water, r is the tube radius, and λ is the angle of contact between the meniscus

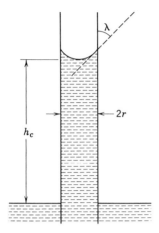

Fig. 2.7 Rise of water in a capillary tube.

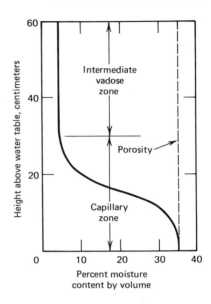

Fig. 2.8 Distribution of water in a coarse sand above the water table after drainage (after Prill[37]).

and the wall of the tube. For pure water in clean glass, $\lambda = 0$, and at 20°C $\tau = 0.074$ g/cm and $\gamma = 1$ g/cm^3, so that the capillary rise approximates

$$h_c = \frac{0.15}{r} \qquad (2.7)$$

It follows from Eq. 2.7 that the thickness of the capillary zone will vary inversely with the pore size of a soil or rock. Measurements of capillary rise in unconsolidated materials shown in Table 2.4 bear out this relationship. Furthermore, for a material containing innumerable pores of a wide range in size, the upper boundary of the zone will form a jagged limit when studied microscopically. Taken macroscopically, however, a gradual decrease in water content results with height. That is, just above the water table almost all pores contain capillary water; higher, only the smaller connected pores contain water; and still higher, only the few smallest connected pores contain water lifted above the water table. This distribution of water above the water table is shown in Fig. 2.8 from a drainage test on a sand. The visual capillary rise is invariably less than the actual capillary zone as defined in Fig. 2.8.

**TABLE 2.4   Capillary Rise in Samples
of Unconsolidated Materials (after Lohman[24])**

Material	Grain Size, mm	Capillary Rise, cm
Fine gravel	5–2	2.5
Very coarse sand	2–1	6.5
Coarse sand	1–0.5	13.5
Medium sand	0.5–0.2	24.6
Fine sand	0.2–0.1	42.8
Silt	0.1–0.05	105.5
Silt	0.05–0.02	200[a]

NOTE: Capillary rise measured after 72 days; all samples have virtually the same porosity of 41 percent.

[a]Still rising after 72 days.

Measurement of Water Content. Determination of the water content of soils can be accomplished by various direct methods based on removal of the water from a sample by evaporation, leaching, or chemical reaction, followed by measurement of the amount removed.[1,17] Thus, the gravimetric method involves weighing a wet soil sample, removing the water by oven-drying it, and reweighing the sample. Indirect methods consist of measuring some soil property affected by soil-water content. Specifically, electrical and thermal conductivity and electrical capacitance of porous materials vary with water content.

Another useful instrument for measuring soil moisture is the neutron probe. When lowered in a small-diameter tube in the ground, determination of soil moisture can be made as a function of depth. The instrument contains a radium-beryllium source of fast neutrons and a detector for slow neutrons. The fast neutrons are slowed by collisions with hydrogen, and because most of the hydrogen in soil is associated with water, the intensity of slow neutrons measured yields, after calibration, the local soil moisture content.[3]

Within the vadose zone a negative-pressure head of water exists, often referred to as *suction,* or *tension* in a positive sense. This tension can be measured by a *tensiometer;* Fig. 2.9 shows a tensiometer installed in a soil column. The depression Δh in water level measures the local soil tension. Such instruments function in the range from atmospheric pressure (near 1000 cm of water) to about 200 cm of water (800 cm water tension). Calibration data for soil suction and water content reveal that the relation between the two variables is not single-valued; instead, soil structure and compaction, as well as effects of wetting or drying, influence the results.[3]

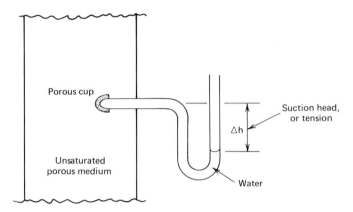

Fig. 2.9 Illustration of a tensiometer for measuring water tension in unsaturated porous media.

Available Water. Soils absorb and retain water, which may be withdrawn by plants during periods between rainfalls or irrigations. This water-holding capacity is defined by the *available water,* which is the range of plant-available water, the moist end being the field capacity and the dry end the wilting point. *Field capacity* can be defined as the amount of water held in a soil after wetting and after subsequent drainage has become negligibly small. The negligible drainage rate is often assumed after two days; however, different soils possess varying drainage rates so that quantitative values may not be comparable. The *wilting point* defines the water content of soils when plants growing in that soil are reduced to a permanent wilted condition. Because factors such as soil type and volume and plant type and age influence wilting point, this moisture content can also be variable.

Zone of Saturation

In the zone of saturation, groundwater fills all of the interstices; hence, the (effective) porosity provides a direct measure of the water contained per unit volume. A portion of the water can be removed from subsurface strata by drainage or by pumping of a well; however, molecular and surface tension forces hold the remainder of the water in place.

Specific Retention. The *specific retention* S_r of a soil or rock is the ratio of the volume of water it will retain after saturation against the force of gravity to its own volume. Thus,

$$S_r = w_r/V \tag{2.8}$$

where w_r is the volume occupied by retained water,* and V is the bulk volume of the soil or rock.

Specific Yield. The *specific yield* S_y of a soil or rock is the ratio of the volume of water that, after saturation, can be drained by gravity to its own volume.[13] Therefore,

$$S_y = w_y/V \qquad (2.9)$$

where w_y is the volume of water drained. Values of S_r and S_y can also be expressed as percentages. Because w_r and w_u constitute the total water volume in a saturated material, it is apparent that

$$\alpha = w_r + w_y \qquad (2.10)$$

where all pores are interconnecting.

Values of specific yield depend on grain size, shape and distribution of pores, compaction of the stratum, and time of drainage.[29] Representative specific yields for various geologic materials are listed in Table 2.5; individual values for a soil or rock can vary considerably from these values. It should be noted that fine-grained materials yield little water, whereas coarse-grained materials permit a substantial release of water—and hence serve as aquifers. In general, specific yields for thick unconsolidated formations tend to fall in the range of 7–15 percent, because of the mixture of grain sizes present in the various strata; furthermore, they normally decrease with depth due to compaction.

Specific yield can be measured by a variety of techniques involving laboratory, field, and estimating techniques.[3,18,19] Methods based on well-pumping tests, described in Chapter 4, generally give the most reliable results for field measurements.

Geologic Formations as Aquifers

A geologic formation that will yield significant quantities of water has been defined as an aquifer. Many types of formations serve as aquifers.[4,27] A key requirement is their ability to store water in the rock pores. Porosity may be derived from intergranular spaces or from fractures. Table 2.6 summarizes the geologic origin of aquifers in terms of type of porosity and rock type. The roles of various geologic formations as aquifers are briefly described in the following subsections.

---

*It should be noted that the terms *field capacity* and *retained water* refer to the same water content but differ by the zone in which they occur.

TABLE 2.5   Representative Values
of Specific Yield (after Johnson[18])

Material	Specific Yield, percent
Gravel, coarse	23
Gravel, medium	24
Gravel, fine	25
Sand, coarse	27
Sand, medium	28
Sand, fine	23
Silt	8
Clay	3
Sandstone, fine-grained	21
Sandstone, medium-grained	27
Limestone	14
Dune sand	38
Loess	18
Peat	44
Schist	26
Siltstone	12
Till, predominantly silt	6
Till, predominantly sand	16
Till, predominantly gravel	16
Tuff	21

Alluvial Deposits.   Probably 90 percent of all developed aquifers consist of unconsolidated rocks, chiefly gravel and sand. These aquifers may be divided into four categories, based on manner of occurrence: water courses, abandoned or buried valleys, plains, and intermontane valleys. Water courses consists of the alluvium that forms and underlies stream channels, as well as forming the adjacent floodplains. Wells located in highly permeable strata bordering streams produce large quantities of water, as infiltration from the streams augments groundwater supplies. Abandoned or buried valleys are valleys no longer occupied by streams that formed them. Although such valleys may resemble water courses in permeability and quantity of groundwater storage, their recharge and perennial yield are usually less. Extensive plains underlain by unconsolidated sediments exist in the United States. In some places gravel and sand beds form important aquifers under these plains; in other places they are relatively thin and have limited productivity. These plains flank highlands or other features that served as the source of the sedimentary deposits. The aquifers are recharged chiefly in areas

TABLE 2.6 Geologic Origin of Aquifers Based on Type of Porosity and Rock Type (after Dept. of Economic and Social Affairs[11])

Type of Porosity	Sedimentary		Igneous and Metamorphic	Volcanic	
	Consolidated	Unconsolidated		Consolidated	Unconsolidated
Intergranular		Gravelly sand Clayey sand Sandy clay	Weathered zone of granite-gneiss	Weathered zone of basalt	Volcanic ejecta, blocks, and fragments Ash
Intergranular and fracture	Breccia Conglomerate Sandstone Slate		Zoogenic limestone Oolitic limestone Calcareous grit	Volcanic tuff Cinder Volcanic breccia Pumice	
Fracture			Limestone Dolomite Dolomitic limestone Granite Gneiss Gabbro Quartzite Diorite Schist Mica schist	Basalt Andesite Rhyolite	

accessible to downward percolation of water from precipitation and from occasional streams. Intermontane valleys are underlain by tremendous volumes of unconsolidated rock materials derived by erosion of bordering mountains. Many of these more or less individual basins, separated by mountain ranges, occur in the western United States. The sand and gravel beds of these aquifers produce large quantities of water, most of which is replenished by seepage from streams into alluvial fans at mouths of mountain canyons.

Limestone. Limestone varies widely in density, porosity, and permeability depending on degree of consolidation and development of permeable zones after deposition. Those most important as aquifers contain sizable proportions of the original rock that have been dissolved and removed.[52] Openings in limestone may range from microscopic original pores to large solution caverns forming subterranean channels sufficiently large to carry the entire flow of a stream.[5,16] The term *lost river* has been applied to a stream that disappears completely underground in a limestone terrane.* Large springs are frequently found in limestone areas.

The solution of calcium carbonate by water causes prevailingly hard groundwater to be found in limestone aquifers; also, by dissolving the rock, water tends to increase the pore space and permeability with time. Solution development of limestone forms a *karst* terrane,[†] characterized by solution channels, closed depressions, subterranean drainage through sinkholes, and caves (see Fig. 2.10). Such regions normally contain large quantities of groundwater.[23] Major limestone aquifers occur in the southeastern United States and in the Mediterranean area.[7,8,21,**]

Volcanic Rock. Volcanic rock can form highly permeable aquifers; basalt flows in particular often display such characteristics.[††] The types of openings contributing to the permeability of basalt

---

*The growing interest of venturesome scuba divers in springs, sinkholes, and caves is yielding new information regarding the hydrogeology of limestone aquifers.

†The term *karst* is derived from the German form of the Slavic word *kras* or *krs*, meaning a black waterless place. Also, it is the German name for a district east of Trieste having such a terrane.[34]

**Awareness of the extensive solution development of limestone along the southern perimeter of Europe continues to increase. Thus, at Gibralter, St. Michael's Cave was only discovered during fortification work in 1942; it is large enough to serve today as an underground amphitheater for concerts.

††An excellent example of a highly permeable volcanic aquifer is in Managua, Nicaragua. Here a circular lake contained in an extinct volcanic crater serves as

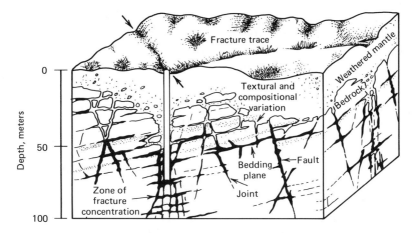

Fig. 2.10 Diagram showing factors influencing cavity distribution in carbonate rocks (after Lattman and Parizek[22]).

aquifers include, in order of importance: interstitial spaces in clinkery lava at the tops of flows, cavities between adjacent lava beds, shrinkage cracks, lava tubes, gas vesicles, fissures resulting from faulting and cracking after rocks have cooled, and holes left by the burning of trees overwhelmed by lava.[27] Most of the largest springs in the United States are associated with basalt deposits. Rhyolites are less permeable than basalt, whereas shallow intrusive rocks can be practically impermeable.

Sandstone.   Sandstone and conglomerate are cemented forms of sand and gravel. As such, their porosity and yield have been reduced by the cement. The best sandstone aquifers yield water through their joints. Conglomerates have limited distribution and are unimportant as aquifers.

Igneous and Metamorphic Rocks.   In solid forms igneous and metamorphic rocks are relatively impermeable and hence serve as poor aquifers. Where such rocks occur near the surface under weathered conditions, however, they have been developed by small wells for domestic water supply.

---

the major municipal water source and yields some 75,000 m³/day. There is no surface inflow, and evaporation exceeds precipitation; hence, the lake, fed entirely by groundwater, acts as a large natural well.

Clay. Clay and coarser materials mixed with clay are generally porous, but their pores are so small that they may be regarded as relatively impermeable. Clayey soils can provide small domestic water supplies from shallow, large-diameter wells.

Types of Aquifers

Most aquifers are of large areal extent and may be visualized as underground storage reservoirs. Water enters a reservoir from natural or artificial recharge; it flows out under the action of gravity or is extracted by wells. Ordinarily, the annual volume of water removed or replaced represents only a small fraction of the total storage capacity. Aquifers may be classed as unconfined or confined, depending on the presence or absence of a water table, while a leaky aquifer represents a combination of the two types.

Unconfined Aquifer. An *unconfined aquifer* is one in which a water table varies in undulating form and in slope, depending on areas of recharge and discharge, pumpage from wells, and permeability. Rises and falls in the water table correspond to changes in the volume of water in storage within an aquifer. Figure 2.5 is an idealized section through an unconfined aquifer; the upper aquifer in Fig. 2.11 is also unconfined. Contour maps and profiles of the water table can be prepared from elevations of water in wells that

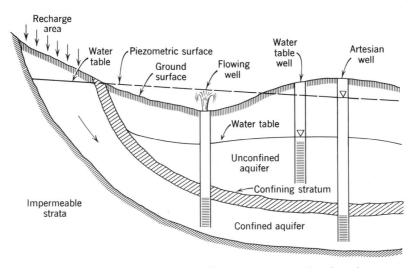

Fig. 2.11 Schematic cross section illustrating unconfined and confined aquifers.

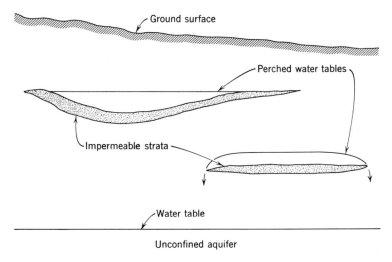

Fig. 2.12   Sketch of perched water tables.

tap the aquifer to determine the quantities of water available and their distribution and movement.

A special case of an unconfined aquifer involves perched water bodies, as illustrated by Fig. 2.12. This occurs wherever a ground-water body is separated from the main groundwater by a relatively impermeable stratum of small areal extent and by the zone of aeration above the main body of groundwater. Clay lenses in sedimentary deposits often have shallow perched water bodies overlying them. Wells tapping these sources yield only temporary or small quantities of water.

Confined Aquifers.   Confined aquifers, also known as *artesian**
or *pressure* aquifers, occur where groundwater is confined under pressure greater than atmospheric by overlying relatively impermeable strata. In a well penetrating such an aquifer, the water level will rise above the bottom of the confining bed, as shown by the artesian and flowing wells of Fig. 2.11. Water enters a confined aquifer in an area where the confining bed rises to the surface; where the confining bed ends underground, the aquifer becomes uncon-

---

*The word *artesian* has an interesting origin. It is derived from the French *artésien,* meaning "of or pertaining to Artois," the northernmost province of France. Here the first deep wells to tap confined aquifers were drilled and investigated, from about 1750. Originally the term referred to a well with freely flowing water, but at present it is applied to any well penetrating a confined aquifer or simply the aquifer itself.

fined. A region supplying water to a confined aquifer is known as a *recharge area;* water may also enter by leakage through a confining bed (see below). Rises and falls of water in wells penetrating confined aquifers result primarily from changes in pressure rather than changes in storage volumes. Hence, confined aquifers display only small changes in storage and serve primarily as conduits for conveying water from recharge areas to locations of natural or artificial discharge.

The *piezometric surface,* or *potentiometric surface,* of a confined aquifer is an imaginary surface coinciding with the hydrostatic pressure level of the water in the aquifer (Fig. 2.11). The water level in a well penetrating a confined aquifer defines the elevation of the piezometric surface at that point. Should the piezometric surface lie above ground surface, a flowing well results. Contour maps and profiles of the piezometric surface can be prepared from well data similar to those for the water table in an unconfined aquifer. It should be noted that a confined aquifer becomes an unconfined aquifer when the piezometric surface falls below the bottom of the upper confining bed. Also, quite commonly an unconfined aquifer exists above a confined one, as shown in Fig. 2.11.

Leaky Aquifer. Aquifers that are completely confined or unconfined occur less frequently than do *leaky,* or *semiconfined, aquifers.* These are a common feature in alluvial valleys, plains, or former lake basins where a permeable stratum is overlain or underlain by

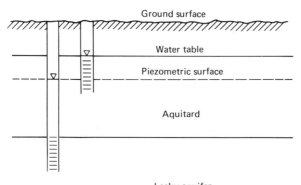

Fig. 2.13 Sketch of a leaky, or semiconfined, aquifer.

a semipervious aquitard, or semiconfining layer (see Fig. 2.13). Pumping from a well in a leaky aquifer removes water in two ways: by horizontal flow within the aquifer and by vertical flow through the aquitard into the aquifer.

Idealized Aquifer. For mathematical calculations of the storage and flow of groundwater, aquifers are frequently assumed to be homogeneous and isotropic. A *homogeneous aquifer* possesses hydrologic properties that are everywhere identical. An *isotropic aquifer* is one with its properties independent of direction. Such idealized aquifers do not exist; however, good quantitative approximations can be obtained by these assumptions, particularly where average aquifer conditions are employed on a large scale. Anisotropic aquifers, which possess directional characteristics, are discussed in Chapter 3.

Storage Coefficient

Water recharged to, or discharged from, an aquifer represents a change in the storage volume within the aquifer. For unconfined aquifers this is simply expressed by the product of the volume of aquifer lying between the water table at the beginning and at the end of a period of time and the average specific yield of the formation. In confined aquifers, however, assuming the aquifer remains saturated, changes in pressure produce only small changes in storage volume. Thus, the hydrostatic pressure within an aquifer partially supports the weight of the overburden while the solid structure of the aquifer provides the remaining support. When the hydrostatic pressure is reduced, such as by pumping water from a well penetrating the aquifer, the aquifer load increases. A compression of the aquifer results that forces some water from it. In addition, lowering of the pressure causes a small expansion and subsequent release of water. The water-yielding capacity of an aquifer can be expressed in terms of its storage coefficient.

A *storage coefficient* (or *storativity*) is defined as the volume of water that an aquifer releases from or takes into storage per unit surface area of aquifer per unit change in the component of head normal to that surface. For a vertical column of unit area extending through a confined aquifer, as in Fig. 2.14a, the storage coefficient S equals the volume of water released from the aquifer when the piezometric surface declines a unit distance. The coefficient is a dimensionless quantity involving a volume of water per volume of aquifer. In most confined aquifers, values fall in the range $0.00005 <$

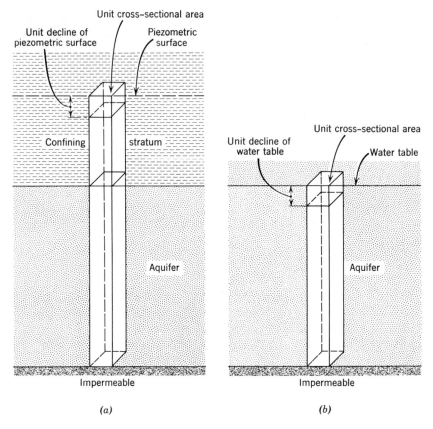

Fig. 2.14 Illustrative sketches for defining storage coefficient of (a) confined and (b) unconfined aquifers.

$S < 0.005$, indicating that large pressure changes over extensive areas are required to produce substantial water yields. Storage coefficients can best be determined from pumping tests of wells (Chapter 4) or from groundwater fluctuations in response to atmospheric pressure or ocean tide variations (see Chapter 6).

The fact that S normally varies directly with aquifer thickness enables the rule-of-thumb relationship[24]

$$S = 3 \times 10^{-6}b \tag{2.10}$$

where b is the saturated aquifer thickness in meters to be applied for estimating purposes.

The storage coefficient for an unconfined aquifer corresponds to its specific yield, as shown in Fig. 2.14b.

Groundwater Basins

A *groundwater basin* may be defined as a hydrogeologic unit containing one large aquifer or several connected and interrelated aquifers.* Such a basin may or may not coincide with a physiographic unit. In a valley between mountain ranges the groundwater basin may occupy only the central portion of the stream drainage basin. In limestone and sandhill areas, drainage and groundwater basins may have entirely different configurations. The concept of a groundwater basin becomes important because of the hydraulic continuity that exists for the contained groundwater resource. In order to ensure continued availability of subsurface water, basin-wide management of groundwater, which is described in Chapter 9, becomes essential.

Springs

A *spring* is a concentrated discharge of groundwater appearing at the ground surface as a current of flowing water. To be distinguished from springs are *seepage areas,*† which indicate a slower movement of groundwater to the ground surface. Water in seepage areas may pond and evaporate or flow, depending on the magnitude of the seepage, the climate, and the topography.

Springs occur in many forms and have been classified as to cause, rock structure, discharge, temperature, and variability. Bryan[6] divided all springs into (1) those resulting from nongravitational forces, and (2) those resulting from gravitational forces. Under the former category are included volcanic springs, associated with volcanic rocks, and fissure springs, resulting from fractures extending to great depths in the earth's crust. Such springs are usually thermal (see following section).

Gravity springs result from water flowing under hydrostatic pressure; the following general types are recognized (see Fig. 2.15):[6]

1. *Depression Springs*—formed where the ground surface intersects the water table.

---

*In practice the term *groundwater basin* is loosely defined; however, it implies an area containing a groundwater reservoir capable of furnishing a substantial water supply.

†*Seepage* is a general term describing the movement of water through the ground or other porous media to the ground surface or surface water bodies. The term is well established in the engineering literature in connection with groundwater movement from and to surface water bodies, particularly where associated with structures such as dams, canals, and levees.

2. Contact Springs—created by a permeable water-bearing formation overlying a less permeable formation that intersects the ground surface.

3. Artesian Springs—resulting from releases of water under pressure from confined aquifers either at an outcrop of the aquifer or through an opening in the confining bed.

4. Impervious Rock Springs—occurring in tubular channels or fractures of impervious rock.

5. Tubular or Fracture Springs—issuing from rounded channels, such as lava tubes or solution channels, or fractures in impermeable rock connecting with groundwater.

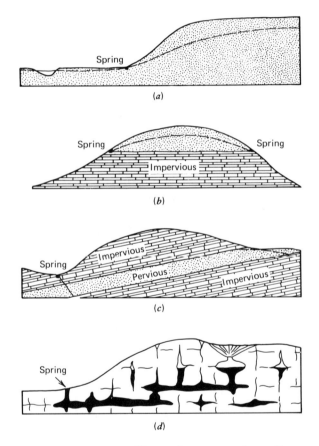

Fig. 2.15 Diagrams illustrating types of gravity springs. (*a*) Depression spring. (*b*) Contact springs. (*c*) Fracture artesian spring. (*d*) Solution tubular spring (after Bryan[6]; copyright © 1919 by the University of Chicago Press).

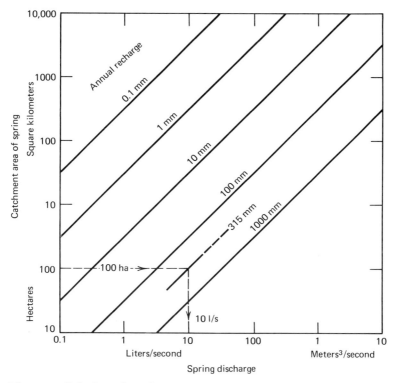

Fig. 2.16 Relation of catchment area and annual recharge to average spring discharge.

To define magnitudes of springs, Meinzer proposed a classification by discharge shown in Table 2.7. Large-magnitude springs occur primarily in volcanic and limestone terranes.[32]

The discharge of a spring depends on the area contributing recharge to the aquifer and the rate of recharge. Fig. 2.16 graphically expresses the relation. It can be noted that adequate spring water to supply the needs of a single family can be obtained from a few hectares, whereas large areas with high rainfalls are necessary to produce a first-magnitude spring.

Most springs fluctuate in their rate of discharge. Fluctuations are in response to variations in rate of recharge with periods ranging from minutes to years, depending on geologic and hydrologic conditions. Perennial springs drain extensive permeable aquifers and discharge throughout the year, whereas intermittent springs discharge only during portions of the year when sufficient groundwater

**TABLE 2.7   Classification of Springs
by Discharge (after Meinzer[30])[a]**

Magnitude	Mean Discharge
First	>10 m^3/s
Second	1–10 m^3/s
Third	0.1–1 m^3/s
Fourth	10–100 l/s
Fifth	1–10 l/s
Sixth	0.1–1 l/s
Seventh	10–100 ml/s
Eighth	<10 ml/s

[a]Another discharge classification of springs, also proposed by Meinzer and based on English units, has been in use for many years in the United States.

is recharged to maintain flow. Areas of volcanic rock and sandhills are noted for their perennial springs of nearly constant discharge. Springs that exhibit more or less regular discharge fluctuations not associated with rainfall or seasonal effects are periodic springs. Such fluctuations may be caused by variations in transpiration, by atmospheric pressure changes, by tides affecting confined aquifers, and by natural siphons acting in underground storage basins.

In coastal areas containing limestone or volcanic rock aquifers, large subsurface channels often discharge groundwater through openings to the sea. Such submarine springs can be found along the borders of the Mediterranean Sea and also in Hawaii. Where the discharge is sufficiently large, potable water can be lifted directly from the sea surface.*

Hydrothermal Phenomena

Thermal Springs.   Thermal springs discharge water having a temperature in excess of the normal local groundwater. The relative terms *warm springs* and *hot springs* are common. Waters of thermal springs are usually highly mineralized and consist for the most part of meteoric water that has been modified in quality by its passage underground.[48]

---

*Lucretius, a Roman poet and philosopher of the first century B.C., described a submarine spring in the Mediterranean Sea in his epic poem *De Rerum Natura*: "In the sea at Arados is a fountain of this kind, which wells up with fresh water and keeps off the salt waters all around it . . . a seasonable help in need to thirsting sailors, vomiting forth fresh waters amid the salt."

Hydrothermal phenomena involving the release of water and steam are nearly always associated with volcanic rocks and tend to be concentrated in regions where large geothermal gradients occur. Also, by implication aquifers must be present that permit water to percolate to great depths—often 1500 to 3000 m. This water, heated from below, forms a large convective current that rises to supply hydrothermal areas (see Fig. 2.17).

A geyser* is a periodic thermal spring resulting from the expansive force of superheated steam within constricted subsurface channels (see Fig. 2.18). Water from surface sources and/or shallow aquifers drains downward into a deep vertical tube where it is heated to above the boiling point. With increasing pressure the steam pushes upward; this releases some water at the surface, which reduces the hydrostatic pressure and causes the deeper superheated water to accelerate upward and to flash into steam. The geyser then surges into full eruption for a short interval until the pressure is dissipated; thereafter, the filling begins again and the cycle is repeated.

Another kind of hot spring, known as a mudpot, results when only a limited supply of water is available. Here water mixes with clay and undissolved particles brought to the surface, forming a muddy suspension by the small amount of water and steam continuing to

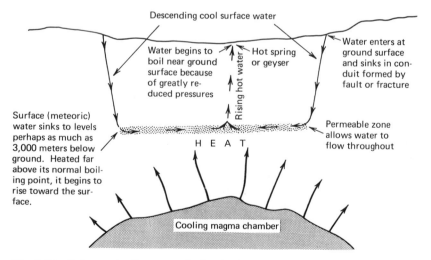

Fig. 2.17 Schematic diagram of a hydrothermal system (after Keefer[20]).

*The word *geyser* is derived from the Icelandic word *geysir,* meaning to gush or rage.

Fig. 2.18 Eruptions of the Midway Geysers in Yellowstone National Park, Wyoming (after Keefer[20]).

bubble to the surface. A fumarole* is an opening through which only steam and other gases such as carbon dioxide and hydrogen sulfide discharge. These features are normally found on hillsides above the level of flowing thermal springs; water can often be heard boiling underground.[36]

Thermal springs of various kinds are found throughout the world;[48] notable areas exist in Iceland, New Zealand, and the Kamchatka Pennisula of the U.S.S.R. Fig. 2.19 shows the regional distribution of thermal springs in the United States. Yellowstone National Park in Wyoming, containing literally thousands of hydrothermal features, possesses the greatest concentration of thermal springs in the world.[20] This area marks the site of an enormous volcanic eruption 600,000 years ago. Today, temperatures of 240°C exist only 300 m below ground surface.

Geothermal Energy Resources. Heat within the earth flows outward at an average rate of 1.5×10^{-6} cal/cm^2/s and creates an average geothermal gradient of 1°C/50 m. But in areas of volcanic and

*The word *fumarole* stems from the Latin *fumus*, meaning smoke.

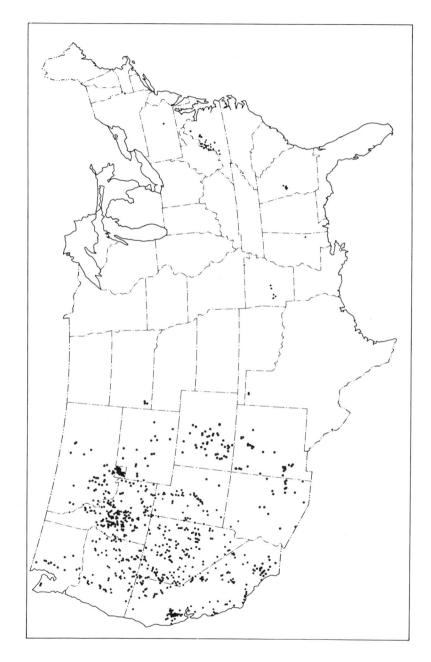

Fig. 2.19 Thermal springs in the United States (after Waring[48]).

tectonic activity, heat flows several orders of magnitude larger than normal have been found. These tremendous reservoirs of heat close to ground surface have been recognized as invaluable sources of energy.[14,42] But most important from a hydrologic standpoint is the fact that utilization of these geothermal resources invariably involves water as the mechanism for extracting heat. Four types of sources are generally recognized:

1. *Dry Steam Field*—permeability is so low that groundwater reaching the heat source is limited and is fully vaporized.

2. *Wet Steam Field*—sufficient groundwater reaches the heat source so that a mixture of water and steam is produced.

3. *Low-Temperature Fields*—temperatures in the range of 50–80°C are found at shallow depths enabling the groundwater to be employed directly for heating purposes.

4. *Dry Rock*—in some areas at depths exceeding 3000 m, temperatures are found in the range of 200–500°C without the presence of groundwater; by injecting water into fractured rock zones through one drill hole, steam can be produced from an adjacent hole.

Geothermal resources exist through the world, but they have been developed in only a few localities.* Most homes in Reykjavik, Iceland, are heated by naturally hot water. Notable power plants generating energy from geothermal resources are located in Italy, New Zealand, and the western United States.

Groundwater in Permafrost Regions

Permafrost, or perennially frozen ground, is defined as unconsolidated deposits or bedrock that continuously have had a temperature below 0°C for two years to thousands of years.[54,55] Fig. 2.20 illustrates the upper and lower limits of permafrost in terms of the depths at which a 0°C ground temperature occurs. Regions of permafrost in the Northern Hemisphere are shown in Fig. 2.21. In the continuous-permafrost zone, permafrost is present everywhere to depths of 150–400 m; while in the more southerly discontinuous-permafrost zone, permafrost is perforated by unfrozen zones that depend on local conditions.

Frozen ground creates an impermeable layer that restricts the movement of groundwater, acts as a confining layer, and limits the volume in which liquid water can be stored. In many areas of frozen

---

*It has been estimated that the total stored heat in the earth to a depth of 3 km amounts to 2×10^{21} cal and that 1 percent, or 2×10^{19} cal, of this can be commercially recovered.[51]

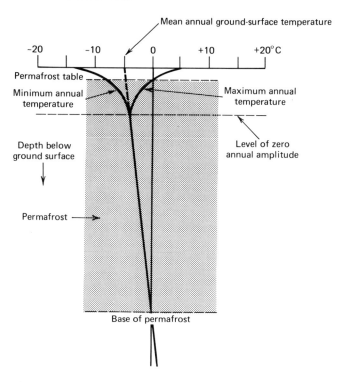

Fig. 2.20 Location of permafrost below ground surface
in relation to ground temperature (after Williams[54]).

ground, shallow aquifers are entirely eliminated, thereby requiring
that wells be drilled deeper than in similar geologic environments
without permafrost. Groundwater can occur above, below, and locally within permafrost.[9] In the continuous-permafrost zone, the
best sources of water are in unfrozen alluvium beneath large lakes,
in major valleys, and adjacent to riverbeds. In the discontinuous-
permafrost zone, groundwater can be produced locally from shallow
aquifers; however, because of potential pollution from ground surface, sources beneath the frozen layer are preferable.

Groundwater in the United States

Productive aquifers and withdrawals from the wells in the United
States are shown on endpaper. This map shows regions in which
moderate to large supplies of water can be obtained from wells. Un-

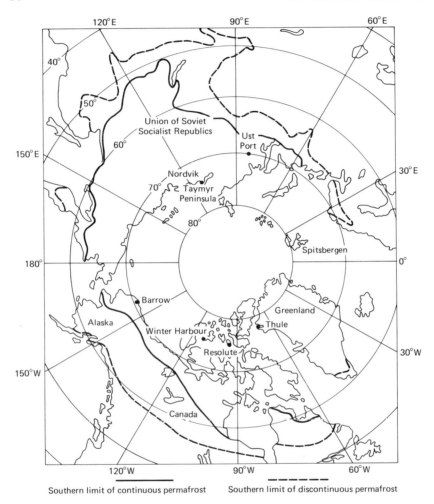

Southern limit of continuous permafrost   Southern limit of discontinuous permafrost

Fig. 2.21 Distribution of permafrost in the Northern Hemisphere (after Williams[54]).

consolidated and consolidated aquifers are indicated in addition to primary geologic formations. Blank areas delineate generally those regions not known to produce yields of more than 3 l/s to a well.

Areal distribution of groundwater in the United States can best be described by dividing the conterminous United States into the 10 groundwater regions shown on endpaper. These regions were developed by Thomas[44] and have been adopted subsequently by others.[12,28] Brief geologic and groundwater summaries for each of the 10 regions plus Alaska and Hawaii follow.*

1. Western Mountain Ranges.   This mountainous region serves as the principal source of water in the western United States because the bulk of precipitation falls here and thereafter supplies streams and aquifers by its runoff. Rocks are mainly hard and dense; they shed water rather than absorbing it, although weathered surficial rock may locally yield limited groundwater. Some aquifers are to be found in alluvium contained in small intermontane valleys. Because of the thinness and rapid draining of the mantle rock, it is not easy to obtain groundwater from wells, as the blank areas on end-paper indicate. Small springs, wells in valleys, and small surface reservoirs meet most domestic water supply needs.

2. Alluvial Basins. The basins in this region consist of vast depressed areas bounded by adjacent highlands. They are partly filled by erosional debris in the form of alluvium and serve as the storage volumes for water flowing from nearby highlands. The alluvial fill functions as an ideal aquifer and creates the opportunity for development of high-yielding wells. Because of the prevailing arid climate, groundwater development for irrigation is much in demand. Replenishment rates usually are far less than withdrawal rates, so groundwater levels decline as storage is depleted. Locally, artificial recharge (see Chapter 13) has helped to alleviate this problem of overproduction.

3. Columbia Lava Plateau.   This area is formed principally by extrusive volcanic rocks, mainly lava flows, interbedded with or overlain by alluvium and lake sediments. Water originates chiefly from mountains on the perimeter of the region. The lava rocks tend to be highly permeable as a result of tubes and shrinkage cracks and thus form highly productive aquifers. The large volumes of groundwater discharge as major springs or as streams with sustained base flows. Because of the great thickness of the lava flows, groundwater is most readily available in valley bottoms; however, in the higher plateau areas, deep wells are required to extract groundwater for irrigation.

---

*An earlier discussion of groundwater in the United States was prepared by Meinzer[31,33] involving 21 groundwater provinces. More recently, a broad-perspective appraisal of groundwater resources in the United States has been undertaken by the U.S. Geological Survey.[45]

4. Colorado Plateaus and Wyoming Basin. This region consists of sedimentary strata, chiefly interbedded sandstone and shale; these are generally horizontal but in places are folded, tilted, or broken by faults. The plateaus are rather high, dry, and deeply dissected by streams. Prospects for large-scale groundwater development are poor; nevertheless, small water supplies for domestic and livestock purposes are widely available. Most aquifers are sandstone beds, although limestone and alluvium yield water in a few places.

5. High Plains. Here alluvium forms a vast plain extending eastward from the Rocky Mountains. The bulk of it is classified as a single stratigraphic unit, the Ogallala Formation, which covers older rocks to thicknesses exceeding 150 m. The sand and the gravel of the formation consitute an aquifer yielding 10 to more than 60 l/s of water from individual wells. The region is generally semiarid so that groundwater recharge from precipitation is extremely small. The productiveness of wells has encouraged pumping of groundwater, especially for irrigation in Texas. This water is derived from storage; as a result, water tables have declined substantially for many years.

6. Unglaciated Central Region. This is a large, complex area characterized by plains and plateaus underlain by consolidated sedimentary rocks. Alluvial deposits of substantial width and thickness form good aquifers but only along major streams. Aquifers in most of the region are dolomitic limestone and sandstone with low to moderate yields. The blank areas on the inside front cover for this region include some of the least productive aquifers in the United States because of low yield, saline water, or both. Wells in some of the karst areas penetrate caverns and exhibit highly variable yields.

7. Glaciated Central Region. Although hydrogeologically similar to the previous region, this area differs fundamentally by the presence of glacial drift deposited by the ice and meltwaters of the continental glaciers. The drift consists mostly of fine-grained rock debris together with beds of water-sorted sand and gravel. In portions of the area the drift is more than 300 m thick and forms an important aquifer. In this glaciated region large-diameter wells will yield sufficient water to meet domestic needs of a family. Excellent aquifers can be found along watercourses where rapidly flowing meltwaters removed fine materials and left behind permeable deposits of sand and gravel.

8. Unglaciated Appalachian Region. This mountainous area consists of varying parallel highlands with differing geologic structures, including crystalline rocks, limestone, dolomite, sandstone, and shale. Groundwater productivity ranges from small to moderate or is erratic. Shallow wells can usually obtain small yields for domestic supplies from the weathered rock. Wells of highest average yield occur in the valleys because rocks are more intensely fractured and because of groundwater drainage from the surrounding hills. In the limestone areas much of the drainage is underground so the chances of obtaining moderately productive wells are good.

9. Glaciated Appalachian Region. Because this region is considerably more rugged than the Glaciated Central Region (No. 7), there is greater local variation in thickness of the glacial drift. Usually, hilltops are covered by a thin mantle of till; slopes contain even less with bedrock outcrops, and valleys are underlain by thicker drift and till. Near the sea glacial material is overlain or interbedded with marine deposits. In most of the area aquifers are not productive. Except for a few areas where bedrock yields moderate quantities of groundwater, the principal groundwater sources are sand and gravel deposits occurring as outwash plains or as channel fillings in the stratified drift.

10. Atlantic and Gulf Coastal Plain. This extensive coastal plain consists of a huge seaward-thickening wedge of generally unconsolidated sedimentary rocks. The sediments are mainly stratified layers of clay, silt, sand, gravel, marl, and limestone. The maximum thickness of deposits beneath the coastline varies from 100 m in the Northeast to more than 10,000 m in the South. Essentially all of the area contains extensive and productive aquifers. The principal aquifers are found in sand and gravel beds, while limestone aquifers are most important on the southeastern portion of the region.[41] Problems of seawater intrusion and land subsidence are locally significant in concentrated pumping localities.

Alaska. The principal aquifers are bodies of water-sorted sand and gravel within glacial drift that covers the uplands and within glacial outwash and other alluvial deposits extending from the uplands into the lowlands. The most productive aquifers are found in the vast central plateau and on the slopes of the southern mountain system. Permafrost predominates in the permeable deposits of the arctic slope so groundwater is available only locally near large bodies of surface water. Along the southeastern coastal area, productive alluvial deposits are scarce.

Hawaii. Basaltic lava flows make up the bulk of the islands and constitute the most important aquifers. Openings within or between the flows are responsible for the generally high permeability. Near eruptive centers vertical dikes of dense impermeable rock may separate high-level groundwater in inland areas from basal groundwater in the coastal lowlands. The basal water forms a lens of fresh groundwater floating on underlying sea water (see Chapter 14). Groundwater is extensively developed for irrigation and municipal purposes on Maui and Oahu.

References

1. Agricultural Research Service, *Field manual for research in agricultural hydrology,* Agric. Handbk. no. 224, U.S. Dept. of Agric., 215 pp., 1962.
2. Baver, L. D., et al., *Soil physics,* 4th ed., John Wiley & Sons, New York, 498 pp., 1972.
3. Black, C. A. (ed.), *Methods of soil analysis,* Part 1, Agronomy Monograph no. 9, Amer. Soc. Agronomy, Madison, Wis., 770 pp., 1965.
4. Blank, N. R., and M. C. Schroeder, Geologic classification of aquifers, *Ground Water,* v. 11, no. 2, pp. 3–5, 1973.
5. Brucker, R. W., et al., Role of vertical shafts in the movement of ground water in carbonate aquifers, *Ground Water,* v. 10, no. 6. pp. 5–13, 1972.
6. Bryan, K., Classification of springs, *Jour. Geology,* v. 27, pp. 522–561, 1919.
7. Burdon, D. J., and N. Papakis, *Handbook of karst hydrogeology with special reference to the carbonate aquifers of the Mediterranean region,* Inst. Geology and Subsfc. Research, United Nations Spec. Fund, Karst Groundwater Inv., Athens, 276 pp., 1963.
8. Burdon, D. J., and C. Safadi, The karst groundwaters of Syria, *Jour. Hydrology,* v. 2. pp. 324–347, 1964.
9. Cederstrom, D. J., et al., Occurrence and development of ground water in permafrost regions, *U.S. Geological Survey Circ. 275,* 30 pp., 1953.
10. Childs, E. C., *An introduction to the physical basis of soil water phenomena,* John Wiley & Sons, London, 493 pp., 1969.
11. Dept. of Economic and Social Affairs, *Ground-water storage and artificial recharge,* Natural Resources, Water Ser. no. 2, United Nations, New York, 270 pp., 1975.
12. Dept. of Economic and Social Affairs, *Ground water in the Western Hemisphere,* Natural Resources, Water Ser. no. 4, United Nations, New York, 337 pp., 1976.

13. Dos Santos, A. G., Jr., and E. G. Youngs, A study of the specific yield in land-drainage situations, *Jour. Hydrology*, v. 8, pp. 59–81, 1969.
14. Dutcher, L. C., Preliminary appraisal of ground water in storage with reference to geothermal resources in the Imperial Valley area, California, *U.S. Geological Survey Circ.* 649, 57 pp., 1972.
15. Feth, J. H., Selected references on saline ground-water resources of the United States, *U.S. Geological Survey Circ.* 499, 30 pp., 1965.
16. Institute of Hydrogeology and Engineering Geology, Chinese Academy of Geological Sciences, *Karst in China*, Shanghai People's Publishing House, 1976.
17. Johnson, A. I., Methods of measuring soil moisture in the field, *U.S. Geological Survey Water-Supply Paper* 1619-U, 25 pp., 1962.
18. Johnson, A. I., Specific yield—compilation of specific yields for various materials, *U.S. Geological Survey Water-Supply Paper* 1662-D, 74 pp., 1967.
19. Jones, O. R., and A. D. Schneider, Determining specific yield of the Ogallala aquifer by the neutron method, *Water Resources Research*, v. 5, pp. 1267–1272, 1969.
20. Keefer, W. R., The geologic story of Yellowstone National Park, *U.S. Geological Survey Bull.* 1347, 92 pp., 1971.
21. LaMoreaux, P. E., et al., Hydrology of limestone terranes, annotated bibliography of carbonate rocks, *Geological Survey Alabama Bull.* 94(A), 242 pp., 1970.
22. Lattman, L. H., and R. R. Parizek, Relationship between fracture traces and the occurrence of ground water in carbonate rocks, *Jour. Hydrology*, v. 2, pp. 73–91, 1964.
23. LeGrand, H. E., and V. T. Stringfield, Karst hydrology—a review, *Jour. Hydrology*, v. 20, pp. 97–120, 1973.
24. Lohman, S. W., Ground-water hydraulics, *U.S. Geological Survey Prof. Paper* 708, 70 pp., 1972.
25. Lohman, S. W., et al., Definitions of selected ground-water terms—revisions and conceptual refinements, *U.S. Geological Survey Water-Supply Paper* 1988, 21 pp., 1972.
26. Manger, G. E., Porosity and bulk density of sedimentary rocks, *U.S. Geological Survey Bull.* 1144-E, 55 pp., 1963.
27. Maxey, G. B., and J. E. Hackett, Applications of geohydrologic concepts in geology, *Jour. Hydrology*, v. 1, pp. 35–46, 1963.
28. McGuinness, C. L., The role of ground water in the national water situation, *U.S. Geological Survey Water-Supply Paper* 1800, 1121 pp., 1963.
29. McQueen, I. S., Evaluating the reliability of specific-yield determinations, *Jour. Research U.S. Geological Survey*, v. 1, pp. 371–376, 1973.
30. Meinzer, O. E., Outline of ground-water hydrology with definitions, *U.S. Geological Survey Water-Supply Paper* 494, 71 pp., 1923.
31. Meinzer, O.E., The occurrence of ground water in the United States, *U.S. Geological Survey Water-Supply Paper* 489, 321 pp., 1923.

32. Meinzer, O. E., Large springs in the United States, *U.S. Geological Survey Water-Supply Paper* 557, 94 pp., 1927.

33. Meinzer, O. E., Ground water in the United States, a summary, *U.S. Geological Survey Water-Supply Paper* 836-D, pp. 157–232, 1939.

34. Monroe, W. H., A glossary of karst terminology, *U.S. Geological Survey Water-Supply Paper* 1899-K, 26 pp., 1970.

35. Morris, D. A., and A. I. Johnson, Summary of hydrologic and physical properties of rock and soil materials, as analyzed by the Hydrologic Laboratory of the U.S. Geological Survey 1948–1960, *U.S. Geological Survey Water-Supply Paper* 1839-D, 42 pp., 1967.

36. Poland, J. F., et al., Glossary of selected terms useful in studies of the mechanics of aquifer systems and land subsidence due to fluid withdrawal, *U.S. Geological Survey Water-Supply Paper* 2025, 9 pp., 1972.

37. Prill, R. C., Specific yield—laboratory experiments showing the effect of time on column drainage, *U.S. Geological Survey Water-Supply Paper* 1662-B, 55 pp., 1965.

38. Smith, D. B., et al., The age of groundwater in the chalk of London Basin, *Water Resources Research,* v. 12, pp. 392–404, 1976.

39. Soil Survey Staff, *Soil survey manual,* U.S. Dept. Agriculture Handbook no. 18, 503 pp., 1951.

40. Stringfield, V. T., Artesian water in Tertiary limestone in the southeastern states, *U.S. Geological Survey Prof. Paper* 517, 226 pp., 1966.

41. Stringfield, V. T., and H. E. LeGrand, Hydrology of carbonate rock terranes—A review with reference to the United States, *Jour. Hydrology,* v. 8, pp. 349–417, 1969.

42. Summers, W. K., *Annotated and indexed bibliography of geothermal phenomena,* New Mexico Bur. Mines and Mineral Resources, Socorro, 665 pp., 1972.

43. Thatcher, L., et al., Dating desert ground water, *Science,* v. 134, no. 3472, pp. 105–106, 1961.

44. Thomas, H. E., *Ground-water regions of the United States—their storage facilities,* v. 3, Interior and Insular Affairs Comm., House of Representatives, U.S. Congress, Washington, D.C., 78 pp., 1952.

45. U.S. Geological Survey, Summary appraisals of the nation's groundwater resources, *Professional Paper* 813, published in sections for various regions of the United States, 1974–.

46. Vogel, J. C., and D. Ehhalt, The use of carbon isotopes in groundwater studies, *Radioisotopes in Hydrology,* Intl. Atomic Energy Agency, Vienna, pp. 383–395, 1963.

47. Walker, E. H., Ground-water resources of the Hopkinsville quadrangle, Kentucky, *U.S. Geological Survey Water-Supply Paper* 1328, 98 pp., 1956.

48. Waring, G. A., Thermal springs of the United States and other countries of the world—a summary, *U.S. Geological Survey Prof. Paper* 492, 383 pp., 1965.

49. White, D. E., Magmatic, connate, and metamorphic waters, *Bull. Geological Soc. Amer.,* v. 68, pp. 1659–1682, 1957.
50. White, D. E., Thermal waters of volcanic origin, *Bull. Geological Soc. Amer.,* v. 68, pp. 1637–1658, 1957.
51. White, D. E., Geothermal energy, *U.S. Geological Survey Circ.* 519, 17 pp., 1965.
52. White, W. B., Conceptual models for carbonate aquifers, *Ground Water,* v. 7, no. 3, pp. 15–21, 1969.
53. Wigley, T. M. L., Carbon 14 dating of groundwater from closed and open systems, *Water Resources Research,* v. 11, pp. 324–328, 1975.
54. Williams, J. R., Ground water in permafrost regions—An annotated bibliography, *U.S. Geological Survey Water-Supply Paper* 1792, 294 pp., 1965.
55. Williams, J. R., Ground water in the permafrost regions of Alaska, *U.S. Geological Survey Prof. Paper* 696, 83 pp., 1970.

Groundwater
Movement

Groundwater in its natural state is invariably moving. This movement is governed by established hydraulic principles. The flow through aquifers, most of which are natural porous media, can be expressed by what is known as Darcy's law. Hydraulic conductivity, which is a measure of the permeability of the media, is an important constant in the flow equation. Determination of hydraulic conductivity can be made by several laboratory or field techniques. Applications of Darcy's law enable groundwater flow rates and directions to be evaluated. The dispersion, or mixing, resulting from flows through porous media produces irregularities of flow that can be studied by tracers. In the zone of aeration the presence of air adds a complicating factor to the flow of water.

Darcy's Law

More than a century ago Henry Darcy,* a French hydraulic engineer, investigated the flow of water through horizontal beds of sand to be used for water filtration. He reported[13] in 1856:

*An interesting summary of the life and accomplishments of Henry Darcy was prepared by Fancher.[18]

I have attempted by precise experiments to determine the law of the flow of water through filters. . . . The experiments demonstrate positively that the volume of water which passes through a bed of sand of a given nature is proportional to the pressure and inversely proportional to the thickness of the bed traversed; thus, in calling s the surface area of a filter, k a coefficient depending on the nature of the sand, e the thickness of the sand bed, $P - H_0$ the pressure below the filtering bed, $P + H$ the atmospheric pressure added to the depth of water on the filter; one has for the flow of this last condition $Q = (ks/e)(H + e + H_0)$, which reduces to $Q = (ks/e)(H + e)$ when $H_0 = 0$, or when the pressure below the filter is equal to the weight of the atmosphere.

This statement, that the flow rate through porous media is proportional to the head loss and inversely proportional to the length of the flow path, is known universally as Darcy's law. It, more than any other contribution, serves as the basis for present-day knowledge of groundwater flow. Analysis and solution of problems relating to groundwater movement and well hydraulics began after Darcy's work.

Experimental Verification. The experimental verification of Darcy's law can be performed with water flowing at a rate Q through a cylinder of cross-sectional area A packed with sand and having piezometers a distance L apart, as shown in Fig. 3.1.[20,29] Total energy heads, or fluid potentials, above a datum plane may be expressed by the Bernoulli equation

$$\frac{p_1}{\gamma} + \frac{v_1{}^2}{2g} + z_1 = \frac{p_2}{\gamma} + \frac{v_2{}^2}{2g} + z_2 + h_L \tag{3.1}$$

where p is pressure, γ is the specific weight of water, v is the velocity of flow, g is the acceleration of gravity, z is elevation, and h_L is head loss. Subscripts refer to points of measurement identified in Fig. 3.1. Because velocities in porous media are usually low, velocity heads may be neglected without appreciable error. Hence, by rewriting, the head loss becomes

$$h_L = \left(\frac{p_1}{\gamma} + z_1\right) - \left(\frac{p_2}{\gamma} + z_2\right) \tag{3.2}$$

Therefore, the resulting head loss is defined as the potential loss within the sand cylinder, this energy being lost by frictional resistance dissipated as heat energy. It follows that the head loss is independent of the inclination of the cylinder.

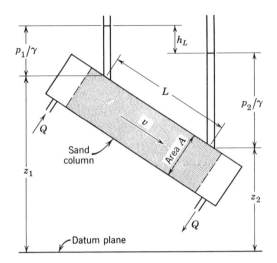

Fig. 3.1 Pressure distribution and head loss in flow through a sand column.

Now, Darcy's measurements showed that the proportionalities $Q \sim h_L$ and $Q \sim 1/L$ exist. Introducing a proportionality constant K leads to the equation

$$Q = -KA\frac{h_L}{L} \qquad (3.3)$$

Expressed in general terms

$$Q = -KA\frac{dh}{dl} \qquad (3.4)$$

or simply

$$v = \frac{Q}{A} = -K\frac{dh}{dl} \qquad (3.5)$$

where v is the *Darcy velocity* or *specific discharge*, K is the *hydraulic conductivity*, a constant that serves as a measure of the permeability of the porous medium, and dh/dl is the hydraulic gradient. The negative sign indicates that the flow of water is in the direction of decreasing head. Equation 3.5 states Darcy's law in its simplest form, that the flow velocity v equals the product of the constant K, known as the hydraulic conductivity, and the hydraulic gradient.

Darcy Velocity. The velocity v in Eq. 3.5 is referred to as the *Darcy velocity* because it assumes that flow occurs through the entire cross section of the material without regard to solids and pores.

Actually, the flow is limited to the pore space only so that the *average interstitial velocity*

$$v_a = \frac{Q}{\alpha A}$$ (3.6)

where α is porosity. This indicates that for a sand with a porosity of 33 percent, $v_a = 3v$. To define the actual flow velocity, one must consider the microstructure of the rock material. In water flowing through a sand, for example, the pore spaces vary continuously with location within the medium. This means that the actual velocity is non-uniform, involving endless accelerations, decelerations, and changes in direction. Thus, the actual velocity depends on specifying a precise point location within the medium. For naturally occurring geologic materials, the microstructure cannot be specified three-dimensionally; hence, actual velocities can only be quantified statistically.

Validity of Darcy's Law. In applying Darcy's law it is important to know the range of validity within which it is applicable. Because velocity in laminar flow, such as water flowing in a capillary tube, is proportional to the first power of the hydraulic gradient (Poiseuille's law), it seems reasonable to believe that Darcy's law applies to laminar flow in porous media. For flow in pipes and other large sections, the Reynolds number, which expresses the dimensionless ratio of inertial to viscous forces, serves as a criterion to distinguish between laminar and turbulent flow. Hence, by analogy, the Reynolds number has been employed to establish the limit of flows described by Darcy's law, corresponding to the value where the linear relationship is no longer valid.

Reynolds number is expressed as

$$N_R = \frac{\rho v D}{\mu}$$ (3.7)

where ρ is the fluid density, v the velocity, D the diameter (of a pipe), and μ the viscosity of the fluid. To adapt this criterion to flow in porous media, the Darcy velocity is employed for v and an effective grain size (d_{10}) is substituted for D. Certainly a grain diameter represents only an approximation of the critical flow dimension for which it is intended; however, measuring pore size distribution is a complex research task.

Experiments show that Darcy's law is valid for $N_R < 1$ and does not depart seriously up to $N_R = 10$.[1] This, then, represents an upper limit to the validity of Darcy's law. A range of values rather than a

unique limit must be stated because as inertial forces increase, turbulence occurs gradually.[49,55,67] The irregular flow paths of eddies
and swirls associated with turbulence occur first in the larger pore
spaces; with increasing velocity they spread to the smaller pores.
For fully developed turbulence the head loss varies approximately
with the second power of the velocity rather than linearly.

Fortunately, most natural underground flow occurs with $N_R < 1$
so Darcy's law is applicable. Deviations from Darcy's law can occur
where steep hydraulic gradients exist, such as near pumped wells;
also, turbulent flow can be found in rocks such as basalt and limestone[10] that contain large underground openings.*

Permeability

Intrinsic Permeability.   The permeability of a rock or soil defines
its ability to transmit a fluid. This is a property of the medium only
and is independent of fluid properties. To avoid confusion with
hydraulic conductivity, which includes the properties of groundwater, an *intrinsic permeability* k may be expressed as

$$K = \frac{K\mu}{\rho g} \tag{3.8}$$

where K is hydraulic conductivity, μ is dynamic viscosity, ρ is fluid
density, and g is acceleration of gravity. Inserting this in Eq. 3.5 yields

$$k = -\frac{\mu v}{\rho g (dh/dl)} \tag{3.9}$$

which has units of

$$k = -\frac{(kg/ms)(m/s)}{(kg/m^3)(m/s^2)(m/m)} = m^2 \tag{3.10}$$

Thus, intrinsic permeability possesses units of area. Because values
of k in Eq. 3.10 are so small, the U.S. Geological Survey expresses
k in square micrometers $(\mu m)^2 = 10^{-12} m^2$.

In the petroleum industry the value of k is measured by a unit
termed the *darcy*, defined as

$$1 \text{ darcy} = \frac{\dfrac{(1 \text{ centipoise})(1 \text{ cm}^3/s)}{1 \text{ cm}^2}}{1 \text{ atmosphere/cm}} \tag{3.11}$$

*It should also be noted that investigations have shown that Darcy's law may not
be valid for very slow water flow through dense clay. Here the effects of electrically
charged clay particles on water in the minute pores produce nonlinearities between
flow rate and hydraulic gradient.[30]

By substitution of appropriate units it can be shown that[41]

$$1 \text{ darcy} = 0.987 \ (\mu m)^2 \qquad (3.12)$$

so the darcy corresponds closely to the intrinsic permeability unit adopted by the U.S. Geological Survey.

Hydraulic Conductivity. For practical work in groundwater hydrology, where water is the prevailing fluid, hydraulic conductivity K is employed. A medium has a unit hydraulic conductivity if it will transmit in unit time a unit volume of groundwater at the prevailing kinematic viscosity* through a cross section of unit area, measured at right angles to the direction of flow, under a unit hydraulic gradient. The units are

$$K = -\frac{v}{dh/dL} = -\frac{m/day}{m/m} = m/day \qquad (3.13)$$

indicating that hydraulic conductivity has units of velocity.

Transmissivity. The term *transmissivity* T is widely employed in groundwater hydraulics. It may be defined as the rate at which water of prevailing kinematic viscosity is transmitted through a unit width of aquifer under a unit hydraulic gradient. It follows that

$$T = Kb = (m/day)(m) = m^2/day \qquad (3.14)$$

where b is the saturated thickness of the aquifer.

Hydraulic Conductivity of Geologic Materials. The hydraulic conductivity of a soil or rock depends on a variety of physical factors, including porosity, particle size and distribution, shape of particles, arrangement of particles, and other factors.[38,47] In general, for unconsolidated porous media, hydraulic conductivity varies with particle size; clayey materials exhibit low values of hydraulic conductivity, whereas sands and gravels display high values.

An interesting illustration of the variation of hydraulic conductivity with particle size is shown by data in Fig. 3.2. Here conductivities were measured for two uniform sieved sands. These two sands were then mixed in varying proportions, and the corresponding hydraulic conductivities were again determined. Results show that any mixture of the two sands displays a conductivity less than a linearly interpolated value. The physical explanation lies in the fact that the smaller grains occupy a larger fraction of the space around larger grains than do uniform grains of either size.

*Kinematic viscosity equals dynamic viscosity divided by fluid density.

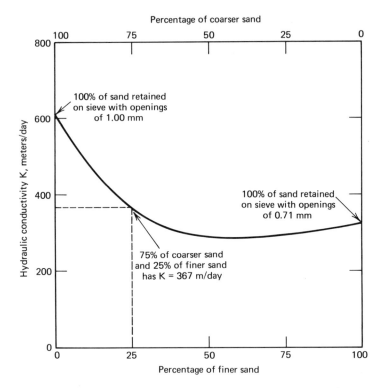

Fig. 3.2 Hydraulic conductivity of various proportions of two uniform sands (courtesy Illinois State Water Survey).

Table 3.1 contains representative hydraulic conductivities for a variety of geologic materials. It should be noted that these are averages of many measurements; clearly, a range of values exists for each rock type depending on factors such as weathering, fracturing, solution channels, and depth of burial.

Magnitudes of hydraulic conductivity for various classes of unconsolidated and consolidated rocks are shown in Fig. 3.3.

Determination of Hydraulic Conductivity

Hydraulic conductivity in saturated zones can be determined by a variety of techniques, including calculation from formulas, laboratory methods, tracer tests, auger hole tests, and pumping tests of wells.

Formulas. Numerous investigators have studied the relationship of permeability or hydraulic conductivity to the properties of porous media. Several formulas have resulted based on analytic or experi-

**TABLE 3.1 Representative Values of Hydraulic Conductivity
(after Morris and Johnson[45])**

Material	Hydraulic Conductivity, m/day	Type of Measurement[a]
Gravel, coarse	150	R
Gravel, medium	270	R
Gravel, fine	450	R
Sand, coarse	45	R
Sand, medium	12	R
Sand, fine	2.5	R
Silt	0.08	H
Clay	0.0002	H
Sandstone, fine-grained	0.2	V
Sandstone, medium-grained	3.1	V
Limestone	0.94	V
Dolomite	0.001	V
Dune sand	20	V
Loess	0.08	V
Peat	5.7	V
Schist	0.2	V
Slate	0.00008	V
Till, predominantly sand	0.49	R
Till, predominantly gravel	30	R
Tuff	0.2	V
Basalt	0.01	V
Gabbro, weathered	0.2	V
Granite, weathered	1.4	V

[a]H is horizontal hydraulic conductivity, R is a repacked sample, and V is vertical hydraulic conductivity.

mental work. Most permeability formulas have the general form

$$k = cd^2 \qquad (3.15)$$

where c is a dimensionless coefficient, or

$$k = f_s f_\alpha d^2 \qquad (3.16)$$

where f_s is a grain (or pore) shape factor, f_α is a porosity factor, and d is characteristic grain diameter.[17,37,43] Few formulas give reliable estimates of results because of the difficulty of including all possible variables in porous media. For an ideal medium, such as an assemblage of spheres of uniform diameter, hydraulic conductivity can be accurately evaluated from known porosity and packing conditions.

Because of the problems inherent in formulas, other techniques for determining hydraulic conductivity are preferable.

Hydraulic conductivity, meters/day									
10^4	10^3	10^2	10^1	1	10^{-1}	10^{-2}	10^{-3}	10^{-4}	10^{-5}
Relative hydraulic conductivity									
Very high		High		Moderate			Low		Very low
REPRESENTATIVE MATERIALS									
Unconsolidated deposits									
Clean gravel	–	Clean sand and sand and gravel	–	Fine sand	–	Silt, clay, and mixtures of sand, silt, and clay		–	Massive clay
Consolidated Rocks									
Vesicular and scoriaceous basalt and cavernous limestone and dolomite		–	Clean sandstone and fractured igneous and metamorphic rocks		–	Laminated sandstone, shale, mudstone	–	Massive igneous and metamorphic rocks	

Fig. 3.3 Hydraulic conductivities for various classes of geologic materials (after Bureau of Reclamation[9]).

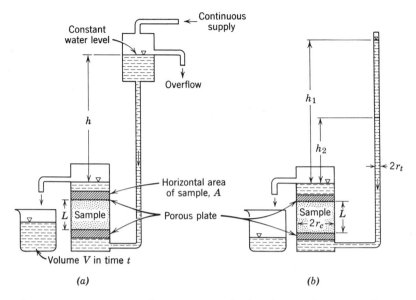

Fig. 3.4 Permeameters for measuring hydraulic conductivity of geologic samples. (*a*) Constant head. (*b*) Falling head.

Laboratory Methods. In the laboratory hydraulic conductivity can be determined by a *permeameter,* in which flow is maintained through a small sample of material while measurements of flow rate and head loss are made.[68] The constant-head and falling-head types of permeameters are simple to operate and widely employed.

The constant-head permeameter shown in Fig. 3.4*a* can measure hydraulic conductivities of consolidated or unconsolidated formations under low heads. Water enters the medium cylinder from the bottom and is collected as overflow after passing upward through the material. From Darcy's law it follows that the hydraulic conductivity can be obtained from

$$K = \frac{VL}{Ath} \tag{3.17}$$

where V is the flow volume in time t, and the other dimensions, A, L, and h, are shown in Fig. 3.4*a*. It is important that the medium be thoroughly saturated to remove entrapped air. Several different heads in a series of tests provide a reliable measurement.

A second procedure utilizes the falling-head permeameter illustrated in Fig. 3.4*b*. Here water is added to the tall tube; it flows upward through the cylindrical sample and is collected as overflow.

The test consists of measuring the rate of fall of the water level in the tube. The hydraulic conductivity can be obtained by noting that the flow rate Q in the tube

$$Q = \pi r_t^2 \, dh/dt \qquad (3.18)$$

must equal that through the sample, which by Darcy's law is

$$Q = \pi r_c^2 K \, h/L \qquad (3.19)$$

After equating and integrating,

$$K = \frac{r_t^2 L}{r_c^2 t} \ln \frac{h_1}{h_2} \qquad (3.20)$$

where L, r_t, and r_c are shown in Fig. 3.4b, and t is the time interval for the water level in the tube to fall from h_1 to h_2.

Permeameter results may bear little relation to actual field hydraulic conductivities. Undisturbed samples of unconsolidated material are difficult to obtain, while disturbed samples experience changes in porosity, packing, and grain orientation, which modify hydraulic conductivities. Then, too, one or even several samples from an aquifer may not represent the overall hydraulic conductivity of an aquifer. Variations of several orders of magnitude frequently occur for different depths and locations in an aquifer. Furthermore, directional properties of hydraulic conductivity may not be recognized.

Tracer Tests. Field determinations of hydraulic conductivity can be made by measuring the time interval for a water tracer to travel between two observation wells or test holes.[11] For the tracer a dye, such as sodium fluorescein, or a salt, such as calcium chloride, is convenient, inexpensive, easy to detect, and safe. Fig. 3.5 shows the cross section of a portion of an unconfined aquifer with groundwater flowing from hole A toward hole B. The tracer is injected as a slug in hole A after which samples of water are taken from hole B to determine the time of passage of the tracer. Because the tracer flows through the aquifer with the average interstitial velocity v_a, then

$$v_a = \frac{K}{\alpha} \frac{h}{L} \qquad (3.21)$$

where K is hydraulic conductivity, α is porosity, and h and L are shown in Fig. 3.5. But v_a also is given by

$$v_a = L/t \qquad (3.22)$$

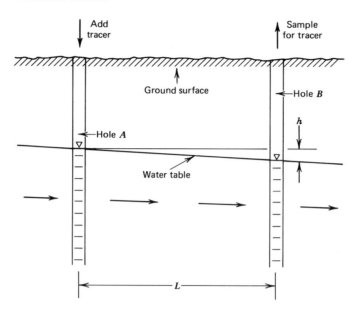

Fig. 3.5 Cross section of an unconfined aquifer illustrating a tracer test for determining hydraulic conductivity.

where *t* is the travel time interval of the tracer between the holes. Equating these and solving for K yields

$$K = \frac{\alpha L^2}{ht} \qquad (3.23)$$

Although this procedure is simple in principle, results are only approximations because of serious limitations in the field.

1. The holes need to be close together; otherwise, the travel time interval can be excessively long.

2. Unless the flow direction is accurately known, the tracer may miss the downstream hole entirely. Multiple sampling holes can help, but these add to the cost and complexity of conducting the test.

3. If the aquifer is stratified with layers having differing hydraulic conductivities, the first arrival of the tracer will result in a conductivity considerably larger than the average for the aquifer.

An alternative tracer technique, which has been successfully applied under field conditions, is the *point dilution method*.[15,24,30] Here a tracer is introduced into an observation well and thoroughly mixed with the contained water. Thereafter, as water flows into and from the well, repeated measurements of tracer concentration are made.

Analysis of the resulting dilution curve defines the groundwater velocity; this, together with the measured water table gradient and Darcy's law, yields a localized estimate of the hydraulic conductivity and also the direction of groundwater flow.[49]

Auger Hole Tests.   The auger hole method involves the measurement of the change in water level after the rapid removal of a volume of water from an unlined cylindrical hole. If the soil is loose, a screen may be necessary to maintain the hole. The method is relatively simple and is most adaptable to shallow water table conditions. The value of K obtained is essentially that for a horizontal direction in the immediate vicinity of the hole.

Fig. 3.6 illustrates an auger hole and the dimensions required for the calculation. It can be shown[6] that hydraulic conductivity is given by

$$K = \frac{C}{864} \frac{dy}{dt} \tag{3.24}$$

where dy/dt is the measured rate of rise in cm/s and the factor 864 yields K values in m/day. The factor C is a dimensionless constant listed in Table 3.2 and governed by the variables shown in Fig. 3.6.

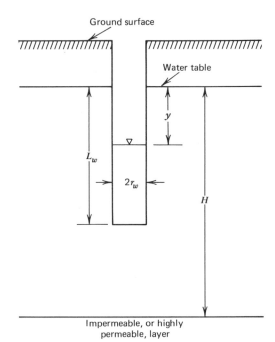

Fig. 3.6 Diagram of an auger hole and dimensions for determining hydraulic conductivity.

TABLE 3.2 Values of the Factor C for the Auger Hole Test to Determine Hydraulic Conductivity (after Boast and Kirkham[6])

L_w/r_w	y/L_w	$(H-L_w)/L_w$ for Impermeable Layer								$H-L_w$	$(H-L_w)/L_w$ for Infinitely Permeable Layer			
		0	0.05	0.1	0.2	0.5	1	2	5	∞	5	2	1	0.5
1	1	447	423	404	375	323	286	264	255	254	252	241	213	166
	0.75	469	450	434	408	360	324	303	292	291	289	278	248	198
	0.5	555	537	522	497	449	411	386	380	379	377	359	324	264
2	1	186	176	167	154	134	123	118	116	115	115	113	106	91
	0.75	196	187	180	168	149	138	133	131	131	130	128	121	106
	0.5	234	225	218	207	188	175	169	167	167	166	164	156	139
5	1	51.9	48.6	46.2	42.8	38.7	36.9	36.1		35.8		35.5	34.6	32.4
	0.75	54.8	52.0	49.9	46.8	42.8	41.0	40.2		40.0		39.6	38.6	36.3
	0.5	66.1	63.4	61.3	58.1	53.9	51.9	51.0		50.7		50.3	49.2	46.6
10	1	18.1	16.9	16.1	15.1	14.1	13.6	13.4		13.4		13.3	13.1	12.6
	0.75	19.1	18.1	17.4	16.5	15.5	15.0	14.8		14.8		14.7	14.5	14.0
	0.5	23.3	22.3	21.5	20.6	19.5	19.0	18.8		18.7		18.6	18.4	17.8
20	1	5.91	5.53	5.30	5.06	4.81	4.70	4.66		4.64		4.62	4.58	4.46
	0.75	6.27	5.94	5.73	5.50	5.25	5.15	5.10		5.08		5.07	5.02	4.89
	0.5	7.67	7.34	7.12	6.88	6.60	6.48	6.43		6.41		6.39	6.34	6.19
50	1	1.25	1.18	1.14	1.11	1.07	1.05			1.04			1.03	1.02
	0.75	1.33	1.27	1.23	1.20	1.16	1.14			1.13			1.12	1.11
	0.5	1.64	1.57	1.54	1.50	1.46	1.44			1.43			1.42	1.39
100	1	0.37	0.35	0.34	0.34	0.33	0.32			0.32			0.32	0.31
	0.75	0.40	0.38	0.37	0.36	0.35	0.35			0.35			0.34	0.34
	0.5	0.49	0.47	0.46	0.45	0.44	0.44			0.44			0.43	0.43

Note that the tabulated values cover the following conditions below the hole: a shallow impermeable layer, an infinite homogeneous stratum, and a shallow, highly permeable (gravel) layer. The value y should correspond to that when dy/dt is measured.

Several other techniques similar to the auger hole test have been developed in which water level changes are measured after an essentially instantaneous removal or addition of a volume of water. With a small-diameter pipe driven into the ground, K can be found by the piezometer, or tube, method.[65] For wells in confined aquifers, the slug method can be employed.[12, 41] Here a known volume of water is suddenly injected or removed from a well after which the decline or recovery of the water level is measured in the ensuing minutes. Where a pump is not available to conduct a pumping test on a well, the slug method serves as an alternative approach.

Pumping Tests of Wells. The most reliable method for estimating aquifer hydraulic conductivity is by pumping tests of wells. Based on observations of water levels near pumping wells, an integrated K value over a sizable aquifer section can be obtained. Then, too, because the aquifer is not disturbed, the reliability of such determinations is superior to laboratory methods. Pump test methods and computations are described in Chapter 4.

Anisotropic Aquifers

The discussion of hydraulic conductivity heretofore assumed that the geologic material was homogeneous and isotropic, implying that the value of K was the same in all directions. In fact, however, this is rarely the case, particularly for undisturbed unconsolidated alluvial materials. Instead, *anisotropy* is the rule where directional properties of hydraulic conductivity exist. In alluvium this results from two conditions. One is that individual particles are seldom spherical so that when deposited underwater they tend to rest with their flat sides down. The second is that alluvium typically consists of layers of different materials, each possessing a unique value of K. If the layers are horizontal, any single layer with a relatively low hydraulic conductivity causes vertical flow to be retarded, but horizontal flow can occur easily through any stratum of relatively high hydraulic conductivity. Thus, the typical field situation in alluvial deposits is to find a hydraulic conductivity K_x in the horizontal direction that will be greater than a value K_z in a vertical direction.

Consider an aquifer consisting of two horizontal layers, each individually isotropic, with different thicknesses and hydraulic con-

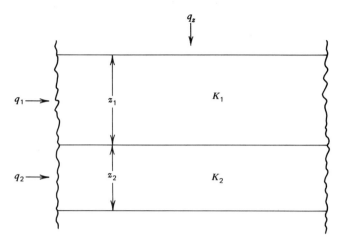

Fig. 3.7 Diagram of two horizontal strata, each isotropic, with different thicknesses and hydraulic conductivities.

ductivities, as shown in Fig. 3.7. For horizontal flow parallel to the layers, the flow q_1 in the upper layer per unit width is

$$q_1 = K_1 i z_1 \qquad (3.25)$$

where i is the hydraulic gradient and K_1 and z_1 are as indicated in Fig. 3.7. Because i must be the same in each layer for horizontal flow, if follows that the total horizontal flow q_x is

$$q_x = q_1 + q_2 = i(K_1 z_1 + K_2 z_2) \qquad (3.26)$$

For a homogeneous system this would be expressed as

$$q_x = K_x i(z_1 + z_2) \qquad (3.27)$$

where K_x is the horizontal hydraulic conductivity for the entire system. Equating these and solving for K_x yields

$$K_x = \frac{K_1 z_1 + K_2 z_2}{z_1 + z_2} \qquad (3.28)$$

which can be generalized for n layers as

$$K_x = \frac{K_1 z_1 + K_2 z_2 + \ldots + K_n z_n}{z_1 + z_2 \quad + \ldots + \quad z_n} \qquad (3.29)$$

This defines the equivalent horizontal hydraulic conductivity for a stratified material.

Now, for vertical flow through the two layers in Fig. 3.7, the flow q_z per unit horizontal area in the upper layer is

$$q_z = K_1 \frac{dh_1}{z_1} \tag{3.30}$$

where dh_1 is the head loss within the first layer. Solving for the head loss

$$dh_1 = \frac{z_1}{K_1} q_z \tag{3.31}$$

From continuity q_z must be the same for the other layer so that the total head loss

$$dh_1 + dh_2 = \left[\frac{z_1}{K_1} + \frac{z_2}{K_2} \right] q_z \tag{3.32}$$

In a homogeneous system

$$q_z = K_z \left[\frac{dh_1 + dh_2}{z_1 + z_2} \right] \tag{3.33}$$

where K_z is the vertical hydraulic conductivity for the entire system. Rearranging,

$$dh_1 + dh_2 = \left[\frac{z_1 + z_2}{K_z} \right] q_z \tag{3.34}$$

and equating with Eq. 3.32,

$$K_z = \frac{z_1 + z_2}{\dfrac{z_1}{K_1} + \dfrac{z_2}{K_2}} \tag{3.35}$$

which can be generalized for n layers as

$$K_z = \frac{z_1 + z_2 + \ldots + z_n}{\dfrac{z_1}{K_1} + \dfrac{z_2}{K_2} + \ldots + \dfrac{z_n}{K_n}} \tag{3.36}$$

This defines the equivalent vertical hydraulic conductivity for a stratified material.

As mentioned earlier, the horizontal hydraulic conductivity in alluvium is normally greater than that in the vertical direction. This observation also follows from the above derivations; thus, if

$$K_x > K_z \tag{3.37}$$

then for the two-layer case from Eqs. 3.28 and 3.35,

$$\frac{K_1 z_1 + K_2 z_2}{z_1 + z_2} > \frac{z_1 + z_2}{\dfrac{z_1}{K_1} + \dfrac{z_2}{K_2}} \tag{3.38}$$

which reduces to[42]

$$\frac{z_1}{z_2}(K_1 - K_2)^2 > 0 \tag{3.39}$$

Because the left side is always positive, it must be greater than zero, thereby confirming that

$$\frac{K_x}{K_z} \geq 1 \tag{3.40}$$

Ratios of K_x/K_z usually fall in the range of 2 to 10 for alluvium,[45] but values up to 100 or more occur where clay layers are present. For consolidated geologic materials, anisotropic conditions are governed by the orientation of strata, fractures, solution openings, or other structural conditions, which do not necessarily possess a horizontal alignment.

In applying Darcy's law to two-dimensional flow in anisotropic media, the appropriate value of K must be selected for the direction of flow. For directions other than horizontal (K_x) and vertical (K_z), the K value can be obtained from

$$\frac{1}{K_\beta} = \frac{\cos^2 \beta}{K_x} + \frac{\sin^2 \beta}{K_z} \tag{3.41}$$

where K_β is the hydraulic conductivity in the direction making an angle β with the horizontal.

Groundwater Flow Rates

From Darcy's law it follows that the rate of groundwater movement is governed by the hydraulic conductivity of an aquifer and the hydraulic gradient. To obtain an idea of the order of magnitude of natural velocities, assume a productive alluvial aquifer with $K = 75$ m/day and a hydraulic gradient $i = 10$ m/1000 m $= 0.01$. Then from Eq. 3.5

$$v = Ki = 75(0.01) = 0.75 \text{ m/day} \tag{3.42}$$

This is approximately equivalent to 0.5 mm/min, which demonstrates the sluggish nature of natural groundwater movement.

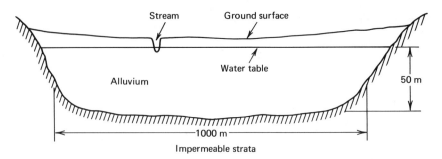

Fig. 3.8 Cross section of a typical alluvial floodplain containing an unconfined aquifer with groundwater flowing perpendicular to the section (not to scale).

If the above flow occurs within and perpendicular to the large alluvial cross section sketched in Fig. 3.8, then the total flow rate

$$Q = Av = (50)(1000)(0.75) = 37,500 \text{ m}^3/\text{day} \tag{3.43}$$

which, when converted to usual streamflow units, amounts to only 0.43 m³/s. Thus, groundwater typically can be conceived of as a massive, slow-moving body of water.

Groundwater velocities vary widely depending on local hydrogeologic conditions; values from 2 m/year to 2 m/day are normal. Usually, velocities tend to decrease with depth as porosities and permeabilities also decrease. Velocities can range from negligible* to those of turbulent streams in underground openings within basalt and limestone. Mechanisms such as wells and drains act to accelerate flows.

An illustration of one-dimensional vertical flow is shown in Fig. 3.9. Here an aquitard separates an overlying unconfined aquifer from an underlying leaky aquifer. The water table stands above the piezometric surface so that water moves vertically downward from the unconfined aquifer, through the aquitard, and into the confined aquifer. For convenience, assume steady-state conditions. Writing Darcy's law from point A to B with the dimensions indicated in Fig. 3.9,

$$v = K \frac{dh}{dl} = 10 \frac{27 - h_B}{27} \tag{3.44}$$

*All groundwater within the hydrologic cycle should be regarded as in continuous motion, although, it must be granted, some of it flows at extremely small rates.

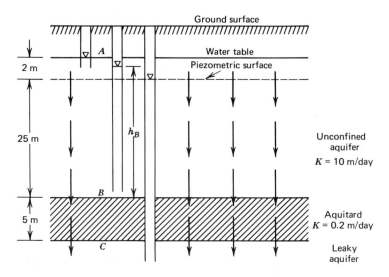

Fig. 3.9 Diagram illustrating application of Darcy's law for vertically downward flow.

and from point B to C,

$$v = K \frac{dh}{dl} = 0.2 \frac{h_B + 5 - 30}{5} \qquad (3.45)$$

Solving these yields $h_B = 26.8$ m* and $v = 0.07$ m/day.

Groundwater Flow Directions

Flow Nets. For specified boundary conditions flow lines[†] and equipotential lines can be mapped in two dimensions to form a flow net. The two sets of lines form an orthogonal pattern of small squares. In a few simplified cases, the differential equation governing flow can be solved to obtain the flow net. Generally, however, graphic solutions based on trial-and-error approximations or laboratory model studies are necessary. Suggestions for graphic construction of flow nets have been given by Cedergren;[11] model studies are described in Chapter 10.

---

*It should be noted that the piezometer open at the bottom of the unconfined aquifer (B) displays a water level below the water table (A) because of the head loss associated with vertical flow through the aquifer.

[†]A flow line is defined here as a line such that the macroscopic velocity vector is everywhere tangent to it.

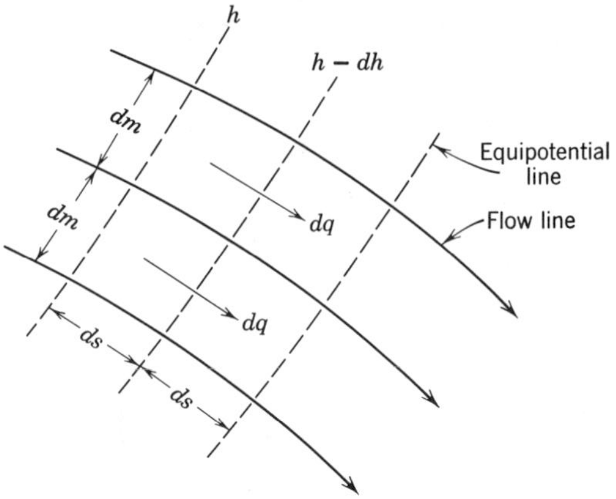

Fig. 3.10 Portion of an orthogonal flow net formed by flow and equipotential lines.

Consider the portion of a flow net shown in Fig. 3.10. The hydraulic gradient i is given by

$$i = \frac{dh}{ds} \tag{3.46}$$

and the constant flow q between two adjacent flow lines by

$$q = K \frac{dh}{ds} dm \tag{3.47}$$

for unit thickness. But for the squares of the flow net, the approximation

$$ds \cong dm \tag{3.48}$$

can be made so that Eq. 3.47 reduces to

$$q = K\, dh \tag{3.49}$$

Applying this to an entire flow net, where the total head loss h is divided into n squares between any two adjacent flow lines, then

$$dh = \frac{h}{n} \tag{3.50}$$

If the flow is divided into m channels by flow lines, then the total flow

$$Q = m\, q = \frac{Kmh}{n} \tag{3.51}$$

Thus, the geometry of the flow net, together with the hydraulic conductivity and head loss, enables the total flow in the section to be computed directly.

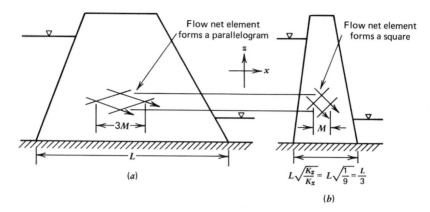

Fig. 3.11 Illustration of flow net analysis for anisotropic hydraulic conductivity in an earth dam. (a) True anisotropic section with $K_x = 9K_z$. (b) Transformed isotropic section with $K_x = K_z$.

In anisotropic media, flow lines and equipotential lines are not orthogonal except when the flow is parallel to one of the principal directions.[5] In order to calculate flows for this situation, the boundaries of a flow section must be transformed so that an isotropic medium is obtained. For the typical alluvial case of $K_x > K_z$, all horizontal dimensions are reduced by the ratio $\sqrt{K_z/K_x}$. This creates a transformed section with an isotropic medium having an equivalent hydraulic conductivity

$$K' = \sqrt{K_x K_z} \tag{3.52}$$

With this transformed section the flow net can be drawn and flow rate determined.

After the flow net has been defined, it can be converted back to the true anisotropic section by multiplying all horizontal dimensions by $\sqrt{K_x/K_z}$. Figure 3.11 illustrates the procedure for an earth dam as well as demonstrating the distortion created by anisotropy in an element of the flow net. The technique can also be extended to anisotropic two-layer systems.[42] Figure 3.12 shows contrasting flow nets for channel seepage through layered anisotropic media.

Flow in Relation to Groundwater Contours. Because no flow crosses an impermeable boundary, flow lines must parallel it. Similarly, if no flow crosses the water table of an unconfined aquifer, it becomes a bounding flow surface. The energy head h_E, or fluid

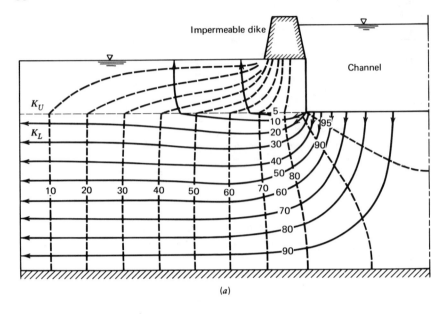

(a)

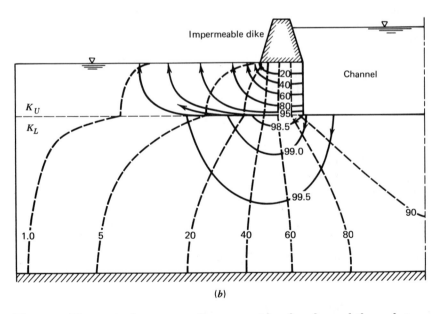

(b)

Fig. 3.12 Flow nets for seepage from one side of a channel through two different anisotropic two-layer systems. (a) $K_u/K_L = 1/50$. (b) $K_u/K_L = 50$. The anisotropy ratio for all layers is $K_x/K_z = 10$ (after Todd and Bear[58]).

potential, from Eq. 3.2 at any point on the water table can be approximated by

$$h_E = \frac{p}{\gamma} + z \qquad (3.53)$$

so that by letting the atmospheric pressure reference be zero, $p = 0$, and $h_E = z$. Therefore, under steady-state conditions, the elevation at any point on the water table equals the energy head and, as a consequence, flow lines lie perpendicular to water table contours. Similarly, flow lines within a confined aquifer are orthogonal to contours of the potentiometric surface.

With only three groundwater elevations known from wells, estimates of local groundwater contours and flow directions can be determined as demonstrated by Fig. 3.13. From field measurements of static water levels in wells within a basin, a water level contour map can be constructed. Flow lines, sketched perpendicular to contours, show directions of movement. An example appears in Fig. 3.14.

Contour maps of groundwater levels, together with flow lines, are useful data for locating new wells. Convex contours indicate regions of groundwater recharge, while concave contours are associated with groundwater discharge. Furthermore, areas of favorable hydraulic conductivity can be ascertained from the spacing of contours. The procedure can be illustrated by treating two adjacent

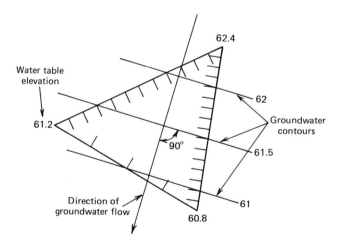

Fig. 3.13 Estimate of groundwater contours and flow direction from water table elevations in three wells.

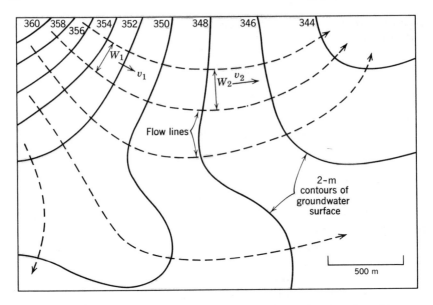

Fig. 3.14 Contour map of a groundwater surface showing flow lines.

flow lines as impermeable boundaries because there can be no flow across a flow line. If the aquifer is uniformly thick, the flow at sections 1 and 2 in Fig. 3.14 equals

$$q = W_1 v_1 = W_2 v_2 \tag{3.54}$$

where v is velocity and W is the width of the flow section perpendicular to the flow. From Darcy's law

$$W_1 K_1 i_1 = W_2 K_2 i_2 \tag{3.55}$$

which can be rewritten

$$\frac{K_1}{K_2} = \frac{W_2 i_2}{W_1 i_1} \tag{3.56}$$

where K is hydraulic conductivity and i is hydraulic gradient. The ratios W_2/W_1 and i_2/i_1 can be estimated from the water level contour map (see Fig. 3.14). For the special case of nearly parallel flow lines, Eq. 3.56 reduces to

$$\frac{K_1}{K_2} = \frac{i_2}{i_1} \tag{3.57}$$

which may be interpreted as indicating for an area of uniform groundwater flow that areas with wide contour spacings (flat gradients) possess higher hydraulic conductivities than those with narrow spacings (steep gradients). Therefore, in Fig. 3.14, prospects for a productive well are better near section 2 than 1.

Where a contour map of groundwater levels contains closed contours around a group of wells of known total discharge Q, the transmissivity of the regional aquifer can be calculated. Figure 3.15 illustrates such a situation resulting from heavy groundwater pumping in and near Savannah, Georgia. If a flow net can be constructed, Eq. 3.51 in the form

$$T = \frac{nQ}{mh} \tag{3.58}$$

can be applied where h represents the difference in elevation between any two selected closed contour lines. The typical irregularity of groundwater contours often makes construction of an accurate

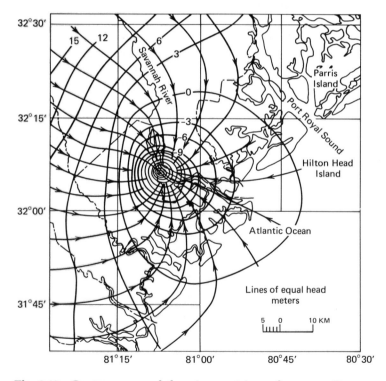

Fig. 3.15 Contour map of the piezometric surface near Savannah, Georgia, 1957, showing closed contours resulting from heavy local groundwater pumping (after USGS Water-Supply Paper 1611).

flow net difficult. As a convenient alternative involving contours but no flow net, Lohman[41] suggested the equation

$$T = \frac{Q}{(L_1 + L_2)\Delta h/\Delta r}$$ (3.59)

where L_1 and L_2 are the lengths of any two concentric closed contours, Δh is the contour interval, and Δr is the average distance between the two closed contours.

Natural permeable boundaries of aquifers include surface water bodies and the ground surface. In a surface water body the energy head is constant everywhere within the water body and equals the elevation of the water surface; consequently, aquifer flow lines must intersect normal to such a bounding surface. For the ground surface this does not apply as only atmospheric pressure exists at the ground surface. Hence, in Eq. 3.53 by letting $p = 0$, $h_E = z$, which is identical to the case for a water table boundary.

Flow Across a Water Table. As long as no flow crosses a water table, it serves as a groundwater boundary; however, if flow, such as percolating water, reaches the water table, flow lines no longer parallel the surface as an impermeable boundary.[29] To illustrate this refraction effect let v_u represent the unsaturated vertical velocity approaching the water table and v_s the saturated velocity below the water table (Fig. 3.16). The head loss dh for flow along the left flow line below the water table occurs in a distance of $b_s \tan(\delta + \varepsilon)$, as defined in Fig. 3.16. Thus

$$v_s = Ki = K\frac{dh}{b_s \tan(\delta + \varepsilon)}$$ (3.60)

but

$$dh = b_u \tan \delta$$ (3.61)

hence

$$v_s = K\frac{v_s \tan \delta}{b_s \tan(\delta + \varepsilon)}$$ (3.62)

From continuity

$$\frac{b_u}{b_s} = \frac{v_s}{v_u}$$ (3.63)

where b_u and b_s are as shown in Fig. 3.16, so that

$$v_s = K\frac{v_s \tan \delta}{v_u \tan(\delta + \varepsilon)}$$ (3.64)

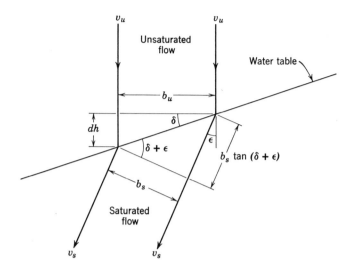

Fig. 3.16 Refraction of flow lines across a water table (after Jacob[33]).

which, when solved for ε, gives

$$\varepsilon = \tan^{-1}\left(\frac{K}{v_u}\tan\delta\right) - \delta \qquad (3.65)$$

This states that for $v_u > 0$, flow lines will form an angle of (90 degrees $- \delta - \varepsilon$) below the water table. For the case of no percolating flow, $v_u = 0$ and $\varepsilon = 90$ degrees $- \delta$, so that v_s parallels the water table.

Flow Across a Hydraulic Conductivity Boundary. Similar to the above analysis, where flow passes from a region of hydraulic conductivity K_1 to one of K_2, a change in flow direction results. The change of direction can be derived from continuity considerations and expressed in terms of the two K values. Visualizing a flow field as shown in Fig. 3.17, it is clear that the normal components of flow approaching and leaving the boundary must be equal; hence, the normal velocities v_n must be such that

$$v_{n_1} = v_{n_2} \qquad (3.66)$$

or

$$K_1 \frac{dh_1}{dL_1}\cos\theta_1 = K_2 \frac{dh_2}{dL_2}\cos\theta_2 \qquad (3.67)$$

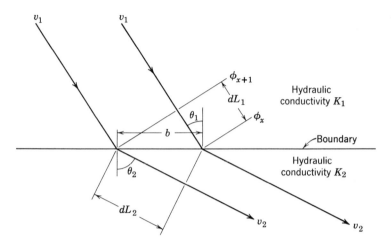

Fig. 3.17 Refraction of flow lines across a boundary between media of different hydraulic conductivities.

where θ_1 and θ_2 are angles with the normal shown in Fig. 3.17. Also, the distance b along the boundary between two adjacent flow lines must be the same on each side of the boundary. From Fig. 3.17 the distance b can be given as

$$b = \frac{dL_1}{\sin \theta_1} = \frac{dL_2}{\sin \theta_2} \tag{3.68}$$

which can be arranged

$$dL_1 \sin \theta_2 = dL_2 \sin \theta_1 \tag{3.69}$$

Dividing this equation by Eq. 3.67, and noting that $dh_1 = dh_2$ between two equipotential lines, gives

$$\frac{K_1}{K_2} = \frac{\tan \theta_1}{\tan \theta_2} \tag{3.70}$$

Thus, for saturated flow passing from a medium of one hydraulic conductivity to that of another, a refraction in flow lines occurs such that the ratio of the K's equals the ratio of the tangents of the angles the flow lines make with the normal to the boundary. Consequences of the relation are illustrated in Fig. 3.18.

   Regional Flow Patterns. Although most groundwater movement in shallow aquifers tends to be nearly horizontal, regional flow patterns can become quite complex. Reasons for this stem from the

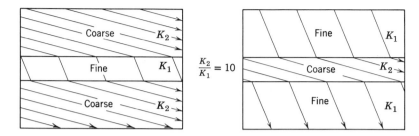

Fig. 3.18 Refraction across layers of coarse and fine sand with a
hydraulic conductivity ratio of 10 (after Hubbert[28]; copyright © 1940
by the University of Chicago Press).

diversity of field parameters: areas and magnitudes of recharge and
discharge, topography, stratigraphy, and anisotropy.[51] Analytic solutions for specified aquifer cross sections by Toth[60,61] demonstrated
that the variability of a water table could produce a variety of flow
patterns. Subsequent work by Freeze[19] extended this approach to
other subsurface boundary conditions. From these contributions it
is clear that accurate evaluation of groundwater flows is contingent
on a detailed knowledge of hydrogeologic conditions.

An illustration of a regional groundwater flow pattern is shown
in Fig. 3.19 for a vertical cross section through an unconfined aquifer.
Flow lines were obtained by computer assuming:

1. A homogeneous and isotropic aquifer with impermeable boundaries
along the sides and the bottom.

2. A rectangular flow region 6100 m wide by 3050 m deep.

3. A sinusoidal potential distribution (equivalent to the water table) with
an amplitude of 15 m and a mean slope of 2 percent on the surface of the
theoretical flow region.

Local, intermediate, and regional systems of flow are indicated in
Fig. 3.19 as well as flows counter to the mean water table gradient.
Although the aquifer portrayed is highly idealized, a similar multi-
formity in flow pattern can be anticipated for actual aquifers where
irregularities of topography, stratigraphy, and anisotropy are introduced.

Dispersion

Concept. In saturated flow through porous media, velocities vary
widely across any single pore, just as in a capillary tube where the
velocity distribution in laminar flow is parabolic. In addition, the
pores possess different sizes, shapes, and orientations. As a result,

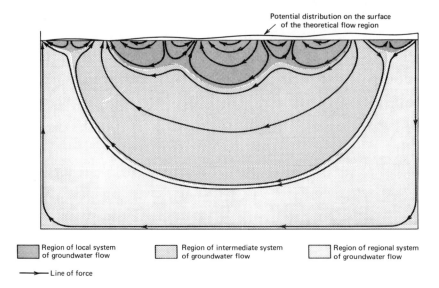

Potential distribution on the surface
of the theoretical flow region

Region of local system   Region of intermediate system   Region of regional system
of groundwater flow   of groundwater flow   of groundwater flow

Line of force

Fig. 3.19 Vertical cross section showing theoretical flow pattern of groundwater through an aquifer with a sloping sinusoidal water table (after Toth[61]).

when a labeled miscible liquid, referred to as a *tracer*, is introduced into a flow system, it spreads gradually to occupy an increasing portion of the flow region. This phenomenon is known as *dispersion* and constitutes a nonsteady, irreversible mixing process by which the tracer disperses within the surrounding water.[48]

In a column packed with sand as in Fig. 3.20a and supplied continuously after a time t_o with water containing a tracer of concentration c_o, dispersion in the longitudinal direction of flow can be measured.[23,52] From samples of water emerging from the column, a tracer concentration c is found. In Fig. 3.20b the solid line shows a typical S-shaped dispersion curve. Besides longitudinal dispersion, lateral dispersion also occurs because water is continually dividing and reuniting as it follows tortuous flow paths around grains of a medium, as illustrated in Fig. 3.21.[14]

The equation for dispersion in homogeneous and isotropic media for the two-dimensional case has the form

$$\frac{\delta c}{\delta t} = D_L \frac{\delta^2 c}{\delta x^2} + D_T \frac{\delta^2 c}{\delta y^2} - v \frac{\delta c}{\delta x} \tag{3.71}$$

where c is the relative tracer concentration ($0 < C < 1$), D_L and D_T are longitudinal and transverse dispersion coefficients, v is fluid

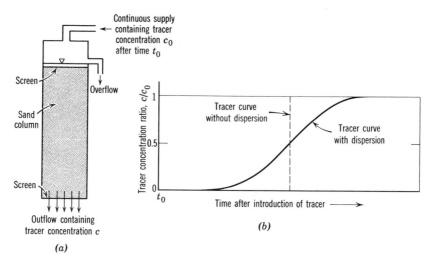

Fig. 3.20 Longitudinal dispersion of a tracer passing through a sand column. (*a*) Sand column. (*b*) Dispersion curve.

velocity, x is the coordinate in the direction of flow, y is the coordinate normal to flow, and t is time. Dimensions of the dispersion coefficients are L^2/T.

Dispersion is essentially a microscopic phenomenon caused by a combination of molecular diffusion and hydrodynamic mixing occurring with laminar flow through porous media. The net result produces a conic form downstream from a continuous point-source

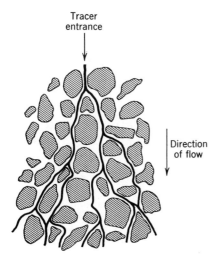

Fig. 3.21 Lateral dispersion of a tracer originating from a point source in a porous medium.

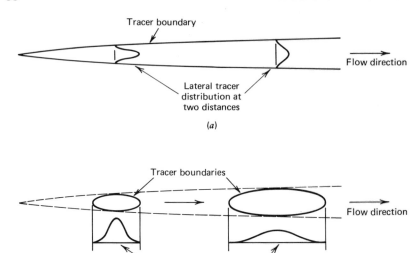

Fig. 3.22 Sketches of tracer distribution resulting from dispersion by flow in porous media. (*a*) Continuous tracer. (*b*) Single slug of tracer.

tracer (Fig. 3.22*a*), and an expanding ellipsoid for a single tracer injection (Fig. 3.22*b*).* In recent years a series of theoretical and experimental studies have provided a better understanding of the phenomenon.[3,16] Most mathematical descriptions of dispersion are based on statistical concepts because of the difficulties in defining the microstructure of porous media as well as the relative roles of molecular diffusion and mechanical dispersion.[40,59] Experimental studies have attempted to establish relations between factors such as velocity, media structural properties, and permeability with dispersion coefficients.[26,27,62] The pattern of a point tracer as it moves downstream from its source tends to a normal (Gaussian) distribution both longitudinally and transversely, as sketched in Fig. 3.22*b*. Furthermore, the longitudinal component is larger than that of the transverse so that the major axis of mixing occurs in the direction of flow.

*One may visualize these two forms, by analogy, as a continuous plume and as a single puff of smoke drifting downwind from a smokestack, respectively, even though the flow in porous media is laminar whereas that in the atmosphere is turbulent.

Dispersion and Groundwater Hydrology. In groundwater hydrology dispersion may be encountered whenever two fluids with different characteristics come into contact. Prime examples of this include tracers for evaluating directions and velocities of groundwater flow, introduction of pollutants into the ground (Chapter 8), artificial recharge of water with one quality into an aquifer containing groundwater of another quality (Chapter 13), and intrusion of saline water into freshwater aquifers (Chapter 14). In general, the magnitude of dispersion for uniform sands can be measured in terms of only a few meters over a travel distance of 10^3 meters.

In addition, the heterogeneity of most geologic materials introduces irregularities of flow with the consequent mixing of a tracer; the effects often far overshadow the effects of microscopic dispersion. The rapid nonsymmetric dispersal of two continuous dye streams in the laboratory demonstration shown in Fig. 3.23 amply illustrates the important role of heterogeneity. In essense, a macroscopic dispersion is superimposed on the flow system. Furthermore, the irregular geometries of groundwater recharge and discharge zones, together with the lack of specific data on aquifer characteristics, preclude quantitative evaluation of dispersion coefficients for most natural groundwater flow situations.

Groundwater Tracers

A variety of tracers have been employed for studying dispersion and also for evaluating directions and rates of groundwater flow under field conditions. An ideal tracer should (1) be susceptible to quantitative determination in minute concentrations, (2) be absent or nearly so from the natural water, (3) not react chemically with the natural water or be absorbed by the porous media, (4) be safe in terms of human health, and (5) be inexpensive and readily available. No tracer completely meets all these requirements, but a reasonably satisfactory tracer can be selected to fit the needs of a particular situation. Possibilities include water-soluble dyes (such as sodium fluorescein), which can be detected by colorimetry; soluble chloride and sulfate salts and sugars, which can be detected chemically; and strong electrolytes, which can be detected by electrical conductivity.

Radioisotopes such as H^3 (tritium), Co^{60}, Rb^{86}, I^{131} have also served as tracers.[22,34,35,57,69,70] Naturally occurring radioisotopes, H^3 and C^{14}, not only have enabled residence times of groundwater to be estimated (see Chapter 2) but also have provided a means for evaluating regional groundwater flow.[4,25,46,66]

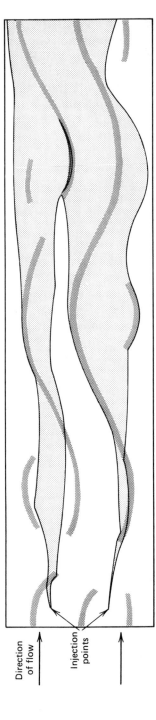

Fig. 3.23 Dispersion of two dye streams in a heterogeneous porous medium. Dotted areas indicate bands of much higher permeability (after Skibitzke and Robinson[53]).

General Flow Equations

Rectangular Coordinates. In general form Darcy's law may be written

$$v = -K\frac{\partial h}{\partial s} \tag{3.72}$$

where v, K, and h are as previously defined, and s is distance along the average direction of flow. For horizontal flow this equation may be generalized by considering the flow through the square element shown in Fig. 3.24. The inflow and outflow components in the x direction can be stated as

$$q_{x,i} = -T_x W\left(\frac{\partial h}{\partial x}\right)_i \quad \text{and} \quad q_{x,o} = -T_x W\left(\frac{\partial h}{\partial h}\right)_o \tag{3.73}$$

where T_x is the transmissivity in the x direction, W is the length of a side of the square, $(\partial h/\partial x)_i$ and $(\partial h/\partial x)_o$ define the hydraulic gradient at the entry and exit faces of the element, respectively. Similar equations can be written for flow in the y direction. The flow rate stored or released in the element as a result of these flows, by continuity, equals

$$(q_{x,i} - q_{x,o}) + (q_{y,i} - q_{y,o}) = -SW^2\frac{\partial h}{\partial t} \tag{3.74}$$

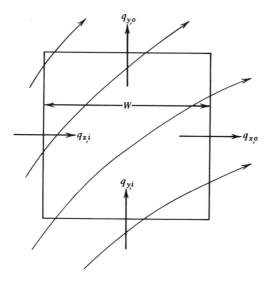

Fig. 3.24 Diagram of horizontal flow through a square element of an aquifer.

where S is the storage coefficient. It follows that

$$-T_x \frac{(\partial h/\partial x)_i - (\partial h/\partial x)_o}{W} - T_y \frac{(\partial h/\partial y)_i - (\partial h/\partial y)_o}{W} = -S \frac{\partial h}{\partial t} \quad (3.75)$$

If the value of W becomes infinitesimally small, the derivatives on the left side become the second derivatives of h, so

$$T_x \frac{\partial^2 h}{\partial x^2} + T_y \frac{\partial^2 h}{\partial y^2} = S \frac{\partial h}{\partial t} \quad (3.76)$$

This is the general partial differential equation for unsteady flow of groundwater in the horizontal direction.

For three dimensions, employing an elemental cube rather than a square, it can be shown that

$$K_x \frac{\partial^2 h}{\partial x^2} + K_y \frac{\partial^2 h}{\partial y^2} + K_z \frac{\partial^2 h}{\partial z^2} = S_s \frac{\partial h}{\partial t} \quad (3.77)$$

where S_s is the *specific storage*, defined as the volume of water a unit volume of saturated aquifer releases from storage for a unit decline in hydraulic head.*

If the flow is steady, $\partial h/\partial t = 0$; therefore,

$$K_x \frac{\partial^2 h}{\partial x^2} + K_y \frac{\partial^2 h}{\partial y^2} + K_z \frac{\partial^2 h}{\partial z^2} = 0 \quad (3.78)$$

and for homogeneous and isotropic aquifers, Eq. 3.78 reduces to

$$\frac{\partial^2 h}{\partial x^2} + \frac{\partial^2 h}{\partial y^2} + \frac{\partial^2 h}{\partial z^2} = 0 \quad (3.79)$$

which is the Laplace equation for potential flow.[28, 29]

Radial Coordinates. For axisymmetric groundwater flow to wells, radial coordinates are preferable. In a homogeneous and isotropic aquifer, it can be shown that Eq. 3.76 is equivalent to

$$\frac{\partial^2 h}{\partial r^2} + \frac{1}{r} \frac{\partial h}{\partial r} = \frac{S}{T} \frac{\partial h}{\partial t} \quad (3.80)$$

where r is the radial coordinate from the well. And for steady flow this reduces to

$$\frac{\partial^2 h}{\partial r^2} + \frac{1}{r} \frac{\partial h}{\partial r} = 0 \quad (3.81)$$

*Note for the particular case of a horizontal confined aquifer of thickness b, $S = S_s b$; therefore, for this condition Eq. 3.77 reduces to Eq. 3.76.

Application to Aquifers. The equations derived above will be applied subsequently to obtain analytic solutions to particular groundwater flow problems. For solution of any problem, idealization of the aquifer and of the boundary conditions of the flow system is necessary. Results may only approximate field conditions; nevertheless, known deviations from assumptions frequently allow analytic solutions to be modified to obtain an answer that otherwise would not have been possible. A common assumption regarding the aquifer is that it is homogeneous and isotropic (see Chapter 2). Often aquifers can be assumed to be infinite in areal extent; if not, boundaries are assumed to be (1) impermeable, such as underlying or overlying rock or clay layers, dikes, faults, or valley walls, or (2) permeable, including surface water bodies in contact with the aquifer, ground surfaces where water emerges from underground, and wells.

Unsaturated Flow

In groundwater hydrology, unsaturated flow is important for downward vertical flow (natural and artificial recharge), upward vertical flow (evaporation and transpiration), movement of pollutants from ground surface, and horizontal flow in the capillary zone above the water table.[54] A large amount of literature exists on unsaturated flow, most of it contributed by soil scientists. Significant summaries of the subject are available in several references.[36,56,65]

A typical distribution of water content above a water table is sketched in Fig. 3.25a. If the water table is lowered Δh, then the capillary zone shifts downward. The crosshatched area serves as a measure of the volume of water drained from above the water table. This water is released by vertical percolation; consequently, specific yield becomes an asymptotic function of time (Fig. 3.25b).

Unsaturated Hydraulic Conductivity. Unsaturated flow in the zone of aeration can be analyzed by Darcy's law; however, the unsaturated hydraulic conductivity K_u is a function of the water content as well as the negative pressure head (tension).[2,8,20,63] Because part of the pore space is filled with air, the available cross-sectional area available for water flow is reduced; consequently, K_u is always less than the saturated value K.

Although there are hysteresis effects present in the relations of K_u with water content and negative pressure, approximations based on empirical evidence can be stated. Water content data fit the form[31]

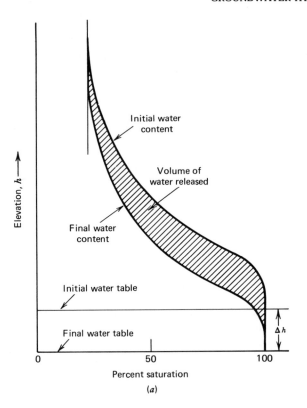

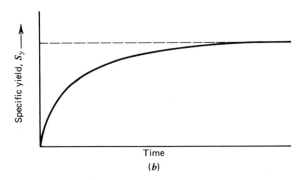

Fig. 3.25 Movement of water in the zone of aeration by lowering the water table. (a) Water content above the water table. (b) Specific yield as a function of the time of drainage.

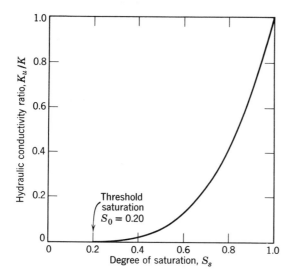

Fig. 3.26 Ratio of unsaturated to saturated hydraulic conductivity as a function of saturation (after Irmay[31]).

$$\frac{K_u}{K} = \left(\frac{S_s - S_o}{1 - S_o}\right)^3 \qquad (3.82)$$

where S_s is the degree of saturation and S_o is the threshold saturation—that part of the voids filled with nonmoving water held primarily by capillary forces. Eq. 3.82 is plotted in Fig. 3.26; note that K_u ranges from zero at $S_s = S_o$ to $K_u = K$ at $S_s = 1$, which is saturation.

For hydraulic conductivity and negative pressure, S-shaped relations as indicated in Fig. 3.27 are generally applicable.[64,65] These can be approximated by a step function or by

$$\frac{K_u}{K} = \frac{a}{\frac{a}{b}(-h)^n + a} \qquad (3.83)$$

where a, b, and n are constants that vary with particle sizes of unconsolidated material and h is the pressure head measured in centimeters. It can be seen that when $h = 0$, which occurs at atmospheric pressure, $K_u = K$. Orders of magnitude of the constants in Eq. 3.83 for different soils are as follows:[7]

Material	a	b	n
Medium sands	5×10^9	10^7	5
Fine sands, sandy loams	5×10^6	10^5	3
Loams and clays	5×10^3	5×10^3	2

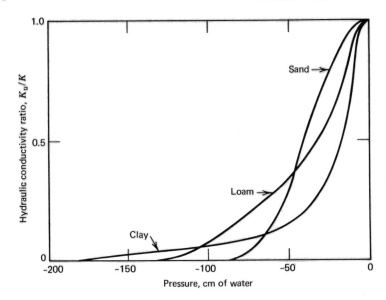

Fig. 3.27   Median relationships between hydraulic conductivity and soil-water pressure (tension) (after Bouwer[7]).

Vertical and Horizontal Flows.   Illustrative of the vertical drainage of water in unsaturated soils are the data shown in Fig. 3.28. An initially saturated sand 180 cm in depth was allowed to drain for 1 day and was then subjected to infiltration for 4 min under a surface head of 3 cm. The resulting depth profiles of water content and pressure head are indicated for various time intervals after the end of the infiltration interval. Note that at the start of the subsurface redistribution of water ($t = 0$), saturation (water content = 0.29 cm^3/cm^3) extends to a depth of about 30 cm and corresponds to a linear pressure gradient. For field conditions where stratifications in soil texture can be expected, the complexity of such downward water migration, as from rainfall or irrigation, can be appreciable.[21, 65]

Lateral flow occurs above the water table in the capillary zone. The flow rate decreases with the degree of saturation, and hence the hydraulic conductivity (see Fig. 3.26). The fraction of flow above the water table can be calculated from an equivalent saturated thickness.[44] For aquifers of substantial depth, the flow component above the water table is negligible, but it can be significant in shallow unconfined aquifers.

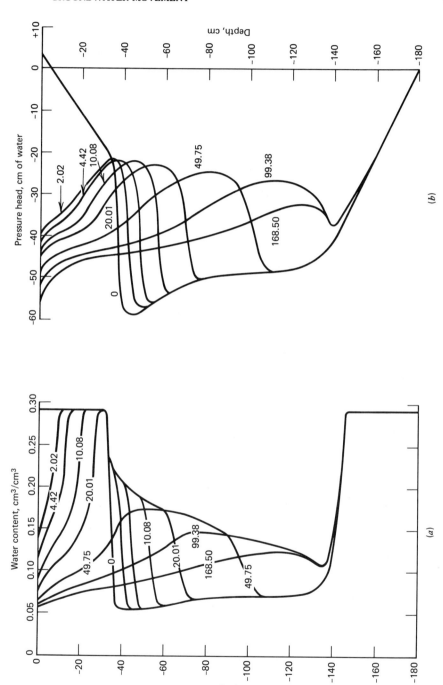

Fig. 3.28 Profiles of water content and pressure head in a sand column following infiltration of water. Numbers on the curves represent time in minutes after start of infiltration. (a) Water content. (b) Pressure head. Data are computer-based numerical results (after Watson in van Schilfgaarde[65]; reproduced from *Drainage for Agriculture*, ASA Monograph No. 17, pp. 368–369, 1974 by permission of the American Society of Agronomy).

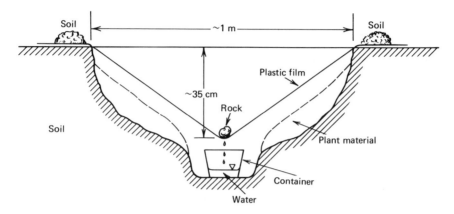

Fig. 3.29 Diagram of a survival still for producing water from soil moisture under desert conditions (after Jackson and van Bavel[32]; copyright © 1965 by the American Association for the Advancement of Science).

Application for Desert Survival. A unique and extremely practical application of unsaturated flow for survival under desert conditions has been demonstrated by Jackson and van Bavel.[32] Water is obtained from the soil-water zone by a simple distillation technique. A small hemispheric hole dug in the soil is covered with a plastic film and held in a conical shape by a rock placed in the center (see Fig. 3.29). Solar energy passes through the plastic and is absorbed by the soil, resulting in evaporation of soil moisture and followed by condensation on the cooler plastic. The condensed water runs down to the point of the cone where it drains into a container for collection and use. Yields of about 1.0 l/day from a single survival still can be obtained under desert conditions, while adding cut plant material (especially cactus) as indicated in Fig. 3.29 can increase yields to near 1.5 l/day.

References

1. Ahmed, N., and D. K. Sunada, Nonlinear flow in porous media, *Jour. Hydraulics Div.*, Amer. Soc. Civil Engrs., v. 95, no. HY6, pp. 1847–1857, 1969.
2. Amer. Soc. Testing Matls., Permeability and capillarity of soils, *ASTM Spl. Tech. Publ. no. 417*, Philadelphia, 210 pp., 1967.
3. Bachmat, Y., and J. Bear, The general equations of hydrodynamic dispersion in homogeneous, isotropic porous mediums, *Jour. Geophysical Research*, v. 69, pp. 2561–2567, 1964.

4. Back, W., et al., Carbon-14 ages related to ocurrence of salt water, *Jour. Hydraulics Div.*, Amer. Soc. Civil Engrs., v. 96, no. HY11, pp. 2325–2336, 1970.
5. Bear, J., and G. Dagan, The relationship between solutions of flow problems in isotropic and anisotropic soils, *Jour. Hydrology*, v. 3, pp. 88–96, 1965.
6. Boast, C. W., and D. Kirkham, Auger hole seepage theory, *Soil Sci. Soc. Amer. Proc.*, v. 35, pp. 365–373, 1971.
7. Bouwer, H., Unsaturated flow in ground-water hydraulics, *Jour. Hydraulics Div.*, Amer. Soc. Civil Engrs., v. 90, no. HY5, pp. 121–144, 1964.
8. Brooks, R. H., and A. T. Corey, Properties of porous media affecting fluid flow, *Jour. Irrig. Drain. Div.*, Amer. Soc. Civil Engrs., v. 92, no. IR2, pp. 61–88, 1966.
9. Bureau of Reclamation, *Ground water manual*, U.S. Dept. Interior, 480 pp., 1977.
10. Castillo, E., et al., Unconfined flow through jointed rock, *Water Resources Bull.*, v. 8, pp. 266–281, 1972.
11. Cedergren, H. R., *Seepage, drainage, and flow nets*, 2nd ed., John Wiley & Sons, New York, 534 pp., 1977.
12. Cooper, H. H., Jr., et al., Response of a finite-diameter well to an instantaneous charge of water, *Water Resources Research*, v. 3, pp. 263–269, 1967.
13. Darcy, H., *Les fontaines publiques de la ville de Dijon*, V. Dalmont, Paris, 647 pp., 1856.
14. de Josselin de Jong, G., Longitudinal and transverse diffusion in granular deposits, *Trans. Amer. Geophysical Union*, v. 39, pp. 67–74, 1958.
15. Drost, W., et al., Point dilution methods of investigating ground water flow by means of radioisotopes, *Water Resources Research*, v. 4, pp. 125–146, 1968.
16. Eldor, M., and G. Dagan, Solutions of hydrodynamic dispersion in porous media, *Water Resources Research*, v. 8, pp. 1316–1331, 1972.
17. Fair, G. M., and L. P. Hatch, Fundamental factors governing the streamline flow of water through sand, *Jour. Amer. Water Works Assoc.*, v. 25, pp. 1551–1565, 1933.
18. Fancher, G., Henry Darcy—engineer and benefactor of mankind, *Jour. Petr. Tech.*, v. 8, pp. 12–14, Oct. 1956.
19. Freeze, R. A., *Theoretical analysis of regional groundwater flow*, Scientific Ser. no. 3, Inland Waters Branch, Dept. Energy, Mines and Resources, Ottawa, Canada, 202 pp., 1969.
20. Freeze, R. A., Three-dimensional, transient, saturated-unsaturated flow in a groundwater basin, *Water Resources Research*, v. 7, pp. 347–366, 1971.
21. Freeze, R. A., and J. Banner, The mechanism of natural ground-water recharge and discharge, *Water Resources Research*, v. 5, pp. 153–171, 1969; vol. 6, pp. 138–155, 1970.

22. Gaspar, E., and M. Oncescu, *Radioactive tracers in hydrology*, Elsevier, Amsterdam, 342 pp., 1972.

23. Gelhar, L. W., and M. A. Collins, General analysis of longitudinal dispersion in nonuniform flow, *Water Resources Research*, v. 7, pp. 1511–1521, 1971.

24. Halevy, E., et al., Borehole dilution techniques: A critical review, *Isotopes in Hydrology*, Intl. Atomic Energy Agency, Vienna, pp. 531–564, 1967.

25. Hanshaw, B. B., et al., Carbonate equilibria and radiocarbon distribution related to groundwater flow in the Floridan limestone aquifer, U.S.A., *Intl. Assoc. Sci. Hydrology Publ.* 74, pp. 601–614, 1965.

26. Harleman, D. R. F., et al., Dispersion-permeability correlation in porous media, *Jour. Hydraulics Div.*, Amer. Soc. Civil Engrs., v. 89, no. HY2, pp. 67–85, 1963.

27. Hoopes, J. A., and D. R. F. Harleman, Dispersion in radial flow from a recharge well, *Jour. Geophysical Research*, v. 72, pp. 3595–3607, 1967.

28. Hubbert, M. K., The theory of ground-water motion, *Jour. Geol.*, v. 48, pp. 785–944, 1940.

29. Hubbert, M. K., Darcy's law and the field equations of the flow of underground fluids, *Trans. Amer. Inst. Min. and Metal. Engrs.*, v. 207, pp. 222–239, 1956.

30. Intl. Assoc. Hydraulic Research, *Fundamentals of transport phenomena in porous media*, Elsevier, Amsterdam, 392 pp., 1972.

31. Irmay, S., On the hydraulic conductivity of unsaturated soils, *Trans. Amer. Geophysical Union*, v. 35, pp. 463–467, 1954.

32. Jackson, R. D., and van Bavel, C. H. M., Solar distillation of water from soil and plant materials: a simple desert survival technique, *Science*, v. 149, pp. 1377–1379, 1965.

33. Jacob, C. E., Flow of ground water, *in Engineering hydraulics* (H. Rouse, ed.), John Wiley and Sons, New York, pp. 321–386, 1950.

34. Kaufman, W. J., and D. K. Todd, Application of tritium tracer to canal seepage measurements, *Tritium in the Physical and Biological Sciences*, Intl. Atomic Energy Agency, Vienna, pp. 83–94, 1962.

35. Keeley, J. W., and M. R. Scalf, Aquifer storage determination by radiotracer techniques, *Ground Water*, v. 7, pp. 17–22, 1969.

36. Kirkham, D., and W. L. Powers, *Advanced soil physics*, John Wiley & Sons, New York, 534 pp., 1972.

37. Krumbein, W. C., and G. D. Monk, Permeability as a function of the size parameters of unconsolidated sand, *Trans. Amer. Inst. Min. and Met. Engrs.*, v. 151, p. 153–163, 1943.

38. LeGrand, H. E., and V. T. Stringfield, Development and distribution of permeability in carbonate aquifers, *Water Resources Research*, v. 7, pp. 1284–1294, 1971.

39. Lewis, D. C., et al., Tracer dilution sampling technique to determine

hydraulic conductivity of fractured rock, *Water Resources Research,* v. 2, pp. 533–542, 1966.

40. Li, W. H., and G. T. Yeh, Dispersion at the interface of miscible liquids in a soil, *Water Resources Research,* v. 4, pp. 369–378, 1968.

41. Lohman, S. W., Ground-water hydraulics, *U.S. Geological Survey Prof. Paper* 708, 70 pp., 1972.

42. Luthin, J. N. (ed.), *Drainage of agricultural lands,* Agronomy Monograph no. 7, Amer. Soc. Agronomy, Madison, Wis., 620 pp., 1957.

43. Masch, F. D., and K. J. Denny, Grain size distribution and its effect on the permeability of unconsolidated sands, *Water Resources Research,* v. 2, pp. 665–677, 1966.

44. Mobasheri, F., and M. Shahbazi, Steady-state lateral movement of water through the unsaturated zone of an unconfined aquifer, *Ground Water,* v. 7, no. 6, pp. 28–34, 1969.

45. Morris, D. A., and A. I. Johnson, Summary of hydrologic and physical properties of rock and soil materials, as analyzed by the Hydrologic Laboratory of the U.S. Geological Survey 1948–60, *U.S. Geological Survey Water-Supply Paper* 1839-D, 42 pp., 1967.

46. Pearson, F. J., Jr., and D. E. White, Carbon 14 ages and flow rates of water in Carrizo Sand, Atascosa County, Texas, *Water Resources Research,* v. 3, pp. 251–261, 1967.

47. Rasmussen, W. C., Permeability and storage of heterogeneous aquifers in the United States, *Intl. Assoc. Sci. Hydrology Publ.* 64, pp. 317–325, 1964.

48. Rumer, R. R., Longitudinal dispersion in steady and unsteady flow, *Jour. Hydraulics Div.,* Amer. Soc. Civil Engrs., v. 88, no. HY4, pp. 147–172, 1962.

49. Rumer, R. R., and P. A. Drinker, Resistance to laminer flow through porous media, *Jour. Hydraulics Div.,* Amer. Soc. Civil Engrs., v. 92, no. HY5, pp. 155–163, 1966.

50. Saleem, M., An inexpensive method of determining the direction of natural flow of ground-water, *Jour. Hydrology,* v. 9, pp. 73–89, 1969.

51. Shahbazi, M., et al., Effect of topography on ground water flow, *Intl. Assoc. Sci. Hydrology Publ.* 77, pp. 314–319, 1968.

52. Shamir, U. Y., and D. R. F. Harleman, Numerical solutions for dispersion in porous mediums, *Water Resources Research,* v. 3, pp. 557–581, 1967.

53. Skibitzke, H. E., and G. M. Robinson, Dispersion in ground water flowing through heterogeneous materials, *U.S. Geological Survey Prof. Paper* 386-B, 3 pp., 1963.

54. Smith, W. O., Infiltration in sands and its relation to groundwater recharge, *Water Resources Research,* v. 3, pp. 539–555, 1967.

55. Smith, W. O., and A. N. Sayre, Turbulence in ground-water flow, *U.S. Geological Survey Prof. Paper* 402-E, 9 pp., 1964.

56. Stallman, R. W., Flow in the zone of aeration, *in* Chow, V. T. (ed.), *Advances in Hydroscience*, v. 4, pp. 151–195, Academic Press, New York, 1967.
57. Stout, G. E. (ed.), *Isotope techniques in the hydrologic cycle*, Geophysical Monograph Ser. no. 11, Amer. Geophysical Union, 199 pp., 1967.
58. Todd, D. K., and J. Bear, Seepage through layered anisotropic porous media, *Jour. Hydraulics Div.*, Amer. Soc. Civil Engrs., v. 87, no. HY3, pp. 31–57, 1961.
59. Todorovic, P., A stochastic model of longitudinal diffusion in porous media, *Water Resources Research*, v. 6, pp. 211–222, 1970.
60. Toth, J., A theory of groundwater motion in small drainage basins in Central Alberta, Canada, *Jour. Geophysical Research*, v. 67, pp. 4375–4387, 1962.
61. Toth, J., A theoretical analysis of groundwater flow in small drainage basins, *Jour. Geophysical Research*, v. 68, pp. 4795–4812, 1963.
62. U.S. Geological Survey, Fluid movement in earth materials, *Prof. Paper* 411, Chaps. A to I, 1961–1970.
63. Vachaud, G., Determination of the hydraulic conductivity of unsaturated soils from an analysis of transient flow data, *Water Resources Research*, v. 3, pp. 697–705, 1967.
64. Vachaud, G., and J. L. Thony, Hysteresis during infiltration and redistribution in a soil column at different water contents, *Water Resources Research*, v. 7, pp. 111–127, 1971.
65. van Schilfgaarde, J. (ed.), *Drainage for agriculture*, Agronomy Monograph no. 17, Amer. Soc. Agronomy, Madison, Wis., 700 pp., 1974.
66. Vogel, J. C., Carbon-14 dating of groundwater, *Isotope Hydrology* 1970, Intl. Atomic Energy Agency, Vienna, pp. 225–239, 1970.
67. Ward, J., Turbulent flow in porous media, *Jour. Hydraulics Div.*, Amer. Soc. Civil Engrs., v. 90, no HY5, pp. 1–12, 1964.
68. Wenzel, L. K., Methods for determining permeability of water-bearing materials with special reference to discharging-well methods, *U.S. Geological Survey Water-Supply Paper* 887, 192 pp., 1942.
69. Wiebenga, W. A., et al., Radioisotopes as groundwater tracers, *Jour. Geophysical Research*, v. 72, pp. 4081–4091, 1967.
70. Working Group on Nuclear Techniques in Hydrology, *Guidebook on nuclear techniques in hydrology*, Tech. Rept. Ser. no. 91, Intl. Atomic Energy Agency, Vienna, 214 pp., 1968.

Groundwater
and Well
Hydraulics

Darcy's law and the fundamental equations governing groundwater movement can now be applied to particular situations. Solutions of groundwater flow to wells rank highest in importance. By pumping tests of wells, storage coefficients and transmissivities of aquifers can be determined; furthermore, with these aquifer characteristics known, future declines in groundwater levels associated with pumpage can be calculated. Well flow equations have been developed for steady and unsteady flows, for various types of aquifers, and for several special boundary conditions. For practical application most solutions have been reduced to convenient graphic or mathematical form.[1,3,4,16]

Steady Undirectional Flow

Steady flow implies that no change occurs with time. Flow conditions differ for confined and unconfined aquifers and hence need to be considered separately, beginning with flow in one direction.

Confined Aquifer. Let groundwater flow with a velocity v in the x-direction of a confined aquifer of uniform thickness. Then for steady flow, Eq. 3.77 reduces to

111

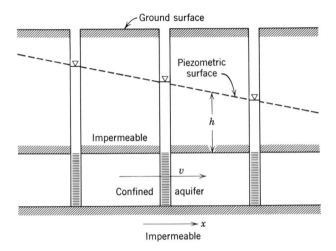

Fig. 4.1 Steady unidirectional flow in a confined aquifer of uniform thickness.

$$\frac{\partial^2 h}{\partial x^2} = 0 \qquad\qquad (4.1)$$

which has for its solution

$$h = C_1 x + C_2 \qquad\qquad (4.2)$$

where h is the head above a given datum and C_1 and C_2 are constants of integration. Assuming $h = 0$ when $x = 0$ and $\partial h/\partial x = -(v/K)$ from Darcy's law, then

$$h = -\frac{vx}{K} \qquad\qquad (4.3)$$

This states that the head decreases linearly, as sketched in Fig. 4.1, with flow in the x-direction.

Unconfined Aquifer. For the similar flow situation in an unconfined aquifer, direct analytic solution of the Laplace equation is not possible. The difficulty arises from the fact that the water table in the two-dimensional case represents a flow line. The shape of the water table determines the flow distribution, but at the same time the flow distribution governs the water table shape. To obtain a solution Dupuit[14] assumed (1) the velocity of the flow to be proportional to the tangent of the hydraulic gradient instead of the sine as defined in Eq. 3.72, and (2) the flow to be horizontal and uniform everywhere in a vertical section. These assumptions, although permitting a solution to be obtained, limit the application of the results. For unidirectional flow, as sketched in Fig. 4.2, the discharge per unit width q at any vertical section can be given as

$$q = -Kh\frac{dh}{dx} \qquad (4.4)$$

where K is hydraulic conductivity, h is the height of the water table above an impervious base, and x is the direction of flow. Integrating,

$$qx = -\frac{K}{2}h^2 + C \qquad (4.5)$$

and, if $h = h_0$ where $x = 0$, then the Dupuit equation

$$q = \frac{K}{2x}(h_0{}^2 - h^2) \qquad (4.6)$$

results, which indicates that the water table is parabolic in form.

For flow between two fixed bodies of water of constant heads h_0 and h_1 as in Fig. 4.2, the water table slope at the upstream boundary of the aquifer (neglecting the capillary zone)

$$\frac{dh}{dx} = -\frac{q}{Kh_0} \qquad (4.7)$$

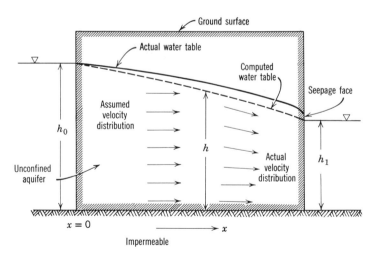

Fig. 4.2 Steady flow in an unconfined aquifer between two water bodies with vertical boundaries.

But the boundary $h = h_0$ is an equipotential line because the fluid potential in a water body is constant; consequently, the water table must be horizontal at this section, which is inconsistent with Eq. 4.7. In the direction of the flow, the parabolic water table described by Eq. 4.6 increases in slope. By so doing, the two Dupuit assumptions, previously stated, become increasingly poor approximations to the actual flow; therefore, the actual water table deviates more and more from the computed position in the direction of flow, as indicated in Fig. 4.2. The fact that the actual water table lies above the computed one can be explained by the fact that the Dupuit flows are all assumed horizontal, whereas the actual velocities of the same magnitude have a downward vertical component so that a greater saturated thickness is required for the same discharge. At the downstream boundary a discontinuity in flow forms because no consistent flow pattern can connect a water table directly to a downstream free-water surface. The water table actually approaches the boundary tangentially above the water body surface and forms a *seepage face*.

The above discrepancies indicate that the water table does not follow the parabolic form of Eq. 4.6; nevertheless, for flat slopes, where the sine and tangent are nearly equal, it closely predicts the water table position except near the outflow. The equation, however, accurately determines q or K for given boundary heads.[48]

Base Flow to a Stream. Estimates of the base flow to streams (see Chapter 6) or average groundwater recharge can be computed by applying the above analysis of one-directional flow in an unconfined aquifer. For example, picture the idealized boundaries shown in Fig. 4.3 of two long parallel streams completely penetrating an unconfined aquifer with a continuous recharge rate W occurring uniformly over the aquifer. With the Dupuit assumptions the flow per unit thickness

$$q = -Kh\frac{dh}{dx}$$ (4.8)

and by continuity

$$q = Wx$$ (4.9)

Combining these equations leads to the result

$$h^2 = h_a{}^2 + \frac{W}{K}(a^2 - x^2)$$ (4.10)

where h, h_a, a, and x are as defined in Fig. 4.3, and K is hydraulic conductivity. From symmetry and continuity

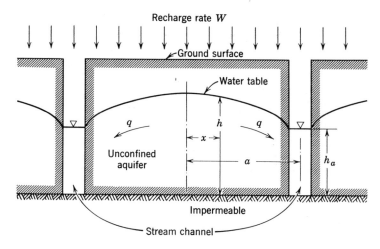

Fig. 4.3 Steady flow to two parallel streams from a uniformly recharged unconfined aquifer.

$$Q_b = 2aW \qquad\qquad (4.11)$$

where Q_b is the base flow entering each stream per unit length of stream channel. If h is known at any point, Q_b or W can be computed provided K is known.

Extensions of this analysis have been applied to design the spacing of parallel drains on agricultural land for specified soil, crop, and irrigation conditions.

Steady Radial Flow to a Well

When a well is pumped, water is removed from the aquifer surrounding the well, and the water table or piezometric surface, depending on the type of aquifer, is lowered. The *drawdown* at a given point is the distance the water level is lowered. A *drawdown curve* shows the variation of drawdown with distance from the well (see Fig. 4.4). In three dimensions the drawdown curve describes a conic shape known as the *cone of depression*. Also, the outer limit of the cone of depression defines the *area of influence* of the well.

Confined Aquifer. To derive the radial flow equation (which relates the well discharge to drawdown) for a well completely penetrating a confined aquifer, reference to Fig. 4.4 will prove helpful. The flow is assumed two-dimensional to a well centered on a circular island and penetrating a homogeneous and isotropic aquifer.

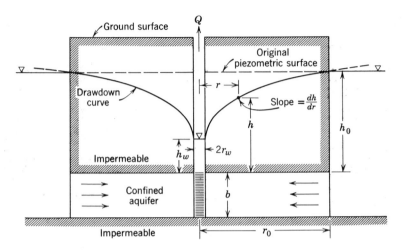

Fig. 4.4 Steady radial flow to a well penetrating a confined aquifer on an island.

Because the flow is everywhere horizontal, the Dupuit assumptions apply without error. Using plane polar coordinates with the well as the origin, the well discharge Q at any distance r equals

$$Q = Av = -2\pi rbK \frac{dh}{dr} \qquad (4.12)$$

for steady radial flow to the well. Rearranging and integrating for the boundary conditions at the well, $h = h_w$ and $r = r_w$, and at the edge of the island, $h = h_0$ and $r = r_0$, yields

$$h_0 - h_w = \frac{Q}{2\pi Kb} \ln \frac{r_0}{r_w} \qquad (4.13)$$

or

$$Q = 2\pi Kb \frac{h_0 - h_w}{\ln(r_0/r_w)} \qquad (4.14)$$

with the negative sign neglected.

In the more general case of a well penetrating an extensive confined aquifer, as in Fig. 4.5, there is no external limit for r. From the above derivation at any given value of r,

$$Q = 2\pi Kb \frac{h - h_w}{\ln(r/r_w)} \qquad (4.15)$$

which shows that h increases indefinitely with increasing r. Yet, the maximum h is the initial uniform head h_0. Thus, from a theoretical

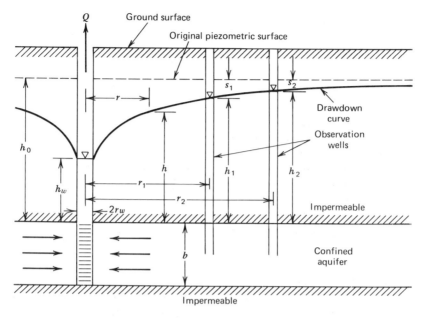

Fig. 4.5 Radial flow to a well penetrating an extensive confined aquifer.

aspect, steady radial flow in an extensive aquifer does not exist because the cone of depression must expand indefinitely with time. However, from a practical standpoint, h approaches h_0 with distance from the well, and the drawdown varies with the logarithm of the distance from the well.

Equation 4.15, known as the *equilibrium*, or *Thiem*,[68] *equation*, enables the hydraulic conductivity or the transmissivity of a confined aquifer to be determined from a pumped well. Because any two points define the logarithmic drawdown curve, the method consists of measuring drawdowns in two observation wells at different distances from a well pumped at a constant rate. Theoretically, h_w at the pumped well can serve as one measurement point; however, well losses caused by flow through the well screen and inside the well introduce errors so that h_w should be avoided. The transmissivity is given by

$$T = Kb = \frac{Q}{2\pi(h_2 - h_1)} \ln \frac{r_2}{r_1} \qquad (4.16)$$

where r_1 and r_2 are the distances and h_1 and h_2 are the heads of the respective observation wells.

From a practical standpoint, the drawdown s rather then the head h is measured so that Eq. 4.16 can be rewritten

$$T = \frac{Q}{2\pi(s_1 - s_2)} \ln \frac{r_2}{r_1} \tag{4.17}$$

where s_1 and s_2 are shown in Fig. 4.5. In order to apply Eq. 4.17, pumping must continue at a uniform rate for a sufficient time to approach a steady-state condition—that is, one in which the drawdown changes negligibly with time.* The observation wells should be located close enough to the pumping well so that their drawdowns are appreciable and can be readily measured. The derivation assumes that the aquifer is homogeneous and isotropic, is of uniform thickness, and is of infinite areal extent; that the well penetrates the entire aquifer; and that initially the piezometric surface is nearly horizontal.

Unconfined Aquifer. An equation for steady radial flow to a well in an unconfined aquifer also can be derived with the help of the Dupuit assumptions. As shown in Fig. 4.6, the well completely penetrates the aquifer to the horizontal base and a concentric boundary of constant head surrounds the well. The well discharge is

$$Q = -2\pi r K h \frac{dh}{dr} \tag{4.18}$$

which, when integrated between the limits $h = h_w$ at $r = r_w$ and $h = h_0$ at $r = r_0$, yields

$$Q = \pi K \frac{h_0^2 - h_w^2}{\ln(r_0/r_w)} \tag{4.19}$$

Converting to heads and radii at two observation wells (see Fig. 4.6),

$$Q = \pi K \frac{h_2^2 - h_1^2}{\ln(r_2/r_1)} \tag{4.20}$$

and rearranging to solve for the hydraulic conductivity

$$K = \frac{Q}{\pi(h_2^2 - h_1^2)} \ln \frac{r_2}{r_1} \tag{4.21}$$

This equation fails to describe accurately the drawdown curve near the well because the large vertical flow components contradict the

---

*In fact, the difference in drawdowns $(s_1 - s_2)$ becomes essentially constant while both values are still increasing so that Eq. 4.17 generally gives good results after only a few days of pumping.

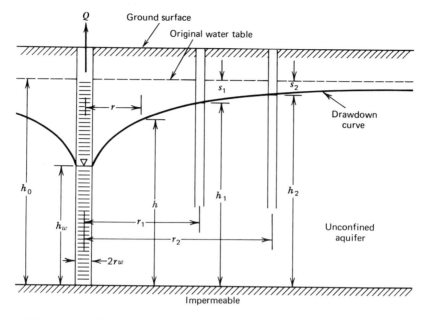

Fig. 4.6 Radial flow to a well penetrating an unconfined aquifer.

Dupuit assumptions; however, estimates of K for given heads are good. In practice, drawdowns should be small in relation to the saturated thickness of the unconfined aquifer.

The transmissivity can be approximated from Eq. 4.21 by

$$T \cong K\frac{h_1 + h_2}{2} \tag{4.22}$$

Where drawdowns are appreciable, the heads h_1 and h_2 in Eq. 4.21 can be replaced by $(h_0 - s_1)$ and $(h_0 - s_2)$, respectively, as shown in Fig. 4.6. Then the transmissivity for the full thickness becomes[4,43]

$$T = Kh_0 = \frac{Q}{2\pi \left[\left(s_1 - \dfrac{s_1^2}{2h_0}\right) - \left(s_2 - \dfrac{s_2^2}{2h_0}\right)\right]} \ln \frac{r_2}{r_1} \tag{4.23}$$

Unconfined Aquifer with Uniform Recharge. Figure 4.7 shows a well penetrating an unconfined aquifer that is recharged uniformly at rate W from rainfall, excess irrigation water, or other surface-water sources. The flow Q toward the well increases as the well is approached, reaching a maximum of Q_w at the well. The increase in flow dQ through a cylinder of thickness dr and radius r comes from the recharged water entering the cylinder from above; hence,

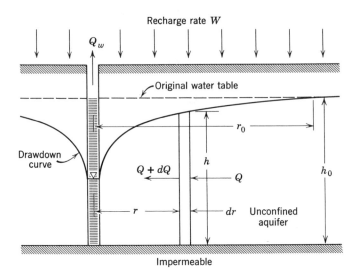

Fig. 4.7 Steady flow to a well penetrating a uniformly recharged unconfined aquifer.

$$dQ = -2\pi r \, dr \, W \qquad\qquad (4.24)$$

Integrating,

$$Q = -\pi r^2 W + C \qquad\qquad (4.25)$$

but at the well $r \to 0$ and $Q = Q_w$, so that

$$Q = -\pi r^2 W + Q_w \qquad\qquad (4.26)$$

Substituting this flow in the equation for flow to the well (Eq. 4.18) gives

$$-2\pi r K h \frac{dh}{dr} = -\pi r^2 W + Q_w \qquad\qquad (4.27)$$

Integrating, and noting that $h = h_0$ at $r = r_0$, yields the equation for the drawdown curve

$$h_0^2 - h^2 = \frac{W}{2K}(r^2 - r_0^2) + \frac{Q_w}{\pi K}\ln\frac{r_0}{r} \qquad\qquad (4.28)$$

By comparing Eq. 4.28 with Eq. 4.19, the effect of the vertical recharge becomes apparent.

It follows that when $r = r_0$, $Q = 0$, so that from Eq. 4.26

$$Q_w = \pi r_0^2 W \qquad\qquad (4.29)$$

Thus, total flow of the well equals the recharge within the circle defined by the radius of influence; conversely, the radius of influence is a function of the well pumpage and the recharge rate only. This results in a steady-state drawdown; however, the analysis assumes an idealized circular outer boundary with a constant head and no flow—conditions that rarely occur in the field.

Well in a Uniform Flow

Drawdown curves for well flow presented heretofore have assumed an initially horizontal groundwater surface. A practical situation is that of a well pumping from an aquifer having a uniform flow field, as indicated by a uniformly sloping piezometric surface or water table. Figure 4.8 shows sectional and plan views of a well penetrating a confined aquifer with a sloping piezometric surface. It is apparent that the circular area of influence associated with a radial flow pattern becomes distorted; however, for most relatively flat natural slopes the Dupuit radial flow equation can be applied without appreciable error.

For wells pumping from an area with a sloping hydraulic gradient, the hydraulic conductivity can be determined from Eq. 4.18 by inserting average heads and hydraulic gradients. The resulting expression has the form

$$K = \frac{2Q}{\pi r(h_u + h_d)(i_u + i_d)} \tag{4.30}$$

for an unconfined aquifer where Q is the pumping rate, h_u and h_d are the saturated thicknesses, and i_u and i_d are the water table slopes at distance r upstream and downstream, respectively, from the well. For a confined aquifer, piezometric slopes replace water table slopes, and $(h_u + h_d)$ is replaced by $2b$ where b is the aquifer thickness.

In Fig. 4.8 the groundwater divide marking the boundary of the region producing inflow to the well is shown. For a well pumping an infinite time, the boundary would extend up to the limit of the aquifer. The expression for the boundary of the region producing inflow can be derived by superposition of radial and one-dimensional flow fields to yield

$$-\frac{y}{x} = \tan\left(\frac{2\pi Kbi}{Q}y\right) \tag{4.31}$$

where the rectangular coordinates are as shown in Fig. 4.8 with the origin at the well, b is the aquifer thickness, Q is the discharge rate, i is the natural piezometric slope, and K is hydraulic conduc-

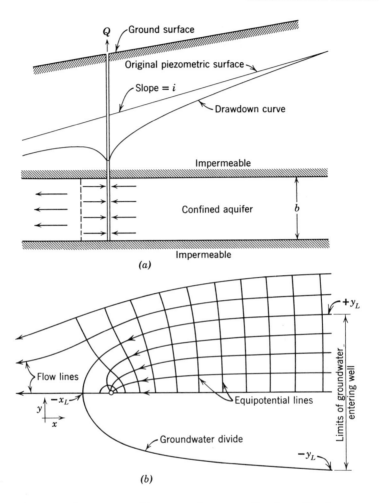

Fig. 4.8 Flow to a well penetrating a confined aquifer having a sloping plane piezometric surface. (a) Vertical section. (b) Plan view.

tivity. From Eq. 4.31 the boundary asymptotically approaches the finite limits

$$y_L = \pm \frac{Q}{2Kbi} \tag{4.32}$$

as $x \to \infty$. The boundary of the contributing area extends downstream to a stagnation point where

$$x_L = -\frac{Q}{2\pi K b i} \qquad (4.33)$$

It follows that the upstream inflow zone equals $2\pi x_L$.

Equations 4.31 to 4.33 also apply to unconfined aquifers by replacing b by the uniform saturated aquifer thickness h_0, providing the drawdown is small in relation to the aquifer thickness. An important practical application of these equations concerns determining whether an upstream pollution source will affect a nearby pumping well (see Chapter 8).

Unsteady Radial Flow in a Confined Aquifer

When a well penetrating an extensive confined aquifer is pumped at a constant rate, the influence of the discharge extends outward with time. The rate of decline of head times the storage coefficient summed over the area of influence equals the discharge. Because the water must come from a reduction of storage within the aquifer, the head will continue to decline as long as the aquifer is effectively infinite; therefore, unsteady, or transient, flow exists. The rate of decline, however, decreases continuously as the area of influence expands.

The applicable differential equation (see Eq. 3.80) in plane polar coordinates is

$$\frac{\partial^2 h}{\partial r^2} + \frac{1}{r}\frac{\partial h}{\partial r} = \frac{S}{T}\frac{\partial h}{\partial t} \qquad (4.34)$$

where h is head, r is radial distance from the pumped well, S is storage coefficient, T is transmissivity, and t is the time since beginning of pumping. Theis[67] obtained a solution for Eq. 4.34 based on the analogy between groundwater flow and heat conduction. By assuming that the well is replaced by a mathematical sink of constant strength and imposing the boundary conditions $h = h_0$ for $t = 0$, and $h \rightarrow h_0$ as $r \rightarrow \infty$ for $t \geq 0$, the solution

$$s = \frac{Q}{4\pi T}\int_u^\infty \frac{e^{-u}\,du}{u} \qquad (4.35)$$

is obtained, where s is drawdown, Q is the constant well discharge, and

$$u = \frac{r^2 S}{4Tt} \qquad (4.36)$$

Equation 4.35 is known as the *nonequilibrium*, or *Theis, equation*.

The integral is a function of the lower limit u and is known as an exponential integral. It can be expanded as a convergent series so that Eq. 4.35 becomes

$$s = \frac{Q}{4\pi T}\left[-0.5772 - \ln u + u - \frac{u^2}{2 \cdot 2!} + \frac{u^3}{3 \cdot 3!} - \frac{u^4}{4 \cdot 4!} + \cdots\right] \qquad (4.37)$$

The nonequilibrium equation permits determination of the formation constants S and T by means of pumping tests of wells. The equation is widely applied in practice and is preferred over the equilibrium equation because: (1) a value for S can be determined, (2) only one observation well is required, (3) a shorter period of pumping is generally necessary, and (4) no assumption of steady-state flow conditions is required.

The assumptions inherent in Eq. 4.35 should be emphasized because they are often overlooked in applying the nonequilibrium equation and thereby can lead to erroneous results. The assumptions include:

1. The aquifer is homogeneous, isotropic, of uniform thickness, and of infinite areal extent.
2. Before pumping, the piezometric surface is horizontal.
3. The well is pumped at a constant discharge rate.
4. The pumped well penetrates the entire aquifer, and flow is everywhere horizontal within the aquifer to the well.
5. The well diameter is infinitesimal so that storage within the well can be neglected.
6. Water removed from storage is discharged instantaneously with decline of head.

Seldom, if ever, are these assumptions strictly satisfied, but recognition of them can create an awareness of the approximations involved for employing the nonequilibrium equation under field conditions. Average values of S and T can be obtained in the vicinity of a pumped well by measuring in one or more observation wells the change in drawdown with time under the influence of a constant pumping rate. Because of the mathematical difficulties encountered in applying Eq. 4.35, or its equivalent Eq. 4.37, several investigators have developed simpler approximate solutions that can be readily applied for field purposes. Three methods, by Theis,[67] Cooper and Jacob,[12] and Chow,[10] are described in the following sections with the necessary tables and/or graphs. An illustrative example accompanies each method.

Theis Method of Solution. Equation 4.35 may be simplified to

$$s = \left(\frac{Q}{4\pi T}\right) W(u) \tag{4.38}$$

where $W(u)$, termed the *well function*, is a convenient symbolic form of the exponential integral. Rewriting Eq. 4.36 as

$$\frac{r^2}{t} = \left(\frac{4T}{S}\right) u \tag{4.39}$$

it can be seen that the relation between $W(u)$ and u must be similar to that between s and r^2/t because the terms in parentheses in the two equations are constants. Given this similarity Theis[67] suggested an approximate solution for S and T based on a graphic method of superposition.

A plot on logarithmic paper of $W(u)$ versus u, known as a *type curve*, is prepared. Table 4.1 gives values of $W(u)$ for a wide range of u. Values of drawdowns are plotted against values of r^2/t on logarithmic paper of the same size as for the type curve. The observed time-drawdown data are superimposed on the type curve, keeping the coordinate axes of the two curves parallel, and adjusted until a position is found by trial whereby most of the plotted points of the observed data fall on a segment of the type curve. Any convenient point is then selected, and the coordinates of this match point are recorded. With values of $W(u)$, u, s, and r^2/t thus determined, S and T can be obtained from Eqs. 4.38 and 4.39.*

In areas where several wells exist near a well being test-pumped, simultaneous readings of s in the wells enable distance-drawdown data to be fitted to a type curve in a manner identical to that for time-drawdown data.

Example of Theis Method. A well penetrating a confined aquifer is pumped at a uniform rate of 2500 m³/day. Drawdowns during the pumping period are measured in an observation well 60 m away; observations of t and s are listed in Table 4.2. Values of r^2/t in m²/min are computed and appear in the right column of Table 4.2. Values of s and r^2/t are plotted on logarithmic paper. Values of $W(u)$ and u from Table 4.1 are plotted on another sheet of logarith-

---

*The computation of r^2/t values can be avoided by plotting s versus t rather than r^2/t. In this case, the type curve must be turned over to obtain coincidence and a match point, but results are identical.

TABLE 4.1 Values of $W(u)$ for Values of u

u	1.0	2.0	3.0	4.0	5.0	6.0	7.0	8.0	9.0
$\times 1$	0.219	0.049	0.013	0.0038	0.0011	0.00036	0.00012	0.000038	0.000012
$\times 10^{-1}$	1.82	1.22	0.91	0.70	0.56	0.45	0.37	0.31	0.26
$\times 10^{-2}$	4.04	3.35	2.96	2.68	2.47	2.30	2.15	2.03	1.92
$\times 10^{-3}$	6.33	5.64	5.23	4.95	4.73	4.54	4.39	4.26	4.14
$\times 10^{-4}$	8.63	7.94	7.53	7.25	7.02	6.84	6.69	6.55	6.44
$\times 10^{-5}$	10.94	10.24	9.84	9.55	9.33	9.14	8.99	8.86	8.74
$\times 10^{-6}$	13.24	12.55	12.14	11.85	11.63	11.45	11.29	11.16	11.04
$\times 10^{-7}$	15.54	14.85	14.44	14.15	13.93	13.75	13.60	13.46	13.34
$\times 10^{-8}$	17.84	17.15	16.74	16.46	16.23	16.05	15.90	15.76	15.65
$\times 10^{-9}$	20.15	19.45	19.05	18.76	18.54	18.35	18.20	18.07	17.95
$\times 10^{-10}$	22.45	21.76	21.35	21.06	20.84	20.66	20.50	20.37	20.25
$\times 10^{-11}$	24.75	24.06	23.65	23.36	23.14	22.96	22.81	22.67	22.55
$\times 10^{-12}$	27.05	26.36	25.96	25.67	25.44	25.26	25.11	24.97	24.86
$\times 10^{-13}$	29.36	28.66	28.26	27.97	27.75	27.56	27.41	27.28	27.16
$\times 10^{-14}$	31.66	30.97	30.56	30.27	30.05	29.87	29.71	29.58	29.46
$\times 10^{-15}$	33.96	33.27	32.86	32.58	32.35	32.17	32.02	31.88	31.76

TABLE 4.2 Pumping Test Data

| | (r = 60 m) | |
t, min	s, m	r^2/t, m²/min
0	0	∞
1.0	0.20	3600
1.5	0.27	2400
2.0	0.30	1800
2.5	0.34	1440
3.0	0.37	1200
4	0.41	900
5	0.45	720
6	0.48	600
8	0.53	450
10	0.57	360
12	0.60	300
14	0.63	257
18	0.67	200
24	0.72	150
30	0.76	120
40	0.81	90
50	0.85	72
60	0.90	60
80	0.93	45
100	0.96	36
120	1.00	30
150	1.04	24
180	1.07	20
210	1.10	17
240	1.12	15

mic paper and a curve is drawn through the points. The two sheets are superposed and shifted with coordinate axes parallel until the observational points coincide with the curve, as shown in Fig. 4.9. A convenient match point is selected with $W(u) = 1.00$ and $u = 1 \times 10^{-2}$, so that $s = 0.18$ m and $r^2/t = 150$ m²/min $= 216,000$ m²/day. Thus, from Eq. 4.38,

$$T = \frac{Q}{4\pi s} W(u) = \frac{2500(1.00)}{4\pi(0.18)} = 1110 \text{ m}^2/\text{day}$$

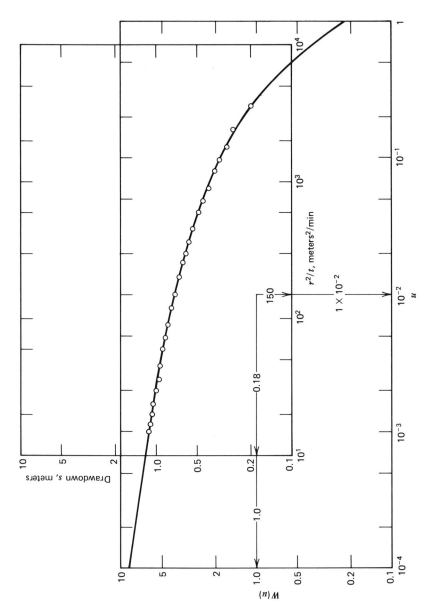

Fig. 4.9 Theis method of superposition for solution of the nonequilibrium equation.

and from Eq. 4.39,

$$S = \frac{4Tu}{r^2/t} = \frac{4(1110)(1 \times 10^{-2})}{216{,}000} = 0.000206$$

Cooper-Jacob Method of Solution. It was noted by Cooper and Jacob[12] that for small values of r and large values of t, u is small, so that the series terms in Eq. 4.37 become negligible after the first two terms. As a result, the drawdown can be expressed by the asymptote

$$s = \frac{Q}{4\pi T}\left(-0.5772 - \ln\frac{r^2S}{4Tt}\right) \qquad (4.40)$$

Rewriting and changing to decimal logarithms, this reduces to

$$s = \frac{2.30Q}{4\pi T}\log\frac{2.25Tt}{r^2S} \qquad (4.41)$$

Therefore, a plot of drawdown s versus the logarithm of t forms a straight line. Projecting this line to $s = 0$, where $t = t_0$ (see Fig. 4.10),

$$0 = \frac{2.30Q}{4\pi T}\log\frac{2.25Tt_0}{r^2S} \qquad (4.42)$$

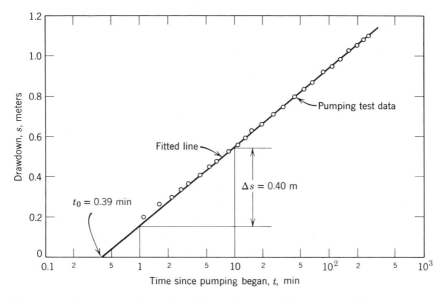

Fig. 4.10 Cooper-Jacob method for solution of the nonequilibrium equation.

and it follows that

$$\frac{2.25Tt_0}{r^2S} = 1 \qquad (4.43)$$

resulting in

$$S = \frac{2.25Tt_0}{r^2} \qquad (4.44)$$

A value for T can be obtained by noting that if $t/t_0 = 10$, then $\log t/t_0 = 1$; therefore, replacing s by Δs, where Δs is the drawdown difference per log cycle of t, Eq. 4.41 becomes[38]

$$T = \frac{2.30Q}{4\pi\Delta s} \qquad (4.45)$$

Thus, the procedure is first to solve for T with Eq. 4.45 and then to solve for S with Eq. 4.44. The straight-line approximation for this method should be restricted to small values of u ($u < 0.01$) to avoid large errors.

Example of Cooper-Jacob Method. From the pumping test data of Table 4.2, s and t are plotted on semilogarithmic paper, as shown in Fig. 4.10. A straight line is fitted through the points, and $\Delta s = 0.40$ m and $t_0 = 0.39$ min $= 2.70 \times 10^{-4}$ day are read. Thus,

$$T = \frac{2.30(2500)}{4\pi(0.40)} = 1090 \text{ m}^2/\text{day}$$

and

$$S = \frac{2.25Tt_0}{r^2} = \frac{2.25(1090)(2.70 \times 10^{-4})}{(60)^2} = 0.000184$$

Chow Method of Solution. Chow[10] developed a method of solution with the advantages of avoiding curve fitting and being unrestricted in its application. Again, measurements of drawdown in an observation well near a pumped well are made. The observational data are plotted on semilogarithmic paper in the same manner as for the Cooper-Jacob method. On the plotted curve, choose an arbitrary point and note the coordinates, t and s. Next, draw a tangent to the curve at the chosen point and determine the drawdown difference Δs, in feet, per log cycle of time. Compute $F(u)$ from

$$F(u) = \frac{s}{\Delta s} \qquad (4.46)$$

and find the corresponding values of $W(u)$ and u from Fig. 4.11.*

*For $F(u) > 2.0$, $W(u) = 2.30F(u)$, and u is obtained from Table 4.1.

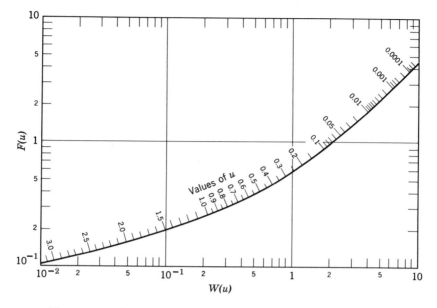

Fig. 4.11 Relation among $F(u)$, $W(u)$, and u (After Chow[10]).

Finally, compute the formation constants T by Eq. 4.38 and S by Eq. 4.39.

Example of Chow Method. In Fig. 4.12 data are plotted from Table 4.2, and point A is selected on the curve where $t = 6$ min $= 4.2 \times 10^{-3}$ day and $s = 0.47$ m. A tangent is constructed as shown; the drawdown difference per log cycle of time is $\Delta s = 0.38$ m. Then $F(u) = 0.47/0.38 = 1.24$, and from Fig. 4.11, $W(u) = 2.75$ and $u = 0.038$. Hence,

$$T = \frac{Q}{4\pi s} W(u) = \frac{2500}{4\pi(0.47)} 2.75 = 1160 \text{ m}^2/\text{day}$$

and

$$S = \frac{4Ttu}{r^2} = \frac{4(1160)(4.2 \times 10^{-3})(0.038)}{(60)^2} = 0.000206$$

Recovery Test. At the end of a pumping test, when the pump is stopped, the water levels in pumping and observation wells will begin to rise. This is referred to as the *recovery* of groundwater levels, while measurements of drawdown below the original static water level (prior to pumping) during the recovery period are known

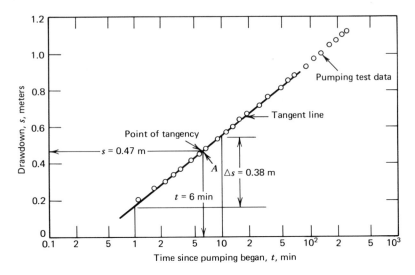

Fig. 4.12 Chow method for solution of the nonequilibrium equation.

as *residual drawdowns*. A schematic diagram of change in water level with time during and after pumping is shown in Fig. 4.13.

It is good practice to measure residual drawdowns because analysis of the data enable transmissivity to be calculated, thereby providing an independent check on pumping test results. Also, costs are nominal in relation to the conduct of a pumping test.* Furthermore, the rate of recharge Q to the well during recovery is assumed constant and equal to the mean pumping rate, whereas pumping rates often vary and are difficult to control accurately in the field.

If a well is pumped for a known period of time and then shut down, the drawdown thereafter will be identically the same as if the discharge had been continued and a hypothetical recharge well with the same flow were superposed on the discharging well at the instant the discharge is shut down. From this principle Theis[67] showed that the residual drawdown s' can be given as

$$s' = \frac{Q}{4\pi T}[W(u) - W(u')] \tag{4.47}$$

where

$$u = \frac{r^2 S}{4Tt} \quad \text{and} \quad u' = \frac{r^2 S}{4Tt'} \tag{4.48}$$

*In addition, it should be noted that measurement of the recovery within a pumped well will provide an estimate of transmissivity even without an observation well.

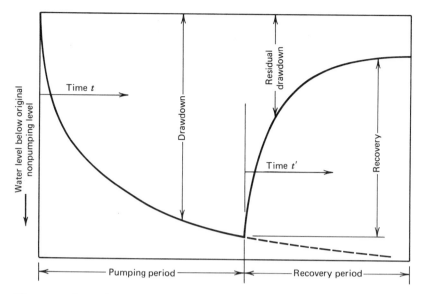

Fig. 4.13 Drawdown and recovery curves in an observation well near a pumping well.

and t and t' are defined in Fig. 4.13. For r small and t' large, the well functions can be approximated by the first two terms of Eq. 4.37 so that Eq. 4.47 can be written as

$$s' = \frac{2.30Q}{4\pi T} \log \frac{t}{t'} \qquad (4.49)$$

Thus, a plot of residual drawdown s' versus the logarithm of t/t' forms a straight line. The slope of the line equals $2.30Q/4\pi T$ so that for $\Delta s'$, the residual drawdown per log cycle of t/t', the transmissivity becomes

$$T = \frac{2.30Q}{4\pi \Delta s'} \qquad (4.50)$$

No comparable value of S can be determined by this recovery test method.

Example of Recovery Test. A well pumping at a uniform rate of 2500 m^3/day was shut down after 240 min; thereafter, measurements of s' and t' tabulated in Table 4.3 were made in an observation well. Values of t/t' are computed, as shown in Table 4.3, and then plotted versus s' on semilogarithmic paper (see Fig. 4.14). A straight line is

TABLE 4.3   Recovery Test Data
(pump shut down at t = 240 min)

t', min	t, min	t/t'	s', m
1.0	241	241	0.89
2.0	242	121	0.81
3.0	243	81	0.76
5	245	49	0.68
7	247	35	0.64
10	250	25	0.56
15	255	17	0.49
20	260	13	0.55
30	270	9	0.38
40	280	7	0.34
60	300	5	0.28
80	320	4	0.24
100	340	3.4	0.21
140	380	2.7	0.17
180	420	2.3	0.14

fitted through the points and $\Delta s' = 0.40$ m is determined; then,

$$T = \frac{2.30Q}{4\pi s'} = \frac{2.30(2500)}{4\pi(0.40)} = 1140 \text{ m}^2/\text{day}$$

Unsteady Radial Flow in an Unconfined Aquifer

The previous solution methods for the nonequilibrium equation applied to pumping tests in confined aquifers can also be applied to unconfined aquifers providing that the basic assumptions are satisfied. In general, if the drawdown is small in relation to the saturated thickness, good approximations are possible.[60]

Where drawdowns are significant, the assumption that water released from storage is discharged instantaneously with decline of head is frequently violated in unconfined aquifers. Pumping test data reveal that as a water table is lowered, gravity drainage of water from the unsaturated zone proceeds at a variable rate, known as *delayed yield*.[6,15,49] In a series of contributions, Boulton[6,7,8] developed special type curves for analyzing pumping test data of unconfined aquifers and for taking account of delayed yield.[71] These time-drawdown curves of delayed yield are shown in Fig. 4.15. The interpretation of any one curve can be considered in three time segments. In the first segment, measured in seconds to a few minutes,

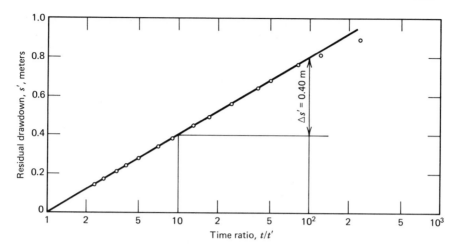

Fig. 4.14 Recovery test method for solution of the nonequilibrium equation.

water is released essentially instantaneously from storage by compaction of the aquifer and by expansion of entrapped air. This portion of the curve can be fitted by a type curve with a storage coefficient equivalent to that of a confined aquifer. The second segment displays a flattening in slope caused by gravity drainage replenishment from the pore space above the cone of depression. Finally, in

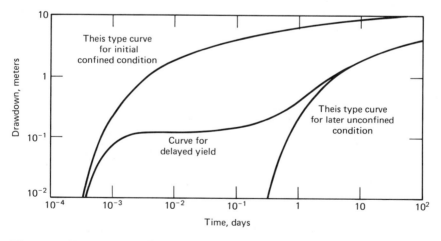

Fig. 4.15 Type curves of drawdown versus time illustrating the effect of delayed yield for pumping tests in unconfined aquifers (after Bureau of Reclamation[9]).

the third segment an equilibrium is approached between gravity drainage and the rate of decline of the water table. This condition occurs after several minutes to several days and can be fitted by a type curve with a storage coefficient for an unconfined aquifer.

From a water production standpoint the storage coefficient obtained from the third segment of the curve in Fig. 4.15, which is the specific yield, is the most reliable and hence most important. For simplicity a pumping test should be continued sufficiently long to define the third segment of the curve; then, by applying one of the solution methods previously described for the nonequilibrium equation, a value for S can be obtained.

The minimum length of pumping test to achieve an accurate estimate of S in an unconfined aquifer depends on the transmissivity of the aquifer. One approach, based on an empirical study of various alluvial aquifer materials, is given by the graphs in Fig. 4.16. The *delay index* t_d is estimated in Fig. 4.16a from the composition of aquifer material. Then knowing the distance r between pumping and observation wells, and estimating S and T, an approximation to the minimum pumping time t_{min} can be calculated from Fig. 4.16b.

Another approach is simply to ensure that the pumping test duration exceeds the following suggested guidelines:[9]

Predominant Aquifer Material	Minimum Pumping Time, hours
Silt and clay	170
Fine sand	30
Medium sand and coarser materials	4

Unsteady Radial Flow in a Leaky Aquifer

When a leaky aquifer, as shown in Fig. 4.17, is pumped, water is withdrawn both from the aquifer and from the saturated portion of the overlying aquitard, or semipervious layer. Lowering the piezometric head in the aquifer by pumping creates a hydraulic gradient within the aquitard; consequently, groundwater migrates vertically downward into the aquifer. The quantity of water moving downward is proportional to the difference between the water table and the piezometric head.[11,33,66]

Steady-state flow is possible to a well in a leaky aquifer because of the recharge through the semipervious layer. The equilibrium will be established when the discharge rate of the pump equals the re-

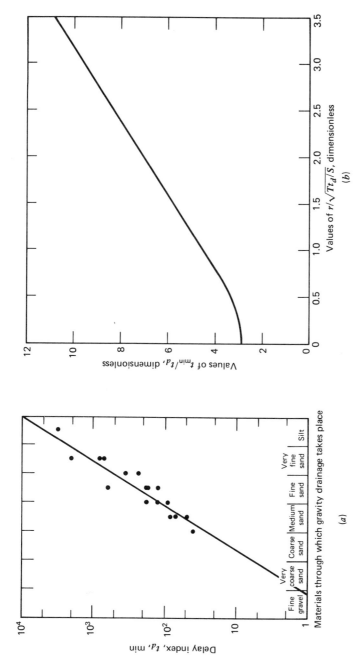

Fig. 4.16 Empirical method for estimating the minimum length of a pumping test in an unconfined aquifer (after Prickett[53]). (*a*) Empirical relation of delay index to character of materials through which gravity drainage occurs. (*b*) Curve for estimating the minimum time t_{min} at which effects of delayed gravity drainage cease to influence drawdown of a pumping well in an unconfined aquifer.

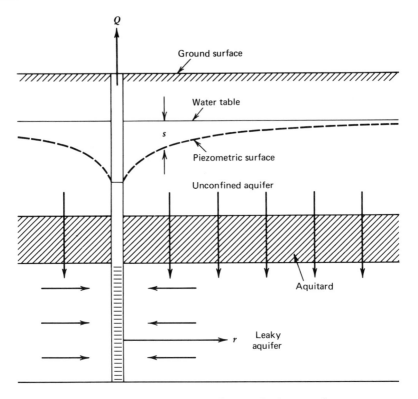

Fig. 4.17 Well pumping from a leaky aquifer.

charge rate of vertical flow into the aquifer, assuming the water table remains constant. Solutions for this special steady-state situation are available,[27,38] but a more general analysis for unsteady flow follows.

When pumping starts from a well in a leaky aquifer, drawdown of the piezometric surface can be given by[21,23,27]

$$s = \frac{Q}{4\pi T} W(u, r/B) \tag{4.51}$$

where s, Q, and r are defined in Fig. 4.17, and again

$$u = \frac{r^2 S}{4Tt} \tag{4.52}$$

The quantity r/B is given by

$$\frac{r}{B} = \frac{r}{\sqrt{T/(K'/b')}} \tag{4.53}$$

where T is transmissivity of the aquifer, K' is the vertical hydraulic conductivity of the aquitard, and b' is the thickness of the saturated semipervious layer (see Fig. 4.17). Values of the function $W(u, r/B)$ were tabulated by Hantush.[21] It can be noted that Eq. 4.51 has the form of the Theis equation (see Eq. 4.38); in fact, for a confined aquifer, $K' \to 0$, so that $B \to \infty$ and $r/B \to 0$, thereby reducing Eq. 4.51 to the Theis equation.

Employing this analogy and the Theis method of solution, Walton[69] prepared a family of type curves for $W(u, r/B)$ as presented in Fig. 4.18. Here values of $W(u, r/B)$ are plotted against $1/u$ for various values of r/B. On another sheet of logarithmic paper of the same scale, s versus t is plotted. Superposing the two sheets while keeping the coordinate axes parallel, a position is found where most of the data points fall on one of the type curves. Selecting any convenient match point, values of $W(u, r/B)$, $1/u$, s, and t are noted. T is then found from Eq. 4.51 and S from Eq. 4.52. Finally, from the value of r/B belonging to the type curve of best fit, it is possible to calculate K'/b' from Eq. 4.53; and if b' is known from field conditions, K' can be evaluated.

Well Flow Near Aquifer Boundaries

Where a well is pumped near an aquifer boundary, the assumption that the aquifer is of infinite areal extent no longer holds. Analysis of this situation involves the principle of superposition by which the drawdown of two or more wells is the sum of the drawdowns of each individual well. By introducing imaginary (or image) wells, an aquifer of finite extent can be transformed into an infinite aquifer so that the solution methods previously described can be applied.

Well Flow Near a Stream. An example of the usefulness of the method of images is the situation of a well near a perennial stream.[20,24,35] It is desired to obtain the head at any point under the influence of pumping at a constant rate Q and to determine what fraction of the pumpage is derived from the stream. Sectional views are shown in Fig. 4.19 of the real system and an equivalent imaginary system. Note in Fig. 4.19b that an imaginary recharge well* has been placed directly opposite and at the same distance from the stream as the real well. This image well operates simultaneously and at the same rate as the real well so that the buildup (increase of head around a recharge well) and drawdown of head along the line of the stream

---

*A recharge well is a well through which water is added to an aquifer; hence, it is the reverse of a pumping well.

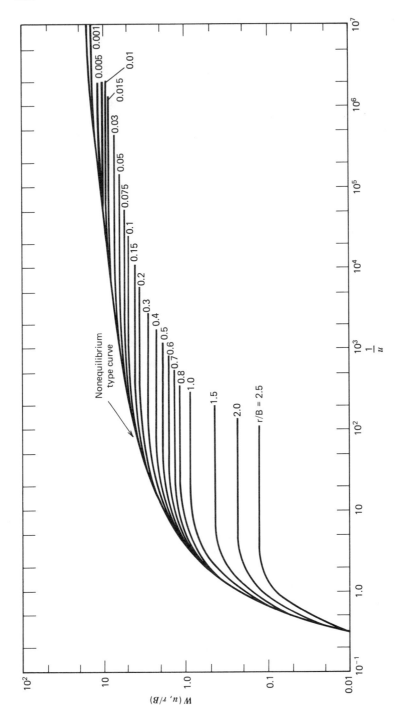

Fig. 4.18 Type curves for analysis of pumping test data to evaluate storage coefficient and transmissivity of leaky aquifers (after Walton[69]).

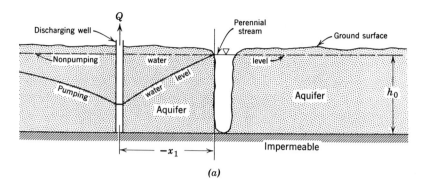

(a)

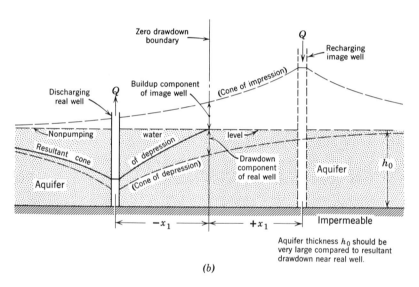

Aquifer thickness h_0 should be
very large compared to resultant
drawdown near real well.

(b)

Fig. 4.19 Sectional views. (a) Discharging well near a perennial
stream. (b) Equivalent hydraulic system in an aquifer of infinite areal
extent (after Ferris, et al.[16]).

exactly cancel. This furnishes a constant head along the stream,
which is equivalent to the constant elevation of the stream forming
the aquifer boundary. Thus, in the plan view of the resulting flow
net, illustrated by Fig. 4.20, a single equipotential line is coincident
with the axis of the stream. The resultant asymmetrical drawdown
of the real well is given at any point by the algebraic sum of the
drawdown of the real well and the buildup of the recharge well, as
if these wells were located in an infinite aquifer.

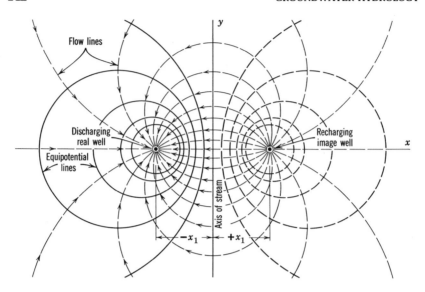

Fig. 4.20 Flow net for discharging real well and recharging image well (after Ferris, et al.[16]).

Examples of hydraulically equivalent aquifer systems bounded by streams with various configurations are shown in Fig. 4.21. Note that combinations of both image recharge and pumping wells are required.[38] For the single stream in Fig. 4.21*a*, the steady-state drawdown at any point (x,y) is given by

$$s = \frac{Q}{4\pi T} \ln \frac{(x + x_w)^2 + (y - y_w)^2}{(x - x_w)^2 + (y - y_w)^2} \tag{4.54}$$

where (x_w, y_w) are the coordinates of the pumped well. Similarly, for the right-angle boundaries of Fig. 4.21*b*,

$$s = \frac{Q}{4\pi T} \ln \frac{[(x - x_w)^2 + (y + y_w)^2][(x + x_w)^2 + (y - y_w)^2]}{[(x - x_w)^2 + (y - y_w)^2][(x + x_w)^2 + (y + y_w)^2]} \tag{4.55}$$

And for the strip aquifer bounded by two straight parallel streams (see Fig. 4.21*c*),

$$s = \frac{Q}{4\pi T} \ln \frac{\cosh \dfrac{\pi(y - y_w)}{2a} + \cos \dfrac{\pi(x + x_w)}{2a}}{\cosh \dfrac{\pi(y - y_w)}{2a} - \cos \dfrac{\pi(x + x_w)}{2a}} \tag{4.56}$$

and the angles are expressed in radians. Actually, in Fig. 4.21*c* the image wells extend to infinity; however, in practice it is only neces-

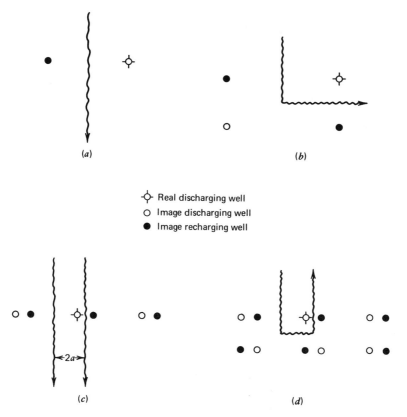

Fig. 4.21 Image well systems for aquifers bounded by streams of various geometries. (*a*). Unidirectional stream. (*b*) Rectangular stream. (*c*) Two parallel streams. (*d*) U-shaped stream. Theoretically, image wells in (*c*) and (*d*) extend left and right to infinity.

sary to include pairs of image wells closest to the real well because others have a negligible influence on the drawdown.

Further analysis of the flow distribution yields the fraction Q_s/Q of well discharge which is obtained from a stream, Q_s being the flow from the stream. The result for nonequilibrium conditions can be expressed as a convergent series or in terms of a probability integral, but the dimensionless graphic solution presented in Fig. 4.22 simplifies computations. Here x is the distance from the well to the stream, T is transmissivity, S is storage coefficient, and t is time from start of pumping. The same result applies to an unconfined aquifer, providing the drawdown is small compared to the saturated thickness.

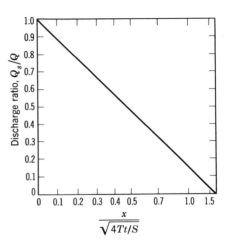

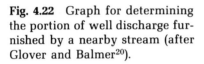

Fig. 4.22 Graph for determining the portion of well discharge furnished by a nearby stream (after Glover and Balmer[20]).

Well Flow Near Other Boundaries. In addition to the previous example, the method of images can be applied to a large number of groundwater boundary problems. As before, actual boundaries are replaced by an equivalent hydraulic system, which includes imaginary wells and permits solutions to be obtained from equations applicable only to extensive aquifers. Three boundary conditions to suggest the adaptability of the method are shown in Figs. 4.23 and and 4.24. Figure 4.23 shows a well pumping near an impermeable boundary. An image discharging well is placed opposite the pumping well with the same rate of discharge and at an equal distance from the boundary; therefore, along the boundary the wells offset one another, causing no flow across the boundary, which is the desired condition. Figure 4.24a shows a discharging well in an aquifer bounded on two sides by impermeable barriers. The image discharge wells I_1 and I_2 provide the required flow but, in addition, a third image well I_3 is necessary to balance drawdowns along the extensions of the boundaries. The resulting system of four discharging wells in an extensive aquifer represents hydraulically the flow system for the physical boundary conditions. Finally, Fig. 4.24b presents the situation of a well near an impermeable boundary and a perennial stream. The image wells required follow from the previous illustrations.

For a wedge-shaped aquifer, such as a valley bounded by two converging impermeable barriers, the drawdown at any location within the aquifer can be calculated by the same method of images.[62] Consider the aquifer formed by two barriers intersecting at an angle of 45 degrees shown in Fig. 4.25. Seven image pumping wells plus the

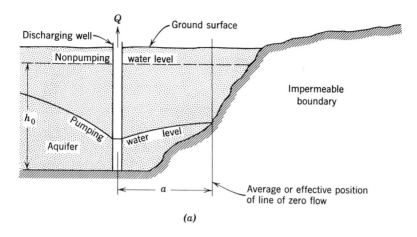

(a)

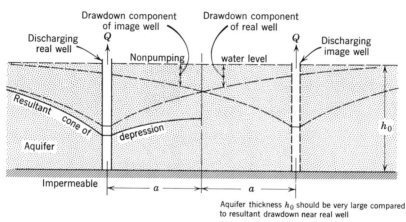

Aquifer thickness h_0 should be very large compared
to resultant drawdown near real well

Fig. 4.23 Sectional views. (*a*) Discharging well near an impermeable
boundary. (*b*) Equivalent hydraulic system in an aquifer of infinite
areal extent (after Ferris, et al. [16]).

single real pumping well form a circle with its center at the wedge
apex; the radius equals the distance from the apex to the real pumping well.[16] The drawdown at any point between the two barriers
can then be calculated by summing the individual drawdowns. In
general, it can be shown that the number of image wells n required
for a wedge angle θ is given by

$$n = \frac{360°}{\theta} - 1 \qquad (4.57)$$

where θ is an aliquot part of 360 degrees.

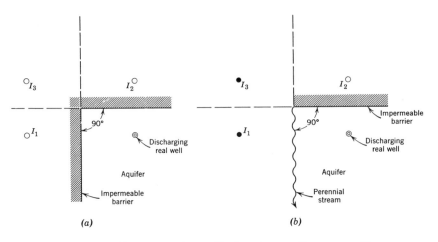

(a) (b)

Fig. 4.24 Image well systems for a discharging well near aquifer boundaries. (a) Aquifer bounded by two impermeable barriers intersecting at right angles. (b) Aquifer bounded by an impermeable barrier intersected at right angles by a perennial stream. Open circles are discharging image wells; filled circles are recharging image wells (after Ferris, et al.[16]).

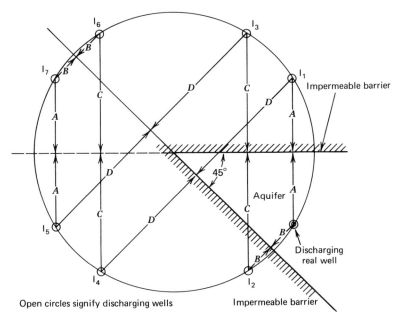

Open circles signify discharging wells

Fig. 4.25 Image well system for a discharging well in an aquifer bounded by two impermeable barriers intersecting at an angle of 45 degrees (after Ferris, et al.[16]).

The above equations involve T for confined aquifers. To adapt them to unconfined aquifers, s should be replaced by $s'' = s - s^2/2h_0$ where h_0 is the initial saturated aquifer thickness. Storage coefficients cannot be calculated from steady-state boundary equations.

Procedures for analyzing unsteady well flows near aquifer boundaries, involving graphic solutions, are available.[16, 23]

Location of Aquifer Boundary. Permeable aquifer boundaries such as streams would normally be visible near a pumping well; however, impermeable subsurface boundaries such as faults or dikes may not be apparent. Where this situation is encountered, the location and orientation of such a barrier can be defined by careful analysis of pumping test data.[9, 16] In the Cooper-Jacob method (see Eq. 4.41) the slope of the straight line on semilogarithmic paper depends only on the pumping rate and the transmissivity. If an impermeable boundary is present, the rate of drawdown in an observation well will double under the influence of an image pumping well (see Fig. 4.26a).* To determine the location of the image well, straight lines are fitted through the two legs of the data. An arbitrary drawdown s_A is selected and a time t_r for this to occur under the influence of the real well is measured (see Fig. 4.26a). Similarly, a time t_i for the same drawdown to be produced by the image well is defined. Then, knowing the distance r_r between the real well and the observation well, the distance r_i to the image well (see Fig. 4.26b) can be found from

$$\frac{r_i{}^2}{t_i} = \frac{r_r{}^2}{t_r} \tag{4.58}$$

The distance r_i defines only the radius of a circle on which the image well lies. It requires measurements in two more observation wells in order to define uniquely by intersection of three arcs the location of the image well (see Fig. 4.26b). The boundary then lies at the midpoint of and perpendicular to a line connecting the real and image wells.

Multiple Well Systems

Where the cones of depression of two nearby pumping wells overlap, one well is said to *interfere* with another because of the increased drawdown and pumping lift created. For a group of wells forming

---

*It should be noted that if the boundary is a stream recharging the aquifer, an image recharge well is introduced. This produces a slope of equal but opposite sign on the drawdown curve, resulting in a horizontal asymptote.

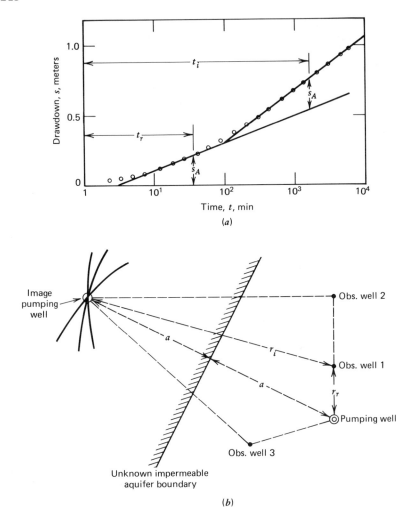

Fig. 4.26 Diagrams illustrating procedure to locate an
unknown impermeable aquifer boundary near a pumping well.
(*a*) Cooper-Jacob drawdown curve showing effect of an impermeable boundary. (*b*) Field situation required to locate an unknown impermeable aquifer boundary.

a well field, the drawdown can be determined at any point if the well
discharges are known, or vice versa. From the principle of superposition, the drawdown at any point in the area of influence caused
by the discharge of several wells is equal to the sum of the drawdowns caused by each well individually. Thus,

$$s_T = s_a + s_b + s_c + \ldots + s_n \qquad (4.59)$$

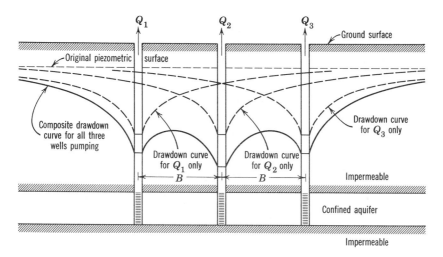

Fig. 4.27 Individual and composite drawdown curves for three wells in a line.

where s_T is the total drawdown at a given point and s_a, s_b, s_c, ..., s_n are the drawdowns at the point caused by the discharge of wells a, b, c, ..., n, respectively. The summation of drawdowns may be illustrated in a simple way by the well line of Fig. 4.27; the individual and composite drawdown curves are given for $Q_1 = Q_2 = Q_3$. Clearly, the number of wells and the geometry of the well field are important in determining drawdowns. Solutions can be based on the equilibrium or nonequilibrium equation. Equations of well discharge for particular well patterns have been developed.[48,54]

In general, wells in a well field designed for water supply should be spaced as far apart as possible so their areas of influence will produce a minimum of interference with each other. On the other hand economic factors such as cost of land or pipelines may lead to a least-cost well layout that includes some interference.[23] For drainage wells designed to control water table elevations, it may be desirable to space wells so that interference increases the drainage effect.

Partially Penetrating Wells

A well whose length of water entry is less than the aquifer it penetrates is known as a *partially penetrating well*. Figure 4.28 illustrates the situation of partially penetrating well in a confined aquifer. The flow pattern to such wells differs from the radial horizontal flow assumed to exist around fully penetrating wells. The average length of a flow line into a partially penetrating well exceeds that into a

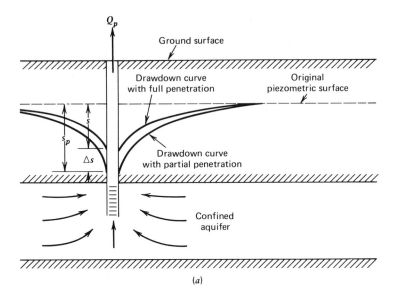

(a)

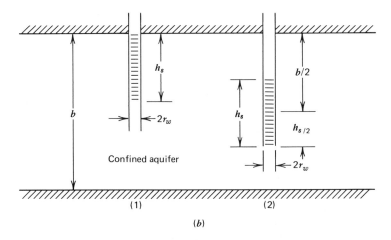

(b)

Fig. 4.28 Partially penetrating wells in a confined aquifer.
(a) Effect of partially penetrating well on drawdown. (b) Two
configurations of partially penetrating wells.

fully penetrating well so a greater resistance to flow is thus encountered. For practical purposes this results in the following relationships between two similar wells, one partially and one fully penetrating the same aquifer: if $Q_p = Q$, then $s_p > s$; and if $s_p = s$, then $Q_p < Q$. Here Q is well discharge, s is drawdown at the well, and the subscript p refers to the partially penetrating well. The effect of partial penetration is negligible on the flow pattern and the drawdown beyond a radial distance larger than 0.5 to 2 times the saturated thickness b, depending on the amount of penetration.

The drawdown s_p at the well face of a partially penetrating well in a confined aquifer (see Fig. 4.28a) can be expressed as

$$s_p = s + \Delta s \tag{4.60}$$

where Δs refers to the additional drawdown resulting from the effect of partial penetration. It can be shown for steady-state conditions and the typical situation* in Fig. 4.28b(1),[30,32]

$$\Delta s = \frac{Q_p}{2\pi T} \frac{1-p}{p} \ln \frac{(1-p)h_s}{r_w} \tag{4.61}$$

where T is transmissivity, p is the penetration fraction ($p = h_s/b$), and h_s and r_w are shown in Fig. 4.28b(1). Equation 4.61 applies where $p > 0.20$.

For the case of a well screen centered in the thickness of the aquifer [see Fig. 4.28b(2)] the value of Δs is given by

$$\Delta s = \frac{Q_p}{2\pi T} \frac{1-p}{p} \ln \frac{(1-p)h_s}{2r_w} \tag{4.62}$$

Equation 4.61 can be modified for a well in an unconfined aquifer by defining

$$\Delta s2h_w = \frac{Q_p}{\pi K} \frac{1-p}{p} \ln \frac{(1-p)h_s}{r_w} \tag{4.63}$$

where h_w is the saturated thickness at the well with full penetration and the hydraulic conductivity $K = T/h_w$. Then

$$s_p{}^2 = s^2 + \Delta s2h_w \tag{4.64}$$

and similarly for Eq. 4.62.

Detailed methods for analyzing effects of partial penetration on well flow for steady and unsteady conditions in confined, unconfined, leaky, and anisotropic aquifers have been outlined by

---

*The drawdown increment is the same whether partial penetration starts from the top or from the bottom of the aquifer.

Hantush[25,29] and others.[37,39,65] Although evaluating the effects is complicated except for the simplest cases, common field situations often reduce the practical importance of partial penetration.* One occurs where an observation well is located more than 1.5 to 2 times the saturated aquifer thickness from a pumping well; in this situation the effect of partial penetration can be neglected for homogeneous and isotropic aquifers. Another applies to many alluvial aquifers with pronounced anisotropy. Here the vertical flow components become small, thereby enabling a pumping well to be approximated as a fully penetrating well in a confined or leaky aquifer with a saturated thickness equal to the length of the well screen.

Well Flow for Special Conditions

A variety of solutions to well flow problems have been derived for special aquifer, pumping, and well conditions.[23,38] Inasmuch as these are of less general application than those outlined heretofore and involve more extensive mathematical treatment, they will be omitted here. It is worth noting, however, that solutions have been obtained for the following special conditions:

1. Constant well drawdown.[2,23]
2. Varying, cyclic, and intermittent well discharges.[22,41,46,58,63,64]
3. Sloping aquifers.[23]
4. Aquifers of variable thickness.[23]
5. Two-layered aquifers.[26,32,51]
6. Anisotropic aquifers.[13,25,29,50,72]
7. Aquifer conditions varying with depth.[47,56]
8. Large-diameter wells.[52,73]
9. Collector wells (see Chapter 5)[28,45]
10. Wells with multiple-sectioned well screens.[57]

Characteristic Well Losses

The drawdown at a well includes not only that of the logarithmic drawdown curve at the well face, but also a *well loss* caused by flow through the well screen and flow inside of the well to the pump intake. Because the well loss is associated with turbulent flow, it may be indicated as being proportional to an nth power of the discharge, as Q^n, where n is a constant greater than one. Jacob[34] suggested that a value $n = 2$ might be reasonably assumed, but Rorabaugh[55] pointed out that n can deviate significantly from 2. An exact value for n

---

*It should be noted that any well with 85 percent or more open or screened hole in the saturated thickness may be considered as fully penetrating.

cannot be stated because of differences of individual wells; detailed investigations of flows inside and outside of wells show that considerable variations occur from assumed flow distributions.

Taking account of the well loss, the total drawdown s_w at the well may be written for the steady-state confined case

$$s_w = \frac{Q}{2\pi T} \ln \frac{r_0}{r_w} + CQ^n \tag{4.65}$$

where C is a constant governed by the radius, construction, and condition of the well. For simplicity let

$$B = \frac{\ln(r_0/r_w)}{2\pi T} \tag{4.66}$$

so that

$$s_w = BQ + CQ^n \tag{4.67}$$

Therefore, as shown in Fig. 4.29, the total drawdown s_w consists of the formation loss BQ and the well loss CQ^n.

Consideration of Eq. 4.67 provides a useful insight to the relation between well discharge and well radius. From Eqs. 4.14 and 4.19 it can be seen that Q varies inversely with $\ln(r_0/r_w)$, if all other variables are held constant. This shows that discharge varies only a small amount with well radius. For example, doubling a well radius

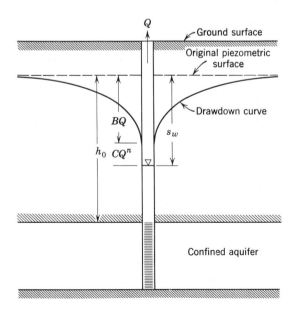

Fig. 4.29 Relation of well loss CQ^n to drawdown for a well penetrating a confined aquifer.

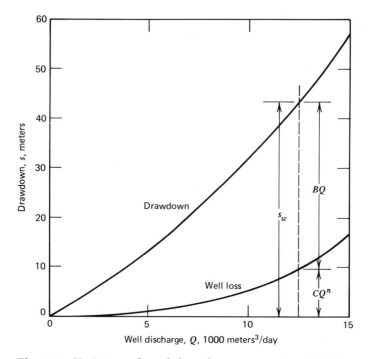

Fig. 4.30 Variation of total drawdown s_w, aquifer loss BQ, and well loss CQ^n with well discharge (after Rorabaugh[55]).

increases the discharge only 10 percent. When the comparison is extended to include well loss, however, the effect is significant. Doubling the well radius doubles the intake area, reduces entrance velocities to almost half, and (if $n = 2$) cuts the frictional loss to less than a third. For axial flow within the well, the area increases four times, reducing this loss an even greater extent.

It is apparent that the well loss can be a substantial fraction of total drawdown when pumping rates are large, as illustrated by Fig. 4.30. With proper design and development of new wells (see Chapter 5), well losses can be minimized. Clogging or deterioration of well screens can increase well losses in old wells.[36] Based on field experience Walton[70] suggested criteria for the well loss coefficient C in Eq. 4.67. These are presented in Table 4.4 to aid in evaluating the condition of a well.

Evaluation of Well Loss. To evaluate well loss a *step-drawdown pumping test* is required. This consists of pumping a well initially at a low rate until the drawdown within the well essentially sta-

**TABLE 4.4 Relation of Well Loss Coefficient
to Well Condition (after Walton[70])**

Well Loss Coefficient C, min^2/m^5	Well Condition
< 0.5	Properly designed and developed
0.5 to 1.0	Mild deterioration or clogging
1.0 to 4.0	Severe deterioration or clogging
> 4.0	Difficult to restore well to original capacity

bilizes.[34,40,59] The discharge is then increased through a successive series of steps as shown by the time-drawdown data in Fig. 4.31a. Incremental drawdowns Δs for each step are determined from approximately equal time intervals. The individual drawdown curves should be extrapolated with a slope proportional to the discharge in order to measure the incremental drawdowns.

From Eq. 4.67 and letting $n = 2$,

$$\frac{s_w}{Q} = B + CQ \tag{4.68}$$

Therefore, by plotting s_w/Q versus CQ (see Fig. 4.31b) and fitting a straight line through the points, the well loss coefficient C is given by the slope of the line and the formation loss coefficient B by the intercept $Q = 0$.

Rorabaugh[55] presented a modification of this graphic analysis to determine n in cases where it deviates significantly from 2.

Specific Capacity

If discharge is divided by drawdown in a pumping well, the *specific capacity* of the well is obtained. This is a measure of the productivity of a well; clearly, the larger the specific capacity, the better the well. Starting from the approximate nonequilibrium equation (Eq. 4.41) and including the well loss,

$$s_w = \frac{2.30Q}{4\pi T} \log \frac{2.25Tt}{r_w^2 S} + CQ^n \tag{4.69}$$

so that the specific capacity

$$\frac{Q}{s_w} = \frac{1}{(2.30/4\pi T)\log(2.25Tt/r_w^2 S) + CQ^{n-1}} \tag{4.70}$$

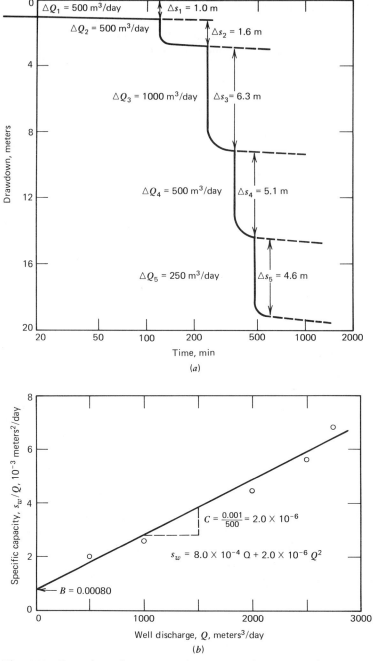

Fig. 4.31 Step-drawdown pumping test analysis to evaluate well loss (after Bierschenk[5]). (a) Time-drawdown data from step-drawdown pumping test. (b Determination of B and C from graph of s_w/Q versus Q.

This indicates that the specific capacity decreases with Q and t; the well data plotted in Fig. 4.32 demonstrate this effect. For a given discharge a well is often assumed to have a constant specific capacity. Although this is not strictly correct, it can be seen that the change with time is minor.

Any significant decline in the specific capacity of a well can be attributed either to a reduction in transmissivity due to a lowering of the groundwater level in an unconfined aquifer or to an increase in well loss associated with clogging or deterioration of the well screen.

If a pumping well is assumed to be 100 percent efficient ($CQ^n = 0$), then the specific capacity from Eq. 4.70 can be presented in the graphic form of Fig. 4.33. Here specific capacity at the end of one day of pumping is plotted as a function of S, T, and a well diameter of 30 cm.[42] This graph provides a convenient means for estimating T from existing pumping wells; any error in S has a small effect on T.[17,31]

Well Efficiency. Figure 4.33 yields a theoretical specific capacity (Q/BQ) for known values of S and T in an aquifer. This computed specific capacity, when compared with one measured in the field (Q/s_w), defines the approximate efficiency of a well.[5] Thus, for a

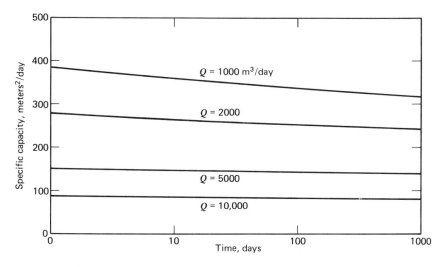

Fig. 4.32 Variation in specific capacity of a pumping well with discharge and time (after Jacob[34]).

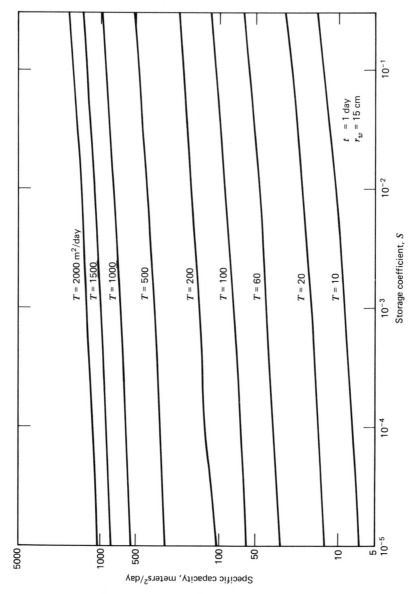

Fig. 4.33 Graph relating specific capacity to transmissivity and storage coefficient from the nonequilibrium equation (after Bentall[4]).

specified duration of pumping, the well efficiency E_w is given as a percentage by

$$E_w = 100\frac{Q/S_w}{Q/BQ} = 100\frac{BQ}{s_w} \tag{4.71}$$

Another method for recognizing an inefficient well is to note its initial recovery rate when pumping is stopped. Where the well loss is large, this drawdown component recovers rapidly by drainage into the well from the surrounding aquifer. A rough rule of thumb for this purpose is: if a pump is shut off after 1 hour of pumping and 90 percent or more of the drawdown is recovered after 5 minutes, it can be concluded that the well is unacceptably inefficient.

References

1. Bennett, G. D., Introduction to ground-water hydraulics, *Techniques of Water-Resources Investigations of the USGS*, Bk. 3, Chap. B2, U.S. Geological Survey, 172 pp., 1976.
2. Bennett, G. D., and E. P. Patten, Jr., Constant-head pumping test of a multiaquifer well to determine characteristics of individual aquifers, *U.S. Geological Survey Water-Supply Paper* 1536-G, pp. 181–203, 1962.
3. Bentall, R., Shortcuts and special problems in aquifer tests, *U.S. Geological Survey Water-Supply Paper* 1545-C, 117 pp., 1963.
4. Bentall, R., Methods for determining permeability, transmissibility, and drawdown, *U.S. Geological Survey Water-Supply Paper* 1536-I, pp. 243–341, 1963.
5. Bierschenk, W. H., Determining well efficiency by multiple step-drawdown tests, *Intl. Assoc. Sci. Hydrology Publ.* 64, pp. 493–507, 1964.
6. Boulton, N. S., The drawdown of the water-table under non-steady conditions near a pumped well in an unconfined formation, *Proc. Inst. Civil Engrs.*, v. 3, pt. III, pp. 564–579, 1954.
7. Boulton, N. S., Analysis of data from non-equilibrium pumping tests allowing for delayed yield from storage, *Proc. Inst. Civil Engrs.*, v. 26, pp. 469–482, 1963.
8. Boulton, N. S., and T. D. Streltsova, New equations for determining the formation constants of an aquifer from pumping test data, *Water Resources Research*, v. 11, pp. 148–153, 1975.
9. Bureau of Reclamation, *Ground water manual*, U.S. Dept. Interior, 480 pp., 1977.
10. Chow, V. T., On the determination of transmissibility and storage coefficients from pumping test data, *Trans. Amer. Geophysical Union*, v. 33, pp. 397–404, 1952.
11. Cooley, R. L., and C. M. Case, Effect of a water table aquitard on drawdown in an underlying pumped aquifer, *Water Resources Research*, v. 9, pp. 434–447, 1973.
12. Cooper, H. H., Jr., and C. E. Jacob, A generalized graphical method for

evaluating formation constants and summarizing well-field history, *Trans. Amer. Geophysical Union*, v. 27, pp. 526–534, 1946.

13. Dagan, G., A method of determining the permeability and effective porosity of unconfined anisotropic aquifers, *Water Resources Research*, v. 3, pp. 1059–1071, 1967.

14. Dupuit, J., *Études théoriques et pratiques sur la mouvement des eaux dans le canaux découverts et à travers les terrains perméables*, 2nd ed., Dunod, Paris, 304 pp., 1863.

15. Ehlig, C., and J. C. Halepaska, A numerical study of confined-unconfined aquifers including effects of delayed yield and leakage, *Water Resources Research*, v. 12, pp. 1175–1183, 1976.

16. Ferris, J. G., et al., Theory of aquifer tests, *U.S. Geological Survey Water-Supply Paper 1536-E*, pp. 69–174, 1962.

17. Gabrysch, R. K., The relation between specific capacity and aquifer transmissibility in the Houston area, Texas, *Ground Water*, v. 6, no. 4, pp. 9–14, 1968.

18. Glover, R. E., *Ground-water movement*, U.S. Bureau of Reclamation Engrng. Monograph no. 31, Denver, 76 pp., 1966.

19. Glover, R. E., *Transient ground water hydraulics*, Dept. of Civil Engrng., Colorado State Univ., Fort Collins, 413 pp., 1974.

20. Glover, R. E., and G. G. Balmer, River depletion resulting from pumping a well near a river, *Trans. Amer. Geophysical Union*, v. 35, pp. 468–470, 1954.

21. Hantush, M. S., Analysis of data from pumping tests in leaky aquifers, *Trans. Amer. Geophysical Union*, v. 37, pp. 702–714, 1956.

22. Hantush, M. S., Drawdown around wells of variable discharge, *Jour. Geophysical Research*, v. 69, pp. 4221–4235, 1964.

23. Hantush, M. S., Hydraulics of wells, *in Advances in hydroscience* (V. T. Chow, ed.), v. 1, Academic Press, New York, pp. 281–432, 1964.

24. Hantush, M. S., Wells near streams with semipervious beds, *Jour. Geophysical Research*, v. 70, pp. 2829–2838, 1965.

25. Hantush, M. S., Wells in homogeneous anisotropic aquifers, *Water Resources Research*, v. 2, pp. 273–279, 1966.

26. Hantush, M. S., Flow to wells in aquifers separated by a semipervious layer, *Jour. Geophysical Research*, v. 72, pp. 1709–1720, 1967.

27. Hantush, M. S., and C. E. Jacob, Non-steady radial flow in an infinite leaky aquifer, *Trans. Amer. Geophysical Union*, v. 36, pp. 95–112, 1955.

28. Hantush, M. S., and I. S. Papadopulos, Flow of ground water to collector wells, *Jour. Hydraulics Div.*, Amer. Soc. Civil Engrs., v. 88. no. HY5, pp. 221–244, 1962.

29. Hantush, M. S., and R. G. Thomas, A method for analyzing a drawdown test in anisotropic aquifers, *Water Resources Research*, v. 2, pp. 281–285, 1966.

30. Huisman, L., *Groundwater recovery*, Winchester, New York, 336 pp., 1972.

31. Hurr, R. T., A new approach for estimating transmissibility from specific capacity, *Water Resources Research*, v. 2, pp. 657–664, 1966.
32. Hydrologisch Colloquium, *Steady flow of ground water towards wells*, Comm. for Hydrological Research, Proc. and Info. no. 10, The Hague, 179 pp., 1964.
33. Jacob, C. E., Radial flow in a leaky artesian aquifer, *Trans. Amer. Geophysical Union*, v. 27, pp. 198–208, 1946.
34. Jacob, C. E., Drawdown test to determine effective radius of artesian well, *Trans. Amer. Soc. Civil Engrs.*, v. 112, pp. 1047–1070, 1947.
35. Jenkins, C. T., Techniques for computing rate and volume of stream depletion by wells, *Ground Water*, v. 6, no. 2, pp. 37–46, 1968.
36. Karanjac, J., Well losses due to reduced formation permeability, *Ground Water*, v. 10, no. 4, pp. 42–49, 1972.
37. Kipp, K. L., Jr., Unsteady flow to a partially penetrating, finite radius well in an unconfined aquifer, *Water Resources Research*, v. 9, pp. 448–462, 1973.
38. Kruseman, G. P., and N. A. de Ridder, *Analysis and evaluation of pumping test data*, Intl. Inst. for Land Reclamation and Improvement, Bull. 11, Wageningen, 200 pp., 1970.
39. Lakshminarayana, V., and S. P. Rajagopalan, Type-curve analysis of time-drawdown data for partially penetrating wells in unconfined anisotropic aquifers, *Ground Water*, v. 16, pp. 328–333, 1978.
40. Lennox, D. H., Analysis and application of step-drawdown test, *Jour. Hydraulics Div.*, Amer. Soc. Civil Engrs., v. 92, no. HY6, pp. 25–48, 1966.
41. Lennox, D. H., and A. Vanden Berg, Drawdowns due to cyclic pumping, *Jour. Hydraulics Div.*, Amer. Soc. Civil Engrs., v. 93, no. HY6, pp. 35–51, 1967.
42. Logan, J., Estimating transmissibility from routine production tests of water wells, *Ground Water*, v. 2, no. 1, pp. 36–37, 1964.
43. Lohman, S. W., Ground-water hydraulics, *U.S. Geological Survey Prof. Paper 708*, 70 pp., 1972.
44. Maasland, D. E. L., and M. W. Bittinger (eds.), *Proceedings of the symposium on transient ground water hydraulics*, Colorado State Univ., Fort Collins, 223 pp., 1963.
45. Milojevic, M., Radial collector wells adjacent to the river bank, *Jour. Hydraulics Div.*, Amer. Soc. Civil Engrs., v. 89, no. HY6, pp. 133–151, 1963.
46. Moench, A., Ground-water fluctuations in response to arbitrary pumpage, *Ground Water*, v. 9, no. 2, pp. 4–8, 1971.
47. Moench, A. F., and T. A. Prickett, Radial flow in an infinite aquifer undergoing conversion from artesian to water table conditions, *Water Resources Research*, v. 8, pp. 494–499, 1972.
48. Muskat, M., *The flow of homogeneous fluids through porous media*, McGraw-Hill, New York, 763 pp., 1937.
49. Neuman, S. P., Theory of flow in unconfined aquifers considering de-

layed response of the water table, *Water Resources Research*, v. 8, pp. 1031–1045, 1972.

50. Neuman, S. P., Analysis of pumping test data from anisotropic unconfined aquifers considering delayed gravity response, *Water Resources Research*, v. 11, pp. 329–342, 1975.

51. Neuman, S. P., and P. A. Witherspoon, Field determination of the hydraulic properties of leaky multiple aquifer systems, *Water Resources Research*, v. 8, pp. 1284–1298, 1972.

52. Papadopulos, I. S., and H. H. Cooper, Jr., Drawdown in a well of large diameter, *Water Resources Research*, v. 3, pp. 241–244, 1967.

53. Prickett, T. A., Type-curve solution to aquifer tests under water-table conditions, *Ground Water*, v. 3, no. 3, pp. 5–14, 1965.

54. Rao, D. B., et al., Drawdown in a well group along a straight line, *Ground Water*, v. 9, no. 4, pp. 12–18, 1971.

55. Rorabaugh, M. I., Graphical and theoretical analysis of step-drawdown test of artesian well, *Proc. Amer. Soc. Civil Engrs.*, v. 79, sep. 362, 23 pp., 1953.

56. Rushton, K. R., and Y. K. Chan, Pumping test analysis when parameters vary with depth, *Ground Water*, v. 14, pp. 82–87, 1976.

57. Selim, S. M., and D. Kirkham, Screen theory for wells and soil drainpipes, *Water Resources Research*, v. 10, pp. 1019–1030, 1974.

58. Sheahan, N. T., Determining transmissibility from cyclic discharge, *Ground Water*, v. 4, no. 3, pp. 33–34, 1966.

59. Sheahan, N. T., Type-curve solution of step-drawdown test, *Ground Water*, v. 9, no. 1, pp. 25–29, 1971.

60. Stallman, R. W., Effects of water table conditions on water level changes near pumping wells, *Water Resources Research*, v. 1, pp. 295–312, 1965.

61. Stallman, R. W., Aquifer test design, observation, and data analysis, *U.S. Geological Survey Techniques of Water Resources Invest.*, Bk. 3, Chap. B1, 26 pp., 1971.

62. Stallman, R. W., and I. S. Papadopulos, Measurement of hydraulic diffusivity of wedge-shaped aquifers drained by streams, *U.S. Geological Survey Prof. Paper* 514, 50 pp., 1966.

63. Sternberg, Y. M., Transmissibility determination from variable discharge pumping tests, *Ground Water*, v. 5, no. 4, pp. 27–29, 1967.

64. Sternberg, Y. M., Simplified solution for variable rate pumping test, *Jour. Hydraulics Div.*, Amer. Soc. Civil Engrs., v. 94, no. HY1, pp. 177–180, 1968.

65. Sternberg, Y. M., Efficiency of partially penetrating wells, *Ground Water*, v. 11, no. 3, pp. 5–8, 1973.

66. Streltsova, T. D., On the leakage assumption applied to equations of groundwater flow, *Jour. Hydrology*, v. 20, pp. 237–253, 1973.

67. Theis, C. V., The relation between the lowering of the piezometric surface and the rate and duration of discharge of a well using ground-water storage, *Trans. Amer. Geophysical Union*, v. 16, pp. 519–524, 1935.

68. Thiem, G. *Hydrologische Methoden*, Gebhardt, Leipzig, 56 pp., 1906.

69. Walton, W. C., *Leaky artesian aquifer conditions in Illinois,* Illinois State Water Survey Rept. Invest. 39, Urbana, 27 pp., 1960.

70. Walton, W. C., *Selected analytical methods for well and aquifer evaluation,* Bull. 49, Illinois State Water Survey, Urbana, 81 pp., 1962.

71. Walton, W. C., Comprehensive analysis of water-table aquifer test data, *Ground Water,* v. 16, pp. 311–317, 1978.

72. Weeks, E. P., Determining the ratio of horizontal to vertical permeability by aquifer-test analysis, *Water Resources Research,* v. 5, pp. 196–214, 1969.

73. Wigley, T. M. L., Flow into a finite well with arbitrary discharge, *Jour. Hydrology,* v. 6, pp. 209–213, 1968.

Water Wells

CHAPTER 5   ···

A water well is a hole or shaft, usually vertical, excavated in the earth for bringing groundwater to the surface. Occasionally wells serve other purposes, such as for subsurface exploration and observation, artificial recharge, and disposal of wastewaters. Many methods exist for constructing wells; selection of a particular method depends on the purpose of the well, the quantity of water required, depth to groundwater, geologic conditions, and economic factors. Shallow wells are dug, bored, driven, or jetted; deep wells are drilled by cable tool or rotary methods. Attention to proper design will ensure efficient and long-lived wells. After a well has been drilled, it should be completed, developed for optimum yield, and tested. Wells should be sealed against entrance of surface pollution and given periodic maintenance. Wells of horizontal extent are constructed where warranted by special groundwater situations.

Test Holes and Well Logs

Before drilling a well in a new area, it is common practice to put down a test hole. The purpose of a test hole is to determine depths to groundwater, quality of water, and physical character and thickness of aquifers without the expense of a regular well, which might prove to be unsuccessful. Diameters seldom exceed 20 cm. Test holes may be put down by any method for well construction; however, cable tool, rotary, and jetting methods are commonly employed. If the test hole appears suitable as a site for a finished well, it can be reamed with hydraulic rotary equipment to convert it to a larger permanent well.

During drilling of a test hole or well, a careful record, or log, is kept of the various geologic formations and the depths at which they are encountered (see Chapter 11). A helpful method is to collect samples of cuttings in containers, labeling each with the depth where obtained. Later these can be studied and analyzed for grain size distribution. Most states require licensed well drillers to submit logs—recording depth, color, character, size of material, and structure of the strata penetrated—for wells they drill. Proper identification of strata in the hydraulic rotary method requires careful analysis because drilling mud is mixed with each sample. A drilling-time log (see Chapter 11) is sometimes helpful in this respect.

Methods for Constructing Shallow Wells

Shallow wells, generally less than 15 m in depth, are constructed by digging, boring, driving, or jetting.[17,49] The methods are briefly described in the following paragraphs; Table 5.1 lists their applications.

Dug Wells. Dating from Biblical times, dug wells have furnished countless water supplies throughout the world.* Depths range up to 20 m or more, depending on the position of the water table, while diameters are usually 1 to 10 m. Dug wells can yield relatively large quantities of water from shallow sources and are most extensively employed for individual water supplies in areas containing unconsolidated glacial and alluvial deposits.[55] Their large diameters permit storage of considerable quantities of water if the wells extend some distance below the water table.

In the past all dug wells were excavated by hand, and even today the same method is widely employed. A typical dug well in underdeveloped portions of the world is often no more than an irregular hole in the ground that intersects the water table (see Fig. 5.1). A pick and shovel are the basic implements. Loose material is hauled to the surface in a container by means of suitable pulleys and lines. Large dug wells can be constructed rapidly with portable excavating equipment such as clamshell and orange-peel buckets. For safety

*See also the description of qanats in Chapter 1.

TABLE 5.1 **Water Well Construction Methods and Applications (after U.S. Soil Conservation Service[50])**

Method	Materials for Which Best Suited	Water Table Depth for Which Best Suited, m	Usual Maximum Depth, m	Usual Diameter Range, cm
Augering Hand auger	Clay, silt, sand, gravel less than 2 cm	2–9	10	5–20
Power auger	Clay, silt, sand, gravel less than 5 cm	2–15	25	15–90
Driven Wells Hand, air hammer	Silt, sand, gravel less than 5 cm	2–5	15	3–10
Jetted Wells Light, portable rig	Silt, sand, gravel less than 2 cm	2–5	15	4–8
Drilled Wells Cable tool	Unconsolidated and consolidated medium hard and hard rock	Any depth	450[b]	8–60
Rotary	Silt, sand, gravel less than 2 cm; soft to hard consolidated rock	Any depth	450[b]	8–45
Reverse-circulation rotary	Silt, sand, gravel, cobble	2–30	60	40–120
Rotary-percussion	Silt, sand, gravel less than 5 cm; soft to hard consolidated rock	Any depth	600[b]	30–50

[a]Yield influenced primarily by geology and availability of groundwater.
[b]Greater depths reached with heavier equipment.

Usual Casing Material	Customary Use	Yield, m^3/day[a]	Remarks
Sheet metal	Domestic, drainage	15–250	Most effective for penetrating and removing clay. Limited by gravel over 2 cm. Casing required if material is loose.
Concrete, steel or wrought-iron pipe	Domestic, irrigation, drainage	15–500	Limited by gravel over 5 cm, otherwise same as for hand auger.
Standard weight pipe	Domestic, drainage	15–200	Limited to shallow water table, no large gravel.
Standard weight pipe	Domestic, drainage	15–150	Limited to shallow water table, no large gravel.
Steel or wrought-iron pipe	All uses	15–15,000	Effective for water exploration. Requires casing in loose materials. Mud-scow and hollow rod bits developed for drilling unconsolidated fine to medium sediments.
Steel or wrought-iron pipe	All uses	15–15,000	Fastest method for all except hardest rock. Casing usually not required during drilling. Effective for gravel envelope wells.
Steel or wrought-iron pipe	Irrigation, industrial, municipal	2500–20,000	Effective for large-diameter holes in unconsolidated and partially consolidated deposits. Requires large volume of water for drilling. Effective for gravel envelope wells.
Steel or wrought-iron pipe	Irrigation, industrial, municipal	2500–15,000	Now used in oil exploration. Very fast drilling. Combines rotary and percussion methods (air drilling) cuttings removed by air. Would be economical for deep water wells.

Fig. 5.1 Women gathering water from a crude dug well in the Shinyanga Region of Tanzania, East Africa (courtesy DHV Consulting Engineers, Amersfoort, The Netherlands).

and to prevent caving, lining (or cribbing) of wood or sheet piling should be placed in the hole to brace the walls.

A modern dug well is permanently lined with a casing (often referred to as a *curb*) of wood staves, brick, rock (Fig. 5.2), concrete, or metal. Curbs should be perforated or contain openings for entry of water and must be firmly seated at the bottom. Dug wells should be deep enough to extend a few meters below the water table. Gravel should be backfilled around the curb and at the bottom of the well to

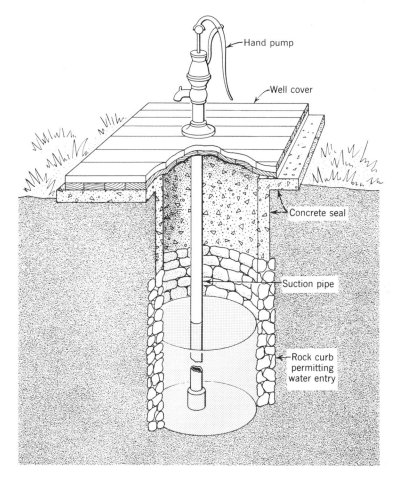

Fig. 5.2 A modern domestic dug well with a rock curb, concrete seal, and hand pump.

control sand entry and possible caving. A properly constructed dug well penetrating a permeable aquifer can yield 2500 to 7500 m³/day, although most domestic dug wells yield less than 500 m³/day.

A serious limitation of large open dug wells involves the ease of their pollution by surface water, airborne material, and objects falling or finding entrance into the wells.

Bored Wells. Where a water table exists at a shallow depth in an unconsolidated aquifer, bored wells can furnish small quantities of water at minimum cost. Bored wells are constructed with hand-operated or power-driven earth augers. Hand augers are available in several shapes and sizes, all operating with cutting blades at the bottom that bore into the ground with a rotary motion. When the blades are full of loose earth, the auger is removed from the hole and emptied; the operation is repeated until the desired hole depth is reached. Hand-bored wells seldom exceed 20 cm in diameter and 15 m in depth. Power-driven augers will bore holes up to 1 m in diameter and, under favorable conditions, to depths exceeding 30 m. The auger consists of a cylindrical steel bucket (Fig. 5.3) with a cutting edge projecting from a slot in the bottom. The bucket is filled by rotating it in the hole by a drive shaft of adjustable length. When full, the auger is hoisted to the surface and the excavated material is removed through hinged openings on the side or bottom of the bucket. Reamers, attached to the top of the bucket, can enlarge holes to diameters exceeding the auger size.

A continuous-flight power auger has a spiral extending from the bottom of the hole to the surface. Cuttings are carried to the surface

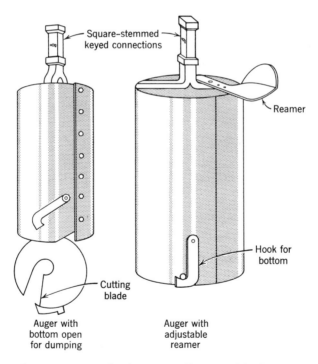

Fig. 5.3 Augers for boring wells. Spiral flight augers are also widely employed for small-diameter holes.

as on a screw conveyor, while sections of the auger may be added as depth increases. The simple equipment, usually truck-mounted, can be operated rapidly by one person and functions to depths exceeding 50 m in unconsolidated formations that do not contain large boulders.

Augers work best in formations that do not cave. Where loose sand and gravel are encountered in a large-diameter hole, or the boring reaches the water table, it may be necessary to lower a concrete or metal casing to the bottom of the hole and continue boring inside. After the desired depth is reached, a permanent well casing and screen are centered in the hole, the outer casing is removed, and the peripheral space is backfilled with gravel. Augers sometimes supplement other well-drilling methods where sticky clay formations are encountered; here augers are more effective than any other penetrating device.

Driven Wells. A driven well consists of a series of connected lengths of pipe driven by repeated impacts into the ground to below the water table. Water enters the well through a *drive* (or *sand*) *point* at the lower end of the well (Fig. 5.4). This consists of a screened cylindrical section protected during driving by a steel cone at the bottom. Diameters of driven wells are small, most falling in the range of 3 to 10 cm. Standard-weight water pipe having threaded couplings serves for casing. Most depths are less than 15 m although a few exceed 20 m. As suction-type pumps extract water from driven wells, the water table must be near the ground surface if a continuous water supply is to be obtained. For best results the water table should be within 3 to 5 m of ground surface in order to provide adequate drawdown without exceeding the suction limit. Yields from driven wells are small, with discharges of about 100–250 m³/day.

Driven wells are best suited for domestic supplies, for temporary water supplies, and for exploration and observation. Batteries of driven wells connected by a suction header to a single pump are effective for localized lowering of the water table. Such installations, known as *well-point systems,* are particularly advantageous for dewatering excavations for foundations and other subsurface construction operations.[34] Figure 5.5 illustrates how a well-point installation reduces the groundwater level to furnish a dry excavation.

Driven wells are limited to unconsolidated formations containing no large gravel or rocks that might damage the drive point. To drive a well, the pipe casing and threads should be protected at the top with a drive cap (see Fig. 5.4). Driving can be done with a maul,

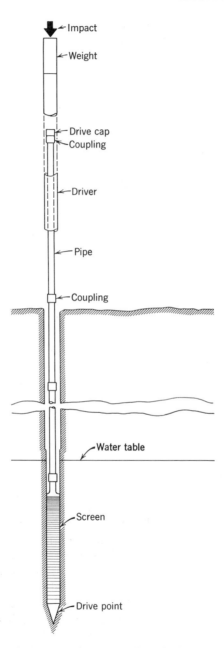

Fig. 5.4 A driven well with driving mechanism.

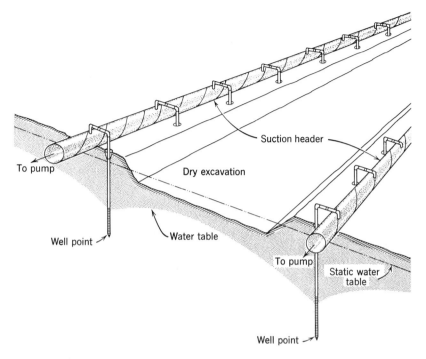

Fig. 5.5 A well-point system to dewater an excavation site.

sledge, drop hammer, or air hammer. It is usually good practice to
place (by boring or driving) an outer protective casing to at least
3 m below ground surface. Screens are available in a variety of
opening sizes, the choice depending on the size of particles in the
water-bearing stratum.

Important advantages of driven wells are that they can be constructed in a short time, at minimum cost, and even by one person.

Jetted Wells. Jetted wells are constructed by the cutting action
of a downward-directed stream of water. The high-velocity stream
washes the earth away, while the casing, which is lowered into the
deepening hole, conducts the water and cuttings up and out of the
well. Small-diameter holes of 3 to 10 cm are formed in this manner
(although the method is capable of producing diameters up to 30 cm
or more) to depths greater than 15 m. Jetted wells have only small
yields and are best adapted to unconsolidated formations. Because
of the speed of jetting a well and the portability of the equipment,

Fig. 5.6   Various designs of jetting drill bits.

jetted wells are useful for exploratory test holes, observation wells, and well-point systems.[9,30]

Various types of jetting drill bits are shown in Fig. 5.6. In penetrating clays and hardpans, the drill pipe is raised and lowered sharply, causing the bit to shatter the formation. During the jetting operation, the drill pipe is turned slowly to ensure a straight hole. To complete a shallow jetted well after the casing extends to below the water table, the well pipe with screen attached is lowered to the bottom of the hole inside the casing. The outer casing is then pulled, gravel is inserted in the outer space, and the well is ready for pumping.

A simplification of the above procedure can be obtained by using a self-jetting well point. This consists of a tube of brass screen ending in a jetting nozzle, which is screwed to the well pipe (Fig. 5.7). As soon as the well point has been jetted to the required depth, the well is completed and ready for pumping. Gravel should be added around the drill pipe for permanent installations.

Methods for Drilling Deep Wells

Most large, deep, high-capacity wells are constructed by drilling. Construction can be accomplished by the cable tool method or by one of several rotary methods.[12,20,23] Each method has particular advantages, so experienced drillers endeavor to have equipment available for a diversity of drilling approaches.[1,2,20] Applications of drilling methods are listed in Table 5.1, while Table 5.2 indicates the performance of the methods in various geologic formations.

Examples of the construction of deep wells in unconsolidated and consolidated formations are shown in Figs. 5.8 and 5.9, respectively, taken from standard specifications for deep wells prepared by the American Water Works Association. The construction procedure

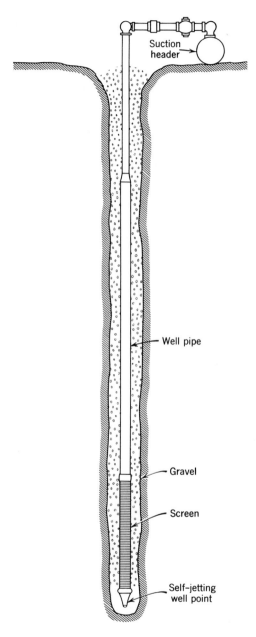

Fig. 5.7 Jetted well with self-jetting well point.

**TABLE 5.2   Performance of Drilling Methods
in Various Types of Geologic Formations
(after Speedstar Div.[40])**

Type of Formation	Drilling Method		
	Cable Tool	Rotary	Rotary Percussion[a]
Dune sand	Difficult	Rapid	NR
Loose sand and gravel	Difficult	Rapid	NR
Quicksand	Difficult, except in thin streaks. Requires a string of drive pipe.	Rapid	NR
Loose boulders in alluvial fans or glacial drift	Difficult; slow but generally can be handled by driving pipe	Difficult, frequently impossible	NR
Clay and silt	Slow	Rapid	NR
Firm shale	Rapid	Rapid	NR
Sticky shale	Slow	Rapid	NR
Brittle shale	Rapid	Rapid	NR
Sandstone, poorly cemented	Slow	Slow	NR
Sandstone, well cemented	Slow	Slow	NR
Chert nodules	Rapid	Slow	NR
Limestone	Rapid	Rapid	Very rapid
Limestone with chert nodules	Rapid	Slow	Very rapid
Limestone with small cracks or fractures			Very rapid
Limestone, cavernous	Rapid	Slow to impossible	Difficult
Dolomite	Rapid	Rapid	Very rapid
Basalts, thin layers in sedimentary rocks	Rapid	Slow	Very rapid
Basalts, thick layers	Slow	Slow	Rapid
Metamorphic rocks	Slow	Slow	Rapid
Granite	Slow	Slow	Rapid

[a]NR: not recommended.

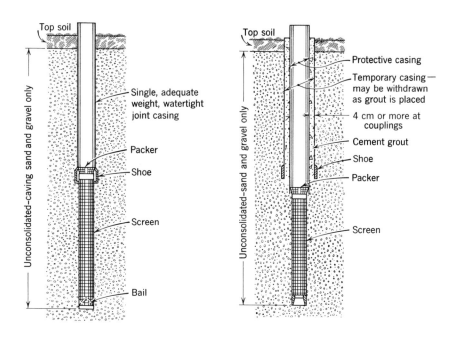

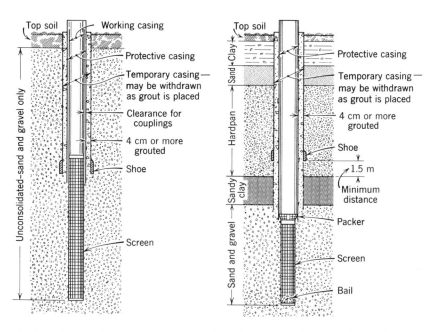

Fig. 5.8 Examples of well construction in unconsolidated formations (reprinted from American Water Works Association *AWWA Standard for Deep Wells*[3], by permission of the Association; copyright © 1967 by American Water Works Association, 6666 West Quincy Avenue, Denver, CO 80235).

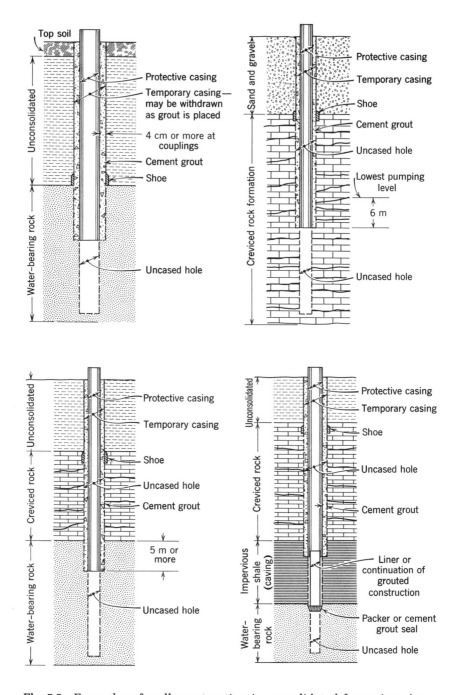

Fig. 5.9 Examples of well construction in consolidated formations (reprinted from American Water Works Association *AWWA Standard for Deep Wells*[3], by permission of the Association; copyright © 1967 by American Water Works Association, 6666 West Quincy Avenue, Denver, CO 80235).

of a successful well is dependent on local conditions encountered in drilling; hence, each well should be treated as an individual project. Construction methods differ regionally within the United States and also from one driller to another. General construction methods are described in the following sections.

Cable Tool Method. Wells drilled by the cable tool (also *percussion* or *standard*) method are constructed with a standard well-drilling rig, percussion tools, and a bailer.[17,19,48] The method is capable of drilling holes of 8 to 60 cm in diameter through consolidated rock materials to depths of 600 m. In unconsolidated sand and gravel, especially quicksand, it is least effective because the loose material slumps and caves around the bit. Drilling is accomplished by regular lifting and dropping of a string of tools. On the lower end, a bit with a relatively sharp chisel edge breaks the rock by impact.

From top to bottom, a string of tools consists of a *swivel socket, a set of jars,* a *drill stem,* and a *drilling bit* (Fig. 5.10). The total weight may amount to several thousand kilograms. Tools are made of steel and are joined with tapered box-and-pin screw joints. The most important part of the string of tools is the bit, which does the actual drilling. Bits are manufactured in lengths of 1 to 3 m and weigh up

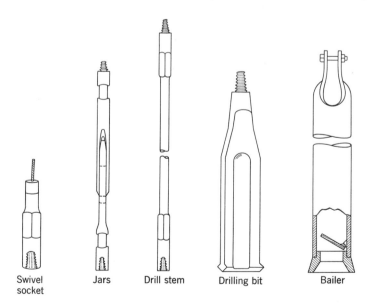

| Swivel socket | Jars | Drill stem | Drilling bit | Bailer |

Fig. 5.10 Basic well drilling tools for the cable tool method.

to 1500 kg. Variously shaped bits are made for drilling in different rock formations. The drill stem is a long steel bar that adds weight and length to the drill so it will cut rapidly and vertically. A set of jars consists of a pair of narrow connecting links. They have no direct effect on the drilling; their purpose is only to loosen the tools should they stick in the hole. Under normal tension on the drilling line the jars remain fully extended. When tools become stuck, the line is slackened to allow the links to open to their full length, whereupon an upstroke of the line will cause the upper section of the jars to impart an upward blow to the tools. The swivel socket attaches the drilling cable to the string of tools.

Drill cuttings are removed from the well by a *bailer* or *sand bucket* (Fig. 5.10). Although several models are manufactured, a bailer consists essentially of a section of pipe with a valve at the bottom and a ring at the top for attachment to the bailer line. When lowered into the well, the valve permits cuttings to enter the bailer but prevents them from escaping. After filling, the bailer is hoisted to the surface and emptied. Bailers are available in a range of diameters, lengths of 3 to 8 m, and capacities up to 0.25 m^3.

The drilling rig for the cable tool method consists of a mast, a multiline hoist, a walking beam, and an engine. In most present-day designs the entire assembly is truck-mounted (Fig. 5.11) for ready portability.

During drilling the tools make 20–40 strokes per minute, ranging from 40 to 100 cm in length. The drilling line is rotated so that the bit forms a round hole, and additional line is let out as needed so that the bit will always strike the bottom of the hole. Water should be added to the hole if none is encountered to form a paste with the cuttings, thereby reducing friction on the falling bit. After the bit has cut 1 or 2 m through a formation, the string of tools is lifted to the surface and the hole is bailed. In unconsolidated formations, casing should be maintained to near the bottom of the hole to avoid caving. Casing is driven down by means of drive clamps fastened to the drill stem; the up and down motion of the tools striking the top of the casing, protected by a *drive head,* sinks the casing. On the bottom of the first section of casing, a *drive shoe* (see Fig. 5.8) with a beveled cutting edge is fastened to protect the casing as it is being driven.

As casing is driven deeper in unconsolidated formations, the vibration causes the sides of the hole to collapse against it. Frictional forces increase until further driving becomes impossible. When this occurs a smaller-diameter casing is telescoped inside of the casing already in the hole; thereafter, drilling is continued using a smaller-

Fig. 5.11 Cable tool drilling rig in operation. The driller holds the cable to sense the progress of the drilling. A bailer is standing to the right of the rig (courtesy Bucyrus-Erie Company).

diameter bit. In deep holes several such reductions in casing may be required.

In drilling any deep well it is important that proper alignment be maintained so as not to interfere with pump installation and operation. The greatest problem occurs in drilling through rock formations. Some drillers have found that holes tending to bend can be corrected by detonating explosives at the bottom. This shatters the surrounding rock and permits drilling to progress vertically.

The cable tool rig is highly versatile in its ability to drill satisfactorily over a wide range of geologic conditions.* Its major draw-

---

*In particular, cable tool rigs can drill through boulders and fractured, fissured, broken, or cavernous rocks, which often are beyond the capabilities of other types of equipment (see Table 5.2).

backs are its slower drilling rate, its depth limitation, the necessity of driving casing coincidentally with drilling in unconsolidated materials, and the difficulty of pulling casing from deep holes. The simplicity of design, ruggedness, and ease of maintenance and repair of the rigs and tools are important advantages in isolated areas.[46] Also, less water is required for drilling than with other methods, a matter of concern in arid and semiarid regions. Furthermore, sampling and formation logging are simpler and more accurate with a cable tool rig.

Rotary Method. A rapid method for drilling in unconsolidated strata is the rotary method.[33] Deep wells up to 45 cm in diameter, and even larger with a reamer, can be constructed. The method operates continuously with a hollow rotating bit through which a mixture of clay and water, or *drilling mud,* is forced. Material loosened by the bit is carried upward in the hole by the rising mud. No casing is ordinarily required during drilling because the mud forms a clay lining, or *mud cake,* on the wall of the well by filtration. This seals the walls, thereby preventing caving, entry of groundwater, and loss of drilling mud.

Drill bits are available in various forms; a group of conical roller gears with teeth that scrape, grind, and fracture the rock is a common design (see Fig. 5.12). The typical string of tools consists of a bit, a drill collar (which adds weight to the bit and aids in maintaining hole alignment), and a drill pipe that extends to the ground surface. The upper end of the drill pipe is attached to the *kelly*—a square

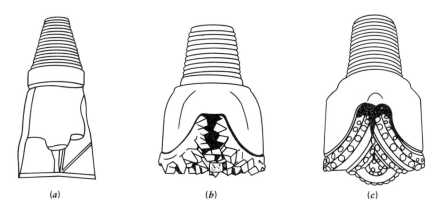

(a)   (b)   (c)

Fig. 5.12 Examples of rotary drill bits. (*a*) Fishtail bit. (*b*) Cone-type rock bit. (*c*) Carbide button bit (after Speedstar Div.[40]).

section of drill rod. The drill is turned by a rotating table that fits closely around the kelly and allows the drill rod to slide downward as the hole deepens. The drilling rig for a rotary outfit consists of a derrick, or mast, a rotating table, a pump for the drilling mud, a hoist, and the engine.

Drilling mud consists of a suspension of water, bentonite, clay, and various organic additives. Maintenance of the correct mud in terms of weight, viscosity, jelling strength, and low percentage of suspended solids is important for trouble-free drilling.[46] Organic additives that degrade with time and thereby cause the mud cake to break down within a few days are a recent innovation. The drilling mud leaves the drill pipe through the bit where it cools and lubricates the cutting surface, entrains drill cuttings, and carries the drill cuttings upward within the annular space between drill pipe and hole wall as the fluid returns to ground surface (see Fig. 5.13). The drilling mud then overflows into a ditch and passes into a settling pit. Here the cuttings settle out; thereafter, the mud is picked up by the pump for recirculation in the hole.

Rotary drilling is employed for oil wells and its application to water-well drilling is steadily increasing. Advantages are the rapid drilling rate, the avoidance of placement of a casing during drilling, and the convenience for electric logging (see Chapter 12). Disadvantages include high equipment cost, more complex operation, the need to remove mud cake during well development, and the problem of lost circulation in highly permeable or cavernous geologic formations.

Air Rotary Method. Rotary drilling can also be accomplished with compressed air in place of drilling mud. The technique is rapid and convenient for small-diameter holes in consolidated formations where a clay lining is unnecessary to support the walls against caving. Larger-diameter holes can be drilled by employing foams and other air additives.[12] Drilling depths can exceed 150 m under favorable circumstances. An important advantage of the air rotary method is its ability to drill through fissured rock formations with little or no water required.

Rotary-Percussion Method. A recently developed rotary-percussion procedure using air as the drilling fluid provides the fastest method for drilling in hard-rock formations. A rotating bit, with the action of a pneumatic hammer, delivers 10 to 15 impacts per second to the bottom of the hole. Penetration rates of as much as

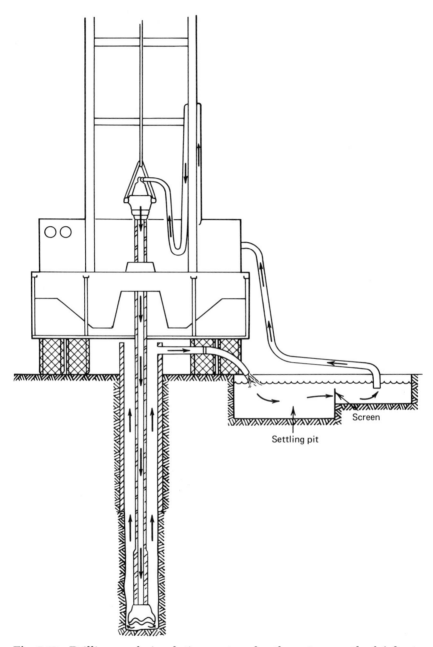

Fig. 5.13   Drilling mud circulation system for the rotary method (after Speedstar Div.[40]).

0.3 m/min have been achieved. Where caving formations or large quantities of water are encountered, a change to conventional rotary drilling with mud usually becomes necessary.

Reverse-Circulation Rotary Method. The reverse-circulation rotary method has become increasingly popular as a means for drilling large-diameter holes in unconsolidated formations. Water is pumped up through the drill pipe employing a large-capacity centrifugal or jet pump similar to those for gravel dredges. Discharge from the hole flows into a large pit where cuttings settle out. Thereafter the water runs through a ditch and back into the hole so that the water level in the hole is maintained at ground surface.

To avoid erosion of the sides of the hole downward velocities must be restricted; therefore, the minimum hole diameter is about 40 cm. Drilling bits range in diameter from 0.4 to 1.8 m. The velocity of water up the drill pipe usually exceeds 2 m/s.

The water table should be 3 to 4 m below ground surface in order to obtain an effective head differential between well and aquifer. With this difference, fine-grained particles suspended in the water aid in stabilizing the walls. Where the water table is closer to the surface, casing can be extended above ground surface to increase the head. On the other hand, where the water table is deep, it may be necessary to place surface casing to minimize water loss.

For unconsolidated formations the reverse-circulation rig is probably the most rapid drilling equipment available. It requires a large volume of readily available water. Such rigs normally can drill to depths of 125 m; modifications with airlift pumping can extend this depth range.* Because reverse-circulation holes have large diameters, completed wells are usually gravel packed.

Well Completion

After a well has been drilled, it must be completed. This can involve placement of casing, cementing of casing, placement of well screens, and gravel packing; however, wells in hard-rock formations can be left as open holes so that these components may not be required.

---

*A versatile modification of reverse rotary drilling is the *dual-tube* method. This involves a rotating outer casing connected to an inner concentric pipe with air serving as the circulating fluid. Rapid drilling rates are possible with diameters up to 40 cm and depths to 400 m. A particular advantage of the method is that continuous and accurate geologic and groundwater quality samples can be obtained as a function of depth.

Well Casings. Well casing serves as a lining to maintain an open hole from ground surface to the aquifer. It seals out surface water and any undesirable groundwater and also provides structural support against caving materials outside the well. Materials commonly employed for well casings are wrought iron, alloyed or unalloyed steel, and ingot iron.[5] Joints normally consist of threaded couplings or are welded, the object being to secure watertightness. In cable-tool drilling, the casing is driven into place; in rotary methods, the casing is smaller than the drilled hole and hence can be lowered into place. Polyvinyl chloride pipe is widely employed as casing for shallow, small-diameter observation wells.

Surface casing is installed from ground surface through upper strata of unstable or fractured materials into a stable and, if possible, relatively impermeable material.[46] Such surface casing serves several purposes, including: (1) supporting unstable materials during drilling, (2) reducing loss of drilling fluids, (3) facilitating installation or removal of other casing, (4) aiding in placing a sanitary seal, and (5) serving as a reservoir for a gravel pack. This casing may be temporary during drilling or it may be permanent. Recommended minimum diameters of surface casing are given in Table 5.3.

Pump chamber casing comprises all casing above the screen in wells of uniform diameter. For telescoping wells it is the casing within which the pump bowls are set. Recommended minimum diameters are listed in Table 5.3. Normally, the pump chamber casing should have a nominal diameter at least 5 cm larger than the nominal diameter of the pump bowls. Nonmetallic pipes are sometimes employed where corrosion or incrustation is a problem. Possibilities include ceramic-clay, concrete, asbestos-cement, plastic, and fiberglass-reinforced plastic pipe; strengths of these materials are not comparable to steel pipe, however.

Cementing. Wells are cemented in the annular space surrounding the casing to prevent entrance of water of unsatisfactory quality, to protect the casing against exterior corrosion, and/or to stabilize caving rock formations. Cement grout, consisting of a mixture of cement and water and sometimes various additives, can be placed by a dump bailer, by a tremie pipe, or by pumping.[5,12] It is important that the grout be introduced at the bottom of the space to be grouted to ensure that the zone is properly sealed.

Screens. In consolidated formations, where the material surrounding the well is stable, groundwater can enter directly into an uncased well. In unconsolidated formations, however, wells are

TABLE 5.3 Recommended Minimum Diameters
for Well Casings and Screens
(after U.S. Bureau of Reclamation[46])

Well Yield, m³/day	Nominal Pump Chamber Casing Diameter, cm	Surface Casing Diameter, cm		Nominal Screen Diameter, cm
		Naturally Developed Wells	Gravel-Packed Wells	
< 270	15	25	45	5
270–680	20	30	50	10
680–1,900	25	35	55	15
1,900–4,400	30	40	60	20
4,400–7,600	35	45	65	25
7,600–14,000	40	50	70	30
14,000–19,000	50	60	80	35
19,000–27,000	60	70	90	40

equipped with screens. These stabilize the sides of the hole, prevent sand movement into the well, and allow a maximum amount of water to enter the well with a minimum of hydraulic resistance.

In the cable-tool method of drilling, screens are normally placed by the *pullback method.* After casing is in place, the screen is lowered inside, and the casing is pulled up to near the top of the screen. A lead packer ring on the top of the screen is flared outward to form a seal between the inside of the casing and the screen. For the rotary method of drilling without casing, screens are lowered into place as drilling mud is diluted and again are sealed by a lead packer to an upper permanent casing. Screens are also sometimes placed by the *bail-down method,* involving bailing out material below the screen until the screen section is lowered to the desired aquifer depth.

In the past, well casings were often perforated in place by a special cutting knife. This practice is now generally discontinued because of the large irregular openings created, the small percentage of open area obtained, and the difficulty of controlling entry of sand with water during pumping. More common is the use of preperforated casing, constructed by sawing, machining, or torch-cutting slots in the casing. Slot openings range from less than 1 to 6 mm; with larger slots the maximum percentage of open area is about 12 percent.[46] Openings by sawing or machining can be properly sized, whereas torch-cut slots tend to be large, irregular, and conducive to sand entry.

A major factor in controlling head loss through a perforated well section is the percentage of open area. For practical purposes a minimum open area of 15 percent is desirable; this value is readily obtained with many commercial screens but not with preperforated casing.[43,46]

Manufactured screens are preferred to preperforated casing because of the ability to tailor opening sizes to aquifer conditions and because of larger percentages of open area that can be achieved. Several types are available: punched, stamped, louvred, wire-wound perforated pipe, and continuous-slot wire-wound screens. The latter type, consisting of a continuous winding of round or specially shaped wire on a cage of vertical rods, is the most efficient, possesses the largest open area, and can be closely matched to aquifer gradations (see Fig. 5.14). Although such screens are more expensive, they may prove to be more economical, especially for thin but highly productive aquifers.

Screens are available in a range of diameters; selection of screen diameter should be made on the basis of the desired well yield and aquifer thickness. Recommended minimum screen diameters are

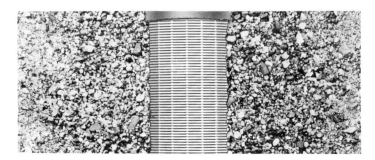

Fig. 5.14 Continuous slot wire-wound well screen in an unconsolidated formation. The grain size distribution around the screen illustrates a properly developed well (courtesy Johnson Div., UOP Inc.).

included in Table 5.3. To minimize well losses and screen clogging, entrance velocities should be kept within specified limits. Because aquifers composed of finer-grained materials tend to clog more easily than do those of coarser materials, field experience has indicated that there is a relationship between hydraulic conductivity of an aquifer and screen entrance velocity, as shown in Table 5.4.

To express the velocities in Table 5.4 in terms of screen size, the following equation can be applied:

$$v_s = \frac{Q}{c\pi d_s L_s P} \tag{5.1}$$

TABLE 5.4 Optimum Entrance Velocity of Water Through a Well Screen (after Walton[52])

Hydraulic Conductivity of Aquifer, m/day	Optimum Screen Entrance Velocity, m/min
>250	3.7
250	3.4
200	3.0
160	2.7
120	2.4
100	2.1
80	1.8
60	1.5
40	1.2
20	0.9
<20	0.6

where v_s is the optimum screen entrance velocity, Q is well discharge, c is a clogging coefficient (estimated at 0.5 on the basis that approximately 50 percent of the open area of a screen will be blocked by aquifer material), d_s is the screen diameter, L_s is the screen length, and P is the percentage of open area in the screen (available from manufacturer's specifications). Thus, for a given aquifer material, aquifer thickness, well yield, and type of screen, the appropriate diameter and length of well screen can be selected.

Screens are made of a variety of metals and metal alloys, plastics, concrete, asbestos-cement, fiberglass-reinforced epoxy, coated base metals, and wood.[46] Because a well screen is particularly susceptible to corrosion and incrustation, nonferrous metals, alloys, and plastics are often selected to prolong well life and efficient operation. Table 5.5 lists the more common metallic screen materials in the order of increasing cost.

A significant characteristic of a well screen is its slot size, which should be determined from mechanical analyses of formation samples obtained during drilling of the well or a pilot hole.* If the uniformity coefficient of an aquifer sample for a naturally developed

TABLE 5.5   Metallic Well Screen
Materials and Their Resistance to Acid and Corrosion
(after U.S. Bureau of Reclamation[46])

Material[a]	Acid Resistance	Corrosion Resistance in Normal Groundwater
Low-carbon steel	Poor	Poor[b,c]
Toncan and Armco iron	Poor	Fair[b,c]
Admiralty red brass	Good	Good[c]
Silicon red brass	Good	Good[c]
304 stainless steel	Good	Very good
Everdure bronze	Very good	Very good[d]
Monel metal	Very good	Very good[d]
Super nickel	Very good	Very good[d]

[a]Materials are listed in order of increasing cost.

[b]Not recommended for permanent installations where incrustation is a serious problem.

[c]Not recommended for permanent installations where sulfate-reducing or similar bacteria are present or where water contains more than 60 mg/l SO_4.

[d]Recommended only in areas where corrosion is very aggressive.

*Screen manufacturers will often recommend the most satisfactory slot size based on the grain size analysis of a given aquifer.

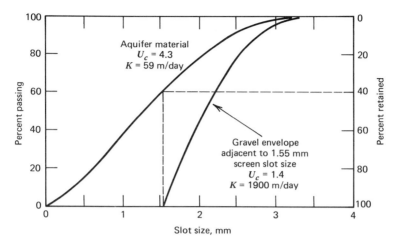

Fig. 5.15 Development of a natural gravel pack around a well screen from a well-graded unconsolidated aquifer (after Mogg[31]).

well (without a gravel pack) is 5 or less, the selected slot size should retain 40 to 50 percent of the aquifer material. For a uniformity coefficient greater than 5, the slot size should retain 30 to 50 percent of the aquifer material.[46] In essence, the screen permits finer material to enter the well and to be removed by bailing during development of the well.[15,21,39,51] But coarser material is retained outside, forming a permeable envelope around the well (see Fig. 5.14). An illustration of how a well screen can form a natural gravel pack around a well is shown in Fig. 5.15. Here, with 40 percent of the aquifer material retained, the resulting gravel envelope has a permeability more than 30 times greater than that of the aquifer.

Where a well screen is to be surrounded by an artificial gravel pack, the size of the screen openings is governed by the size of the gravel (see below).

Gravel Packs. A gravel-packed well is one containing an artificially placed gravel screen or envelope surrounding the well screen (see Fig. 5.16). A gravel pack (1) stabilizes the aquifer, (2) minimizes sand pumping, (3) permits use of a large screen slot with a maximum open area, and (4) provides an annular zone of high permeability, which increases the effective radius and yield of the well. Maximum grain size of a pack should be near 1.0 cm, while the thickness should be in the range of 8 to 15 cm.

Various formulas for relating gravel pack grain-size gradations to aquifer grain-size gradations have been developed.[22,26,31,38] Criteria

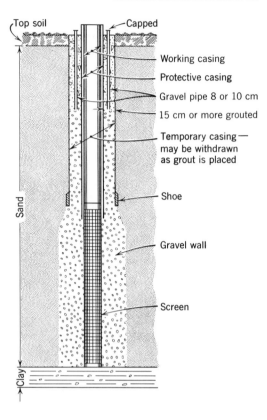

Fig. 5.16 Vertical cross-section of a gravel-packed well (reprinted from American Water Works Association *AWWA Standard for Deep Wells*[3], by permission of the Association; copyright © 1967 by American Water Works Association, 6666 West Quincy Avenue, Denver, CO 80235).

conforming to U.S. Bureau of Reclamation field experience are summarized in Table 5.6. The selected gravel should be washed and screened siliceous material that is rounded, abrasive-resistant, and dense. Gravel should be placed in such a manner as to ensure complete filling of the annular space and to minimize segregation. A common procedure is to extend two tremie pipes to the bottom of the well on opposite sides of the screen. Gravel is poured, washed, or pumped into the tremie pipes; these are then withdrawn in stages as the pack is placed.* In cable-tool holes the inner casing and screen are set inside the blank outer casing, the annular space is filled with

---

*A prudent design is to install permanent tremie pipes that extend from between the surface and pump chamber casings to the aquifer (see Fig. 5.16). In the event of excessive gravel settlement, additional gravel can move downward to maintain the pack.

**TABLE 5.6   Criteria for Selection of Gravel Pack Material
(after U.S. Bureau of Reclamation[46])**

Uniformity Coefficient (U_c) of Aquifer	Gravel Pack Criteria	Screen Slot Size
<2.5	(a) U_c between 1 and 2.5 with the 50% size not greater than 6 times the 50% size of the aquifer (b) If (a) is not available, U_c between 2.5 and 5 with 50% size not greater than 9 times the 50% size of the aquifer	≤10% passing size of the gravel pack
2.5–5	(a) U_c between 1 and 2.5 with the 50% size not greater than 9 times the 50% size of the formation (b) If (a) is not available, U_c between 2.5 and 5 with 50% size not greater than 12 times the 50% size of the aquifer	≤10% passing size of the gravel pack
>5	(a) Multiply the 30% passing size of the aquifer by 6 and 9 and locate the points on the grain-size distribution graph on the same horizontal line (b) Through these points draw two parallel lines representing materials with $U_c \leq 2.5$ (c) Select gravel pack material that falls between the two lines	≤10% passing size of the gravel pack

gravel, and thereafter the outer casing is pulled out of the well. In sandy aquifers, where a gravel pack is most essential, deep wells should be constructed by the rotary or reverse-circulation rotary method. The drilling fluid should be circulated and diluted with water before the gravel is introduced.

Well Development

Following completion, a new well is developed to increase its specific capacity, prevent sanding, and obtain maximum economic well life. These results are accomplished by removing the finer material from the natural formations surrounding the perforated sections of the casing. Where a well has been gravel-packed, much of the same purpose has been accomplished, although development is still beneficial. The importance of developing wells cannot be underestimated; all too often development is not carried out adequately to produce full potential yields.

Development procedures are varied and include pumping, surging, use of compressed air, hydraulic jetting, addition of chemicals, hydraulic fracturing, and use of explosives.[23,46] These are briefly described in subsequent paragraphs.

Pumping. This procedure involves pumping a well in a series of steps from a low discharge to one exceeding the design capacity.* To be most effective the intake area of the pump should extend to near the center of the screened section. At each step the well is pumped until the water clears, after which the power is shut off and water in the pump column surges back into the well. The step is repeated until only clear water appears. The discharge rate is then increased and the procedure repeated until the final rate is the maximum capacity of the pump or well. This irregular and noncontinuous pumping agitates the fine material surrounding the well so that it can be carried into the well and pumped out. The coarser fraction entering the well is removed by a bailer or sand pump from the bottom. This development method by pumping is recommended as a finishing procedure after any of the development techniques described subsequently.

Surging. Another method for developing a well is by the up-and-down motion of a surge block attached to the bottom of a drill stem. Such blocks are particularly applicable with a cable tool rig.[46] Solid, vented, and spring-loaded surge blocks, often constructed by well-drilling contractors, are employed.† The cylindrical block is 2 to

---

*In the field this technique is sometimes referred to as "rawhiding" a well. If a high discharge occurs initially, "bridging" (wedging sand grains around individual perforations formed by the sudden pull on the sand toward the well) can prevent fine material from being removed and reduce the effectiveness of the development process.

†Surging can also be accomplished with a flap-valve bailer, if a close fit exists within the well screen.

5 cm smaller than the well screen and fitted with belting, rubber, or leather that will not damage the screen. As the block is moved up and down in the screen, a surging action is imparted to the water. The downstroke causes backwash to break up any bridging which may occur, while the upstroke pulls dislodged sand grains into the well.

Initially, surging should begin with a slow stroke at the bottom of the screen and progress to the top of the screen. This should then be repeated with increasingly faster strokes. The procedure is completed when material accumulating in the bottom of the well becomes negligible. For wells in rock aquifers, surging can be accomplished in the casing above open holes.

Surging with Air. To develop wells by compressed air, an air compressor is connected to an air pipe into the well. Around the air pipe a discharge pipe is fitted, as shown in Fig. 5.17. Both pipes should be capable of being shifted vertically by clamps. Initially, the pipes extend to near the bottom of the screened section; for efficient operation, the water depth in the discharge pipe should exceed two-thirds the length of the pipe. To begin the development, the air pipe is closed and the air pressure is allowed to build up to 0.7 to 1.0×10^6 Pa, whereupon it is released suddenly into the well by means of a quick-opening valve. The inrush of air creates a powerful surge within the well, first increasing then decreasing the pressure as water is forced up the discharge pipe. The process loosens the fine material surrounding the perforations; the material may then be brought into the well by continuous air injection creating an airlift pump. The operation is repeated at intervals along the screened section until sand accretion becomes negligible.

Backwashing with Air. In the backwashing method the top of the well is fitted with an airtight cover. Discharge and air pipes are installed similar to the previous method, together with a separate short air pipe and a three-way valve, as shown in Fig. 5.18. Compressed air is released through the long air pipe, forcing air and water out of the well through the discharge pipe. After the water clears, the air supply is shut off and the water is allowed to return to its static level. The three-way valve is then turned to admit air into the top of the well through the short air pipe. This backwashes the water from the well through the discharge pipe and at the same time agitates the sand grains surrounding the well. Air is forced into the well until it begins escaping from the discharge pipe, after which

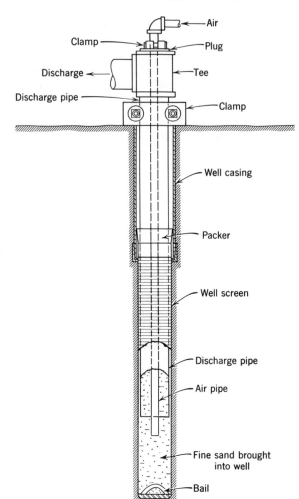

Fig. 5.17 Installation for well development with compressed air.

the three-way valve is turned and the air supply is again directed down the long air pipe to pump the well. Backwashing is repeated until the well is fully developed.

Hydraulic Jetting. Jetting with a high-velocity stream of water is an effective development technique in open rock holes and in wells containing screens with large percentage openings (see Fig. 5.19). The jet nozzle, mounted horizontally, is attached to a string of pipe, which is connected through a swivel and hose to a high-pressure, high-capacity pump. The jet head is slowly rotated and to succes-

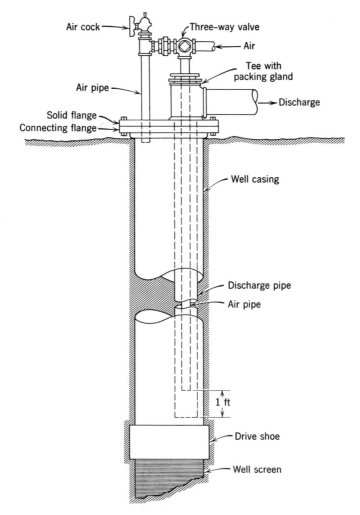

Fig. 5.18   Installation for well development by backwashing with air.

sively higher levels. Fine-grained material from unconsolidated aquifers is carried into the well by the turbulent flow; in addition, the method is particularly effective in developing gravel-packed wells.

Chemicals.   Open-hole wells in limestone or dolomite formations can be developed by adding hydrochloric acid to water in the well. The solvent action removes fine particles and tends to widen fractures leading into the well bore. Normally this procedure would be

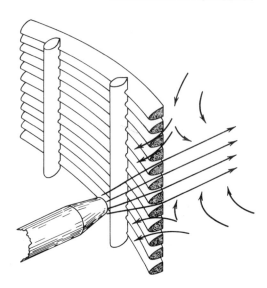

Fig. 5.19 High-velocity hydraulic jetting through a continuous slot wire-wound well screen for well development (after Johnson Div., UOP Inc.[23])

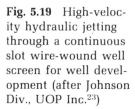

followed by one of the previously described development methods. Hydrofluoric acid can be similarly employed for rocks containing silicates.

For most development methods adding one of the polyphosphates* to water in the well will aid the development process.[46] These compounds act as deflocculants and dispersants of clays and other fine-grained materials, thereby enabling the mud cake on the wall of a hole and the clay fractions in an aquifer to be more readily removed by the development.

Blocks of solid carbon dioxide (dry ice) are sometimes added to a well after acidizing and surging with compressed air to complete well development. The accumulation of gaseous carbon dioxide released by sublimation builds up a pressure within the well; upon release this causes a burst of muddy water (Fig. 5.20) from the well.

Hydraulic Fracturing. Hydraulic fracturing, a technique borrowed from the petroleum industry, is occasionally employed to enhance the yield of open-hole rock wells. Inflatable packers on a pipe extending to ground surface isolate a section of aquifer. After filling the pipe and isolated section with water, pump pressure is applied to fracture the rock. Sand is sometimes pumped into the section to force the grains into the rock fractures so as to maintain the openings.

---

*Common household phosphate-based detergents can serve as a substitute; excessive foaming when the well is pumped may result.

Fig. 5.20 Eruption of mud and water extending 40 m into the air from a well in Utah after development with dry ice (courtesy H.S. Snyder).

Explosives. Detonation of explosives in rock wells often increases yields by enlarging the hole, increasing rock fractures, and removing fine-grained deposits on the face of the well bore.

Testing Wells for Yield

Following development of a new well, it should be tested to determine its yield and drawdown. This information provides a basis for determining the water supply available from the well, for selecting the type of pump, and for estimating the cost of pumping. A test is accomplished by measuring the static water level, after which the well is pumped at a maximum rate until the water level in the well stabilizes. The depth to water is then noted (techniques for measuring water well depths are described in Chapter 12). The difference in depths is the drawdown, and the discharge-drawdown ratio is an estimate of the specific capacity of the well. The discharge can be determined by any of several measuring devices connected to the discharge pipe.

Pumping Equipment

Well pumps produce flow by transforming mechanical energy to hydraulic energy. A wide variety of pumps are produced annually by many manufacturers. The selection of a particular size and type of pump depends on several factors including: (1) pumping capacity, (2) well diameter and depth, (3) depth and variability of pumping level, (4) straightness of the well, (5) sand pumping, (6) total pumping head, (7) duration of pumping, (8) type of power available, and (9) costs.[46] Table 5.7 summarizes characteristics of pumps most frequently employed in wells.

Total Pumping Head. The total pumping head, or total dynamic head, of a pump represents the total vertical lift of the water from the well. As shown in Fig. 5.21, this total head consists of three components: (1) the drawdown inside the well (including aquifer and well losses), (2) the static head, being the difference between the static groundwater level and the static discharge elevation, and (3) friction losses due to flow through the intake and discharge pipes. As indicated in Fig. 5.21, the total pumping head increases with discharge.

Pumps for Shallow Wells. For shallow wells where only small discharges are needed, hand-operated pitcher pumps, turbine pumps, and gear pumps may be installed. Discharges range up to 500 m³/day. Suction lifts should not exceed about 7 m for efficient and continuous service.

Where a larger discharge is required from a shallow well, a centrifugal pump is commonly employed. The assembly may be mounted with a horizontal or a vertical shaft. The horizontal design is efficient, is easy to install and maintain, and is usually connected directly to an electric motor. Because of the low suction head, the pump is often placed a short distance above the water level in large-diameter wells.

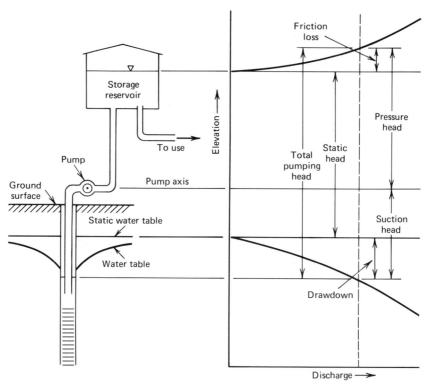

Fig. 5.21 Diagram illustrating total pumping head for a well supplying a storage reservoir. Note the increase in head as a function of well discharge.

Pumps for Deep Wells. In deep wells requiring high lifts, large-capacity pumps serving irrigation, municipal, or industrial water requirements are installed. Several types of pumps are suitable for deep-well operation: plunger, displacement, airlift, jet, and, most importantly, deep-well turbine and submersible.

The deep-well turbine pump has been widely adopted for large, deep, high-capacity wells (see Fig. 5.22a). This type of pump has its impeller suspended vertically on a long drive shaft within a discharge pipe. The bowl of the pump contains the impeller and guide vanes; several bowls connected in series for higher heads form a multiple-stage pump. The pump is usually driven by an electric motor at the ground surface and connected by a long vertical shaft positioned by bearings within the discharge pipe. Deep-well pumps, being submerged, require no priming and are capable of operating under a wide range of water levels without having to be reset.

The submersible pump is simply a deep-well turbine pump close-coupled to a small-diameter submersible electric motor, as shown in

TABLE 5.7 Characteristics of Pumps Frequently Employed in Wells (after U.S. Public Health Service[49])

Type of Pump	Practical Suction Lift[a]	Usual Well-Pumping Depth	Usual Pressure Heads	Advantages	Disadvantages
Reciprocating:					
Shallow well...	6–7 m	6–7 m	30–60 m	Positive action; discharge against variable heads; pumps water containing sand and silt; especially adapted to low capacity and high lifts.	Pulsating discharge; subject to vibration and noise; maintenance cost may be high; may cause destructive pressure if operated against closed valve
Deep well...	6–7 m	Up to 180 m	Up to 180 m above cylinder		
Centrifugal:					
Shallow well straight centrifugal (single stage)	6 m max	3–6 m	30–45 m	Smooth, even flow; pumps water containing sand and silt; pressure on system is even and free from shock; low-starting torque; usually reliable and good service life	Loses prime easily; efficiency depends on operating under design heads and speed
Regenerative vane turbine type (single impeller)	8 m max	8 m	30–60 m	Same as straight centrifugal except not suitable for pumping water containing sand or silt; self-priming	Same as straight centrifugal except maintains priming easily
Deep well Vertical line shaft turbine (multistage)	Impellers submerged	15–90 m	30–250 m	Same as shallow well turbine	Efficiency depends on operating under design head and speed;

Type				Advantages	Remarks
					requires straight well large enough for turbine bowls and housing; lubrication and alignment of shaft critical; abrasion from sand
Submersible turbine (multistage)	Pump and motor submerged	15–120 m	15–120 m	Same as shallow well turbine; easy to frost-proof installation; short pump shaft to motor	Repair to motor or pump requires pulling from well; sealing of electrical equipment from water vapor critical; abrasion from sand
Jet:					
Shallow well...	4-6 m below ejector	Up to 4-6 m below ejector	25–45 m	High capacity at low heads; simple in operation; does not have to be installed over well; no moving parts in well	Capacity reduces as lift increases; air in suction or return line will stop pumping
Deep well.........	4-6 m below ejector	7–35 m 60 m max.	25–45 m	Same as shallow well jet	Same as shallow well jet
Rotary:					
Shallow well... (gear type)	7 m	7 m	15–75 m	Positive action; discharge constant under variable heads; efficient operation	Subject to rapid wear if water contains sand or silt; wear of gears reduces efficiency
Deep well...... (helical rotary type).	Usually submerged	15–150 m	30–150 m	Same as shallow well rotary; only one moving pump device in well	Same as shallow well rotary except no gear wear

[a]Practical suction lift at sea level. Reduce lift 0.3 m for each 300 m above sea level.

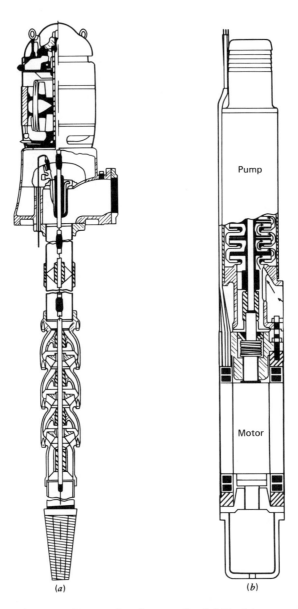

Fig. 5.22 Pumps for deep wells. (*a*) Turbine.
(*b*) Submersible (after Anderson[6]).

Fig. 5.22b. The efficiency of the pump is increased by direct coupling, while effective cooling results from complete immersion. Pump sizes range from small units that fit inside an 8-cm casing to large-capacity units involving numerous stages. An advantage of submersible pumps is that they can lift water from deep wells where long shafts in crooked casings might prohibit installation of a deep-well turbine pump. Other advantages, which account for increasing installations of this type of pump, include ease of maintenance, freedom from noise, protection from weathering and flooding, and avoidance of a large aboveground installation.

Protection of Wells

Sanitary Protection. Wherever groundwater pumped from a well is intended for human consumption, proper sanitary precautions must be taken to protect the water quality. Pollution sources may exist either above or below ground surface (see Chapter 8). Precautions apply equally to springs; Fig. 5.23 shows, for example, a typical method for protecting a spring water supply.

Surface pollution can enter wells either through the annular space outside of the casing or through the top of the well itself. To close avenues of access for undesirable water outside of the casing, the annular space should be filled with cement grout as shown for deep wells in Figs. 5.8 and 5.9. Entry through the top of the well can be avoided by providing a watertight cover to seal the top of the casing.[10,28] Some pumps are available with closed metal bases that provide the necessary closure. For pumps having an open-type base, or where the pump is not placed directly over the well, a seal is required for the annular opening between the discharge pipe and casing.* Seals may be made of metal or lead packing; asphaltic and mastic compounds are also satisfactory. Covers around the well should be made of concrete, should be elevated above the adjacent land level, and should slope away from the well (see Fig. 5.24).

Whenever a new well is completed or an old well repaired, contamination from equipment, well materials, or surface water may be introduced into the well. Addition and agitation of a chlorine compound will disinfect the well. Following disinfection, the well should be pumped to waste until all traces of chlorine are removed. As a final check on the potability of the water, a sample should be collected and sent to a laboratory for bacteriological examination.

*It is desirable to provide a small opening in or below the pump base to allow for periodic water level measurements.

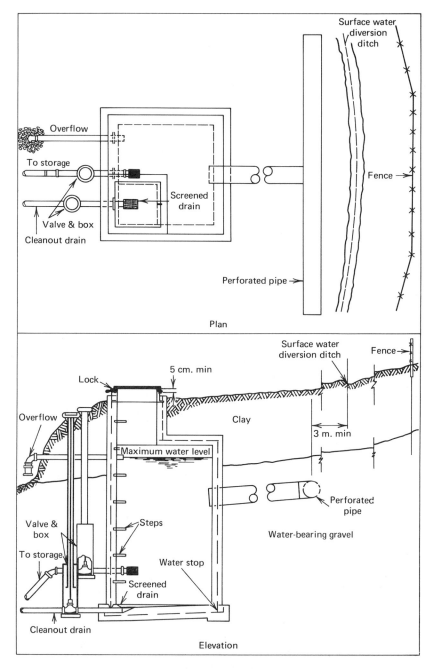

Fig. 5.23 Plan and elevation views of a developed spring showing a typical method for providing sanitary protection (after U.S. Public Health Service [49]).

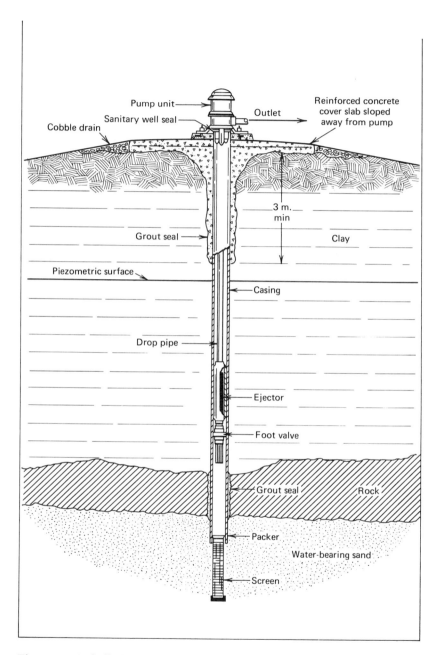

Fig. 5.24   A drilled well showing grout seal, concrete slab, and well seal for sanitary protection (after U.S. Public Health Service[49]).

Frost Protection.   In regions where winter frost occurs, it is important to protect pumps and water lines from freezing. A typical method for frostproofing a domestic well is shown in Fig. 5.25. The pitless adapter, attached to the well casing, provides access to the well, while the discharge pipe runs about 2 m underground to the basement of the house.

Abandonment of Wells.   Whenever a well is abandoned, for whatever reason, it should be sealed by filling it with clay, concrete, or earth. Not only is surface contamination then unable to enter the

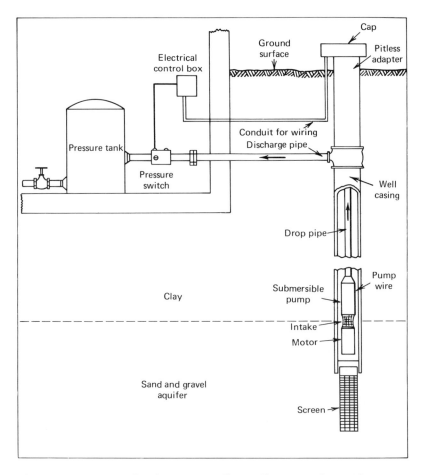

Fig. 5.25   Diagram of a domestic well installation with a pitless adapter to protect the well from frost (after Gibb[16]).

well, but sealing serves other useful purposes: prevents accidents, avoids possible movement of inferior water from one aquifer to another, and conserves water in flowing wells.[11]

Well Rehabilitation

A new well, properly drilled, cased, and developed, will give years of satisfactory service with little attention. Many wells fail, however; that is, they yield decreasing quantities of water with time.* Well rehabilitation refers to the treatment of a production well by mechanical, chemical, or other means to recover as much as possible of the lost production capacity.[24,25] Table 5.8 lists well rehabilitation methods and their applications to various types of aquifers.

One cause of failure is depletion of the groundwater supply. Not a fault of the well, this trouble can sometimes be remedied by decreasing pumping drafts, resetting the pump, or deepening the well. A second cause of well trouble results from faulty well construction. Such items as poor casing connections, improper perforations or screens, incomplete placement of gravel packs, and poorly seated wells are typical of difficulties encountered. Depending on the particular situation as determined from a television or photographic survey of the well (see Chapter 12), it may be possible to repair the well, but sudden failures involving entrance of sand or collapse of a casing often require replacement of the entire well.

The third and most prevalent cause of well failure results from corrosion or incrustation of well screens.[7,27,37] Corrosion may result from direct chemical action of the groundwater or from electrolytic action caused by the presence of two different metals in the well. The effects of corrosion can be minimized by selecting nonmetallic well screens or ones of corrosion-resistant metal (such as nickel, copper, or stainless steel), and by providing cathodic protection.† If the damage is localized it may be possible to insert a liner inside the screen to prevent excessive sand pumping.

Incrustation is caused by precipitation on or near well screens of materials carried in solution by groundwater.[32] The sudden pressure drop associated with water entering a well under heavy pumping can release carbon dioxide and cause precipitation of calcium car-

---

*Frequently the pump rather then the well is at fault; hence, it should be checked before beginning any extensive well repair.

†One method of providing cathodic protection for a well is to introduce a metal low on the electrochemical scale that will be corroded instead of the well casing. Rods of magnesium suspended in the well water serve this purpose.

TABLE 5.8 Rehabilitation Methods and Their Applications to Various Types of Aquifers (after Erickson[13])

Method	Unconsolidated Aquifers	Consolidated Sandstone	Consolidated Limestone
Muriatic acid[a] followed by chlorine	Removes iron, sulfur, and carbonate deposits	Not usually effective	Sometimes beneficial; best results obtained by pressure acidizing
Polyphosphate followed by chlorine	Removes fine silt, clay, colloids, disseminated shale, and soft iron deposits	Not usually effective	Not usually effective
Dynamiting	Not recommended	Effective for all types of well-screen deposits	Effective when very large charges are used
Compressed air	Removes plugging deposits of silt and fine sand in areas adjacent to screens	Not used	Not used
Dry ice	Same as compressed air	Used only rarely, to remove cuttings from the face of a new production well	Not usually effective
Surging	Same as compressed air	Rarely used	Rarely used
Chlorine[b]	Removes iron and slime-forming bacteria	Removes iron and slime-forming bacteria	Removes iron and slime-forming bacteria
Caustic soda	Removes oil scum left by oil-lubricating pumps	Removes oil scum left by oil-lubricating pumps	Removes oil scum left by oil-lubricating pumps

[a]Not to be used with concrete screens.
[b]Usually used in a concentration of 500 mg/l.

bonate. Another cause of incrustation stems from the presence of oxygen in a well; this can change soluble ferrous iron to insoluble ferric hydroxide. Screens can be cleaned by shooting a string of vibratory explosives in the well or by adding hydrochloric acid (HCl) or sulfamic acid (H_2NSO_3H) to the well, followed by agitation and surging. Where slime-forming organisms block screens, particularly in recharge wells, treatment with chlorine gas or hypochlorite solutions[46] can remedy the problem. For improving yields of rock wells, acidizing or shooting with explosives is generally effective.[53]

Horizontal Wells

Subsurface conditions often preclude groundwater development by normal vertical wells. Such conditions may involve aquifers that are thin, poorly permeable, or underlain by permafrost or saline water. In other circumstances, where groundwater is to be derived primarily from infiltration of streamflow, a horizontal well system may be advantageous. Also, in developing areas of the world, the cost of constructing a horizontal well may be far less costly than drilling a vertical well.

Infiltration Galleries. An infiltration gallery is a horizontal conduit for intercepting and collecting groundwater by gravity flow.[8] Qanats, described in Chapter 1, illustrate one type of gallery. Galleries, normally constructed at the water table elevation, discharge into a sump where a pump lifts the water to ground surface for use. In Europe and the United States many infiltration galleries are laid parallel to riverbeds, where with induced infiltration an adequate perennial water supply can be obtained (see Chapter 13). Depending on the type of aquifer penetrated, galleries may be unlined or lined with vitrified clay, brick, concrete, or cast iron.*

In Alaska galleries have been widely employed to obtain water supplies where underlying permafrost (see Chapter 2) would not contribute groundwater.[14] On oceanic islands galleries have the particular advantage of enabling fresh water to be collected with little disturbance of underlying saline water (see Chapter 14). Such installations are well known in Hawaii (see Fig. 5.26), for example, where they are unlined in basalt,[36,54] and in Barbados, where they are unlined in coral limestone.

---

*Deep infiltration galleries exist in the chalk aquifers of southeastern England. Here the permeability of solid chalk is quite low, and unlined horizontal tunnels (or adits), roughly 2 m in diameter and extending for distances up to 2 km, intersect a maximum number of fissures from which most of the water is obtained.

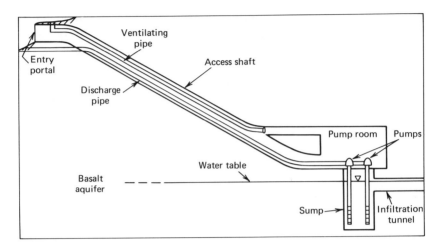

Fig. 5.26 Cross section of an infiltration gallery in Hawaii. These are locally known as Maui-type wells because they were first constructed on the island of Maui to provide water for irrigation of sugar cane (after Watson[54]).

Horizontal Pipes. On sloping ground surfaces small-diameter horizontal holes can be drilled by the rotary method.[45,56] Perforated pipes placed in these holes tap groundwater that would otherwise be discharged by seepage or from small springs. Two examples are shown in Fig. 5.27. Such horizontal pipes provide sanitary and low-cost water, and, in addition, enable flow to be controlled by valves

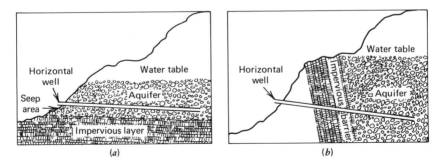

Fig. 5.27 Examples of horizontal wells consisting of small-diameter perforated pipes drilled into hillsides. These wells should have a downward slope into the aquifer to prevent formation of a vacuum inside the pipes (after Welchert and Freeman[56]). (*a*) Contact spring formation. (*b*) Dike spring formation.

at the discharge ends. The same technique has also been widely employed to drain side slopes, such as in highway cuts, in order to reduce the possibility of landslides.

Collector Wells. For cities and industries located near rivers, the problem of obtaining high-quality, low-temperature water at reasonable cost has become increasingly difficult. In Europe and the United States, groundwater pumped from collector wells tapping permeable alluvial aquifers has often proved to be a successful solution.[18,41] If located adjacent to a surface water source, a collector well lowers the water table and thereby induces infiltration of surface water through the bed of the water body to the well (see Chapter 13). In this manner greater supplies of water can be obtained than would be available from groundwater alone. Analytic solutions of the flow to a collector well have been developed (see Chapter 4).

Plan and elevation views of a collector well are shown in Fig. 5.28. The central cylinder, consisting of a monolithic concrete caisson about 5 m in diameter, is sunk into the aquifer by excavating the inside earth material. After the requisite depth is reached, a thick concrete plug is poured to seal the bottom. Perforated pipes, 15 to 20 cm in diameter, are jacked hydraulically into the water-bearing formation through precast portholes in the caisson to form a radial pattern of horizontal pipes. During construction fine-grained material is washed into the caisson so that natural gravel packs form around the perforations (see Fig. 5.28). The number, length, and radial pattern of the collector pipes can be varied to obtain the maximum capacity; usually more pipes are extended toward than away from the surface water source.

The large area of exposed perforations in a collector well causes low inflow velocities, which minimize incrustation, clogging, and sand transport. Polluted river water is filtered by its passage through the unconsolidated aquifer to the well. The initial cost of a collector well exceeds that of a vertical well; however, advantages of large yields, reduced pumping heads, and low maintenance costs are factors to be considered. Yields vary with local conditions; the average for a large number of such wells approximated 27,000 m^3/day. Installations bordering streams with regulated stages may encounter decreases in yield with time because of sediment deposition on streambeds.[18]

Collector wells can also function in permeable aquifers removed from surface water. Several such installations gave an average yield of about 15,000 m^3/day.

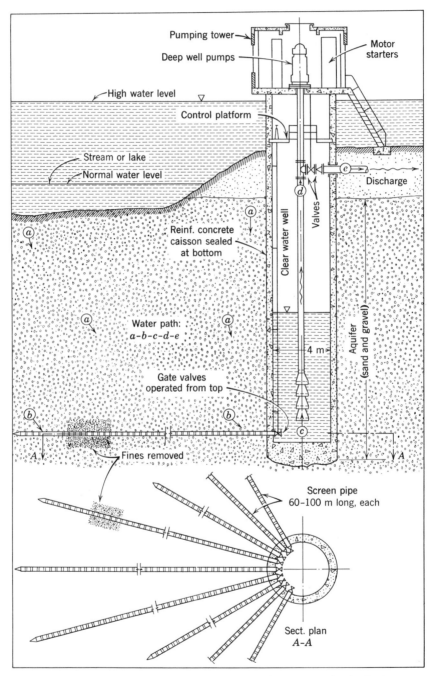

Fig. 5.28 A collector well located near a surface water body (courtesy Ranney Method Water Supplies, Inc.).

References

1. Ahmad, N., *Tubewells, construction and maintenance,* Scientific Research Stores, Lahore, Pakistan, 250 pp., 1969.
2. Ahrens, T. P., Basic considerations of well design, *Water Well Jour.,* v. 24, no. 4, pp. 45–50; no. 5, pp. 49–51; no. 6, pp. 47–51; no. 8, pp. 35–37, 1970.
3. Amer. Water Works Assoc., *AWWA standard for deep wells,* AWWA-A100-66, Denver, Colorado, 57 pp., 1967.
4. Amer. Water Works Assoc., *Getting the most from your well supply,* Proceedings AWWA Seminar, New York, 62 pp., 1972.
5. Amer. Water Works Assoc., *Ground water,* AWWA Manual M21, 130 pp., 1973.
6. Anderson, K. E., *Water well handbook,* 2nd ed., Missouri Water Well & Pump Contractors Assoc., Rolla, 281 pp., 1967.
7. Barnes, I., and F. E. Clarke, Chemical properties of ground water and their corrosion and encrustation effects on wells, *U.S. Geological Survey Prof. Paper* 498-D, 58 pp., 1969.
8. Bennett, T. W., On the design and construction of infiltration galleries, *Ground Water,* v. 8, no. 3, pp. 16–24, 1970.
9. Bentall, R., Methods of collecting and interpreting ground-water data, *U.S. Geological Survey Water-Supply Paper* 1544-H, 97 pp., 1963.
10. Bernhard, A. P., Protection of water-supply wells from contamination by wastewater, *Ground Water,* v. 11, no. 3, pp. 9–15, 1973.
11. California Dept. of Water Resources, *Water well standards: State of California,* Bull. 74, Sacramento, 205 pp., 1968.
12. Campbell, M. D., and J. H. Lehr, *Water well technology,* McGraw-Hill, New York, 681 pp., 1973.
13. Erickson, C. R., Cleaning methods for deep wells and pumps, *Jour. Amer. Water Works Assoc.,* v. 53, pp. 155–162, 1961.
14. Feulner, A. J., Galleries and their use for development of shallow ground-water supplies, with special reference to Alaska, *U.S. Geological Survey Water-Supply Paper* 1809-E, 16 pp., 1964.
15. Garg, S. P., and J. Lal, Rational design of well screens, *Jour. Irrig. Drainage Div.,* Amer. Soc. Civil Engrs., v. 97, no. IR1, pp. 131–147, 1971.
16. Gibb, J. P., *Wells and pumping systems for domestic water supplies,* Illinois State Water Survey Circ. 117, Urbana, 17 pp., 1973.
17. Gibson, U. P., and R. D. Singer, *Water well manual,* Premier Press, Berkeley, Calif., 156 pp., 1971.
18. Gidley, H. K., and J. H. Millar, Performance records of radial collector wells in Ohio River Valley, *Jour. Amer. Water Works Assoc.,* v. 52, pp. 1206–1210, 1960.
19. Gordon, R. W., *Water well drilling with cable tools,* Bucyrus-Erie Co., South Milwaukee, Wis., 230 pp., 1958.
20. Huisman, L., *Groundwater recovery,* Winchester Press, New York, 336 pp., 1972.
21. Hunter Blair, A., Well screens and gravel packs, *Ground Water,* v. 8, no. 1, pp. 10–21, 1970.

22. Johnson, A. I., et al., Laboratory study of aquifer properties and well design for an artificial-recharge site, *U.S. Geological Survey Water-Supply Paper* 1615-H, 42 pp., 1966.

23. Johnson Div., UOP Inc., *Ground water and wells,* 2nd ed., Edward E. Johnson, St. Paul, Minn., 440 pp., 1972.

24. Koenig, L., Survey and analysis of well stimulation performance, *Jour. Amer. Water Works Assoc.,* v. 52, pp. 333–350, 1960.

25. Koenig, L., Relation between aquifer permeability and improvement achieved by well stimulation, *Jour. Amer. Water Works Assoc.,* v. 53, pp. 652–670, 1961.

26. Kruse, G., *Selection of gravel packs for wells in unconsolidated aquifers,* Colorado State Univ. Exp. Sta. Tech. Bull. 66, Fort Collins, 22 pp., 1960.

27. Larson, T. E., *Corrosion by domestic waters,* Bull. 59, Illinois State Water Survey, Urbana, 48 pp., 1975.

28. Moehrl, K. E., Well grounding and well protection, *Jour. Amer. Water Works Assoc.,* v. 56, pp. 423–431, 1964.

29. Mackaness, F. G. (ed.), *Manual of water well construction practices for the State of Oregon,* 2nd ed., Oregon Drilling Assoc., Salem, 84 pp., 1968.

30. Matlock, W. G., Small diameter wells drilled by jet-percussion method, *Ground Water,* v. 8, no. 1, pp. 6–9, 1970.

31. Mogg, J. L., The technical aspects of gravel well construction, *Jour. New England Water Works Assoc.,* v. 77, pp. 155–164, 1963.

32. Mogg, J. L., Practical construction and incrustation guide lines for water wells, *Ground Water,* v. 10, no. 2, pp. 6–11, 1972.

33. Moore, P. L., *Drilling practices manual,* Petroleum Publishing, Tulsa, Okla., 448 pp., 1974.

34. Noble, D. G., Well points for dewatering, *Ground Water,* v. 1, no. 3, pp. 21–26, 1963.

35. Patchick, P. F., Quicksand and water wells, *Ground Water,* v. 4, no. 2, pp. 32–46, 1966.

36. Peterson, F. L., Water development on tropic volcanic islands—type example: Hawaii, *Ground Water,* v. 10, no. 5, pp. 18–23, 1972.

37. Ritchie, E. A., Cathodic protection wells and ground-water pollution, *Ground Water,* v. 14, pp. 146–149, 1976.

38. Schwartz, D. H., Successful sand control design for high rate oil and water wells, *Jour. Petr. Tech.,* v. 21, pp. 1193–1198, 1969.

39. Soliman, M. M., Boundary flow considerations in the design of wells, *Jour. Irrig. Drain. Div.,* Amer. Soc. Civil Engrs., v. 91, no. IR1, pp. 159–177, 1965.

40. Speedstar Div., *Well drilling manual,* Koehring Co., Enid, Okla., 72 pp. (no date).

41. Spiridonoff, S. V., Design and use of radial collector wells, *Jour. Amer. Water Works Assoc.,* v. 56, pp. 689–698, 1964.

42. Stow, G. R. S., Modern water-well drilling techniques in use in the United Kingdom, *Ground Water,* v. 1, no. 3, pp. 3–12, 1963.

43. Stramel, G. J., Maintenance of well efficiency, *Jour. Amer. Water Works Assoc.*, v. 57, pp. 996–1010, 1965.

44. Trescott, P. C., and G. F. Pinder, Air pump for small-diameter piezometers, *Ground Water*, v. 8, no. 3, pp. 10–15, 1970.

45. U.S. Bureau of Mines, *Horizontal boring technology: a state of the art study*, Info. Circ. 8392, 86 pp., 1968.

46. U.S. Bureau of Reclamation, *Ground water manual*, U.S. Dept. of the Interior, 480 pp., 1977.

47. U.S. Dept. of the Army, *Well drilling operations*, Tech. Manual TMS-297, 249 pp., 1965.

48. U.S. Environmental Protection Agency, *Manual of water well construction practices*, Rept. Environmental Protection Agency-570/9-75-001, Washington, D.C., 156 pp., 1976.

49. U.S. Public Health Service, *Manual of individual water supply systems*, Publ. no. 24, 121 pp., 1962.

50. U.S. Soil Conservation Service, *Engineering field manual for conservation practices*, 995 pp., 1969.

51. Vaadia, Y., and V. H. Scott, Hydraulic properties of perforated well casings, *Jour. Irrig. Drainage Div.*, Amer. Soc. Civil Engrs., v. 84, no. IR1, Paper 1505, 26 pp., 1958.

52. Walton, W. C., *Selected analytical methods for well and aquifer evaluation*, Bull. 49, Illinois State Water Survey, Urbana, 81 pp., 1962.

53. Walton, W. C., and Csallany, S., *Yields of deep sandstone wells in Northern Illinois*, Rept. of Inv. 43, Illinois State Water Survey, Urbana, 47 pp., 1962.

54. Watson, L. J., Development of ground water in Hawaii, *Jour. Hydraulics Div.*, Amer. Soc. Civil Engrs., v. 90, no. HY6, pp. 185–202, 1964.

55. Watt, S. B., and W. E. Wood, *Hand dug wells and their construction*, Intermediate Technology, London, 234 pp., 1976.

56. Welchert, W. T., and B. N. Freeman, Horizontal wells, *Jour. Range Management*, v. 26, pp. 253–256, 1973.

Groundwater Levels and Environmental Influences

CHAPTER 6 ·······································

A groundwater level, whether it be the water table of an unconfined aquifer or the piezometric surface of a confined aquifer, indicates the elevation of atmospheric pressure of the aquifer. Any phenomenon that produces a change in pressure on groundwater will cause the groundwater level to vary. Differences between supply and withdrawal of groundwater cause levels to fluctuate. Streamflow variations are closely related to groundwater levels. Other diverse influences on groundwater levels include meteorological and tidal phenomena, urbanization, earthquakes, and external loads. And, finally, subsidence of the land surface can occur due to changes in underlying groundwater conditions.

Time Variations of Levels

Secular Variations. Secular variations of groundwater levels are those extending over periods several years or more. Alternating series of wet and dry years, in which the rainfall is above or below the mean, will produce long-period fluctuations of levels. The long records of rainfall and groundwater levels shown in Fig. 6.1 illustrate this point. Rainfall is not an accurate indicator of groundwater level changes. Recharge is the governing factor (assuming annual with-

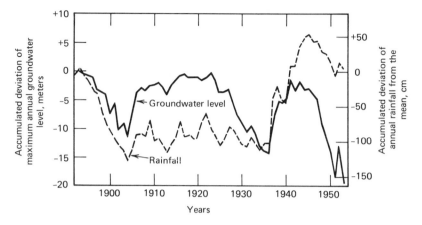

Fig. 6.1 Secular variations of maximum annual groundwater level and annual rainfall in San Bernardino Valley, California.

drawals are constant); it depends on rainfall intensity and distribution and amount of surface runoff.

In other instances pronounced trends may be noted. Thus, in overdeveloped basins where draft exceeds recharge, a downward trend of groundwater levels may continue for many years. Figure 6.2 dramatically illustrates the decline in piezometric surface of a deep sandstone aquifer as a result of nearly a century of intensive pumping in the Chicago metropolitan area. Another example is that of the confined aquifer in Queensland, Australia, shown in Fig. 6.3. Here groundwater is obtained from flowing wells. Over a period of more than 80 years the total number of wells has steadily increased, but the resulting decline of 120 m in pressure head has caused the total flow to decrease gradually since 1914.

Seasonal Variations. Many groundwater levels show a seasonal pattern of fluctuation. This results from influences such as rainfall and irrigation pumping that follow well-defined seasonal cycles. The variations shown in Fig. 6.4 are typical for areas subject to frozen ground in winter. Highest levels occur in late spring and lowest in winter. In irrigated areas where frozen ground is not a factor, lowest levels normally occur during fall at the end of the irrigation season. The amplitude depends on recharge, pumpage, and the type of aquifer; confined aquifers normally display a greater range in levels than do unconfined aquifers.

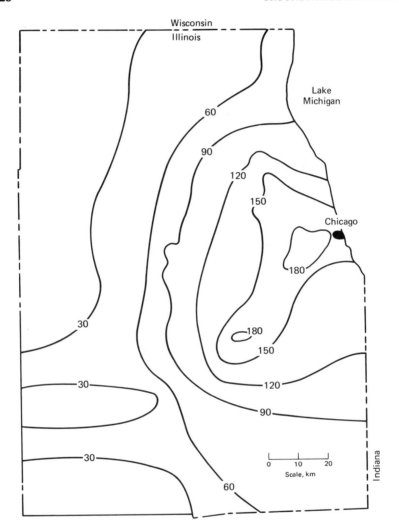

Fig. 6.2 Decline of the piezometric surface in the Chicago metropolitan area due to extended heavy pumping. Contours are lines of equal decline in meters for the period 1864–1958 (after Russell[59]).

Short-term Variations. Groundwater levels often display characteristic short-term fluctuations governed by the primary use of groundwater in a locality. Clearly defined diurnal variations may be associated with municipal water-supply wells. Similarly, weekly patterns occur with pumping for industrial and municipal purposes.

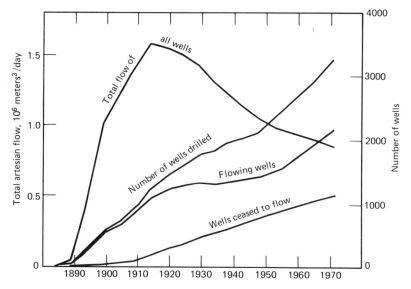

Fig. 6.3 Time variation of the number of flowing wells tapping the confined aquifer in Queensland, Australia, and of the total flow produced by the wells. The discharge of an average flowing well decreased by a factor of 4 between 1914 and 1971 (courtesy Australian Water Resource Council).

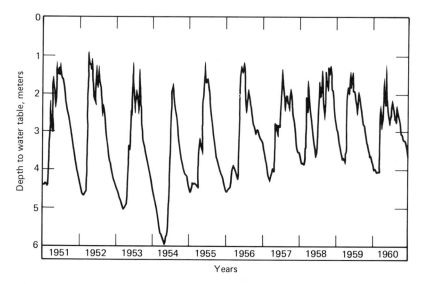

Fig. 6.4 Seasonal fluctuations of the water table in a glacial till aquifer in Ohio. Well depth is 9 m (after Klein and Kaser[33]).

Streamflow and Groundwater Levels

Where a stream channel is in direct contact with an unconfined aquifer, the stream may recharge the groundwater or receive discharge from the groundwater, depending on the relative levels. A *gaining stream* is one receiving groundwater discharge; a *losing stream* is one recharging groundwater (see Fig. 6.5). Often a gaining stream may become a losing one, and conversely, as the stream stage changes.[32,44]

The term *rising water* is applied to marked increases in streamflow in reaches where a subsurface restriction forces groundwater to the surface. Fig. 6.6 illustrates the phenomenon for a situation where a dry stream channel exists above and below the convergent section.

Bank Storage. During a flood period of a stream, groundwater levels are temporarily raised near the channel by inflow from the stream. The volume of water so stored and released after the flood is referred to as *bank storage*. Field data are rarely adequate to evaluate bank storage and its rate of inflow and outflow; therefore, analytic or model approaches are necessary to obtain quantitative estimates for specified boundary conditions.

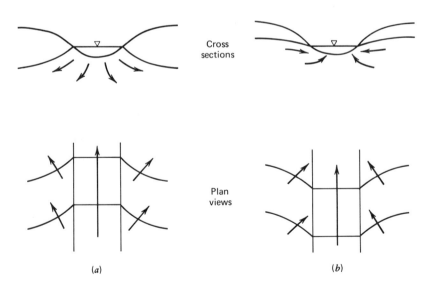

Fig. 6.5 Water table contours and groundwater flow directions in relation to stream stages. (*a*) Losing stream. (*b*) Gaining stream.

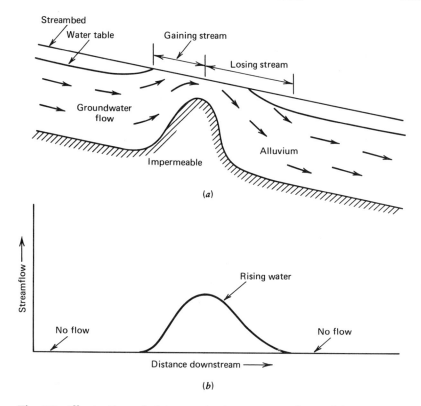

Fig. 6.6 Illustration of rising water in a stream channel from emerging groundwater flow. (a) Cross section along stream channel in an alluvial valley. (b) Streamflow as a function of distance along the stream.

Figure 6.7 illustrates idealized groundwater conditions adjacent to a flooding stream. A flood hydrograph of sinusoidal form (Fig. 6.7a) was superposed on an aquifer and stream situation sketched in Fig. 6.7b. As a result of the flood, the bank storage increased and then decreased; the variation of the volume of water in storage is depicted in Fig. 6.7c. The derivative of the volume curve yields the groundwater flow curve (Fig. 6.7d). From this it can be seen that a stream fluctuation produces large variations in magnitude and direction of local groundwater flow.

Cooper and Rorabaugh[7] derived solutions for changes in groundwater head near the stream, groundwater flow to the stream, and bank storage. Their comprehensive analysis also included a family

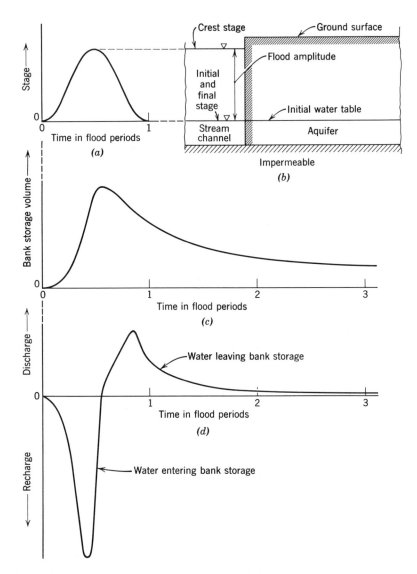

Fig. 6.7 Groundwater in relation to a flooding stream. (a) Flood hydrograph. (b) Vertical cross section of field conditions. (c) Volume of bank storage as a function of time. (d) Groundwater flow to and from bank storage. Results are from a laboratory model (after Todd[67]).

of asymmetric flood-wave stage hydrographs, which facilitate study of the effects of a wide variety of flood shapes on groundwater.

Study of bank storage produced by the spring rise along an 80-km reach of the Columbia River indicates a total storage volume (including both banks of the stream) of some 2.07×10^8 m^3.[43] During a typical 45-day rise of the river to flood peak, an average diversion to bank storage of 4.60×10^6 m^3/day occurs; in the subsequent 165-day decline, the return flow averages 1.25×10^6 m^3/day.

Base Flow.   Streamflow originating from groundwater discharge is referred to as *groundwater runoff* or *base flow.* During periods of precipitation streamflow is derived primarily from surface runoff, whereas during extended dry periods all streamflow may be contributed by base flow.[23,45] Typically, base flow is not subject to wide fluctuations and is indicative of aquifer characteristics within a basin.[31,41,42]

To estimate base flow a rating curve of groundwater runoff can be prepared by plotting mean groundwater stage (water table level) within the basin against streamflow during periods when all flow is contributed by groundwater.[61] Figure 6.8a shows rating curves for a small drainage basin in Illinois. Data were fitted by two curves: one for the April–October period when evapotranspiration from groundwater is significant, and the other for the November–March period when evapotranspiration is minimal. With these rating curves and the mean groundwater stages for one year (see Fig. 6.8b), the separation of surface runoff and base flow hydrographs shown in Fig. 6.8c could be achieved. It can be noted that frozen ground impeded groundwater recharge during February and March and that base flows were largest during the spring and summer months. Groundwater contributed 33 percent of total streamflow for the year.

Streamflow at any instant contains groundwater contributed at previous times and different locations within the drainage area. During and after a storm period in a small drainage basin, the water table will rise, causing the base flow to increase also (Fig. 6.8a). But superimposed on this will be the bank storage fluctuation (Fig. 6.7d). The effects of these two variations are shown schematically in Fig. 6.9.

An alternative approach to determining the separation of total streamflow into surface-runoff and groundwater components during flood periods can be accomplished from measurements of chemical concentrations.[36,75] Total dissolved solids or any major ion will serve the purpose with the equation

$$C_{TR}Q_{TR} = C_{GW}Q_{GW} + C_{SR}Q_{SR} \tag{6.1}$$

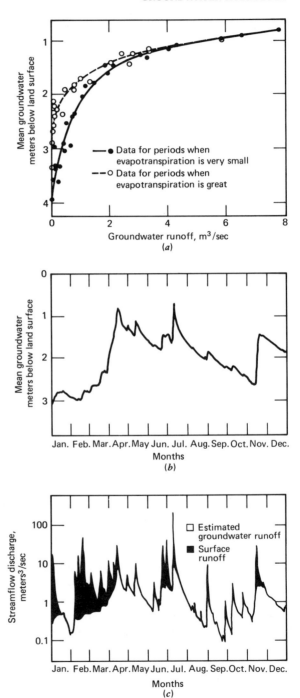

Fig. 6.8 Estimaof base flow for Panther Creek near Peoria, Illinois (drainage area: 246 km²). (*a*) Rating curves of mean groundwater stage versus groundwater runoff (base flow). (*b*) Mean groundwater stage for 1951. (*c*) Streamflow hydrograph for 1951 showing surface runoff and base flow components (after Schicht and Walton[60]).

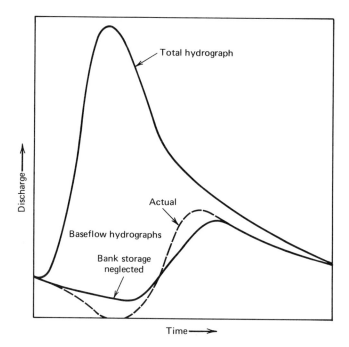

Fig. 6.9 Schematic diagram of the variation of base flow during a flood hydrograph with and without effects of bank storage (after Singh[62]).

where C is ionic concentration, Q is streamflow, TR is total runoff, GW is groundwater contribution (base flow), and SR is surface run-off. Solving for the base flow,

$$Q_{GW} = [(C_{TR} - C_{SR})/(C_{GW} - C_{SR})]Q_{TR} \qquad (6.2)$$

where

$$Q_{TR} = Q_{GW} + Q_{SR} \qquad (6.3)$$

Values of C_{GW} are measured during rainless periods, C_{SR} is measured in small tributary streams during storm events, and C_{TR} is measured during the peak flow period in the main stream. Measurements by this method for three small basins (6 to 13 km²) showed that groundwater contributed from 32 to 42 percent of the total flow at peak discharge.[49]

Base Flow Recession Curve. A *recession curve* shows the variation of base flow with time during periods of little or no rainfall over a drainage basin. In essence it is a measure of the drainage rate of groundwater storage from the basin.[16,80] If large, highly permeable aquifers are contained within a drainage area, the base flow will be sustained even through prolonged droughts; if the aquifers are small and of low permeability, the base flow will decrease relatively rapidly and may even cease.[11,69]

Analyses of streamflow hydrographs show that the recession curve can often be fitted by the equation

$$Q = Q_o K^t \tag{6.4}$$

where Q is streamflow at time t in days after a given discharge Q_o, and K is a recession constant governed by the hydrogeologic characteristics of the basin.[2,34] The value of K can be empirically determined from the slope of a straight line fitted to a series of consecutive discharges plotted on semilogarithmic paper, as shown in Fig. 6.10. Typical values lie in the range of 0.89 and 0.95. Thus, prior knowledge of the shape of the recession curve enables future estimates to be made of streamflow during rainless periods.

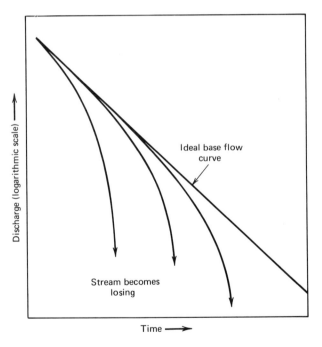

Fig. 6.10 Base flow recession curves of streamflow for varying magnitudes of evapotranspiration losses from groundwater.

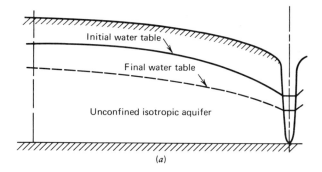

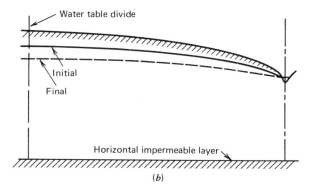

Fig. 6.11 Water table and stream channel conditions affecting base flow. (*a*) Fully penetrating stream. (*b*) Partially penetrating stream (after Singh[62]).

An analytic study of base flow by Singh[62] demonstrated that base flow recession curves depend on the degree to which a stream channel is entrenched in an aquifer. For a fully penetrating stream (see Fig. 6.11*a*), recession curves do not plot as straight lines on semilogarithmic paper; instead, the recession rate continuously decreases with time, forming a concave curve. But for deep aquifers and partially penetrating streams (see Fig. 6.11*b*), the straight-line approximation is generally applicable. The value of K in Eq. 6.4 varies directly with the degree of stream entrenchment.

These approaches to base flow assume that groundwater drains only toward the stream channel. Groundwater also can flow downward to an underlying leaky aquifer and can be lost by evapotranspiration to the atmosphere.[62] Where these diversions are significant, the recession curve will be deflected downward. In semiarid regions where streamflow is intermittent, evapotranspiration losses become significant; this causes the recession curve to steepen (see Fig. 6.10) until streamflow finally ceases.

Fluctuations Due to Evapotranspiration

Unconfined aquifers with water tables near ground surface frequently exhibit diurnal fluctuations that can be ascribed to evaporation and/or transpiration. Both processes cause a discharge of groundwater into the atmosphere and have nearly the same diurnal variation because of their high correlation with temperature.

Evaporation Effects. Evaporation from groundwater increases as the water table approaches ground surface. The rate also depends on the soil structure, which controls the capillary tension above the water table and hence its hydraulic conductivity (see Chapter 3). Computation of actual evaporation from bare soil is complicated by variations in external evaporative conditions at the soil surface.[19,25,74] For isothermal conditions, upward movement is essentially all in the liquid phase, but a soil may have a high surface temperature, causing it to dry out and establishing upward vapor movement in response to a vapor pressure gradient.[56]

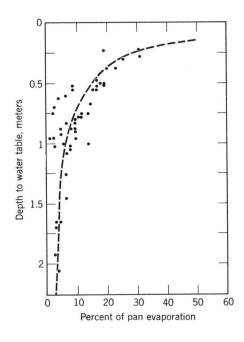

Fig. 6.12 Groundwater evaporation, expressed as a percentage of pan evaporation, as a function of depth to water table (after White[81]).

Field measurements of groundwater evaporation from tanks filled with soil (lysimeters) have been made. Water tables were maintained at prescribed depths below ground surface. Results, expressed as a percentage of pan evaporation at ground surface, are shown in Fig. 6.12. For water tables within one meter of ground surface, evaporation is largely controlled by atmospheric conditions, but below this soil properties become limiting and the rate decreases markedly with depth.

Transpiration Effects. Where the root zone of vegetation reaches the saturated stratum, the uptake of water by roots equals (for practical purposes) the transpiration rate. Figure 6.13 shows water level variations measured in a well in a thicket of willows. Rapid foliage growth during August (Fig. 6.13a) caused daily fluctuations averaging about 10 cm with the water table between 1.5 and 1.8 m below ground surface. Heavy frosts occurred in early October and most

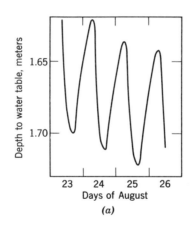

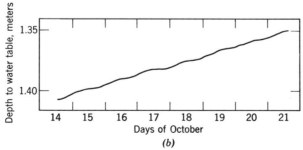

Fig. 6.13 Effect of transpiration discharge on groundwater levels near Milford, Utah. (a) In summer. (b) After frost (after White[81]).

leaves had fallen by mid-October; thereafter, diurnal fluctuations were negligible (Fig. 6.13b) with the vegetation dormant.

Magnitudes of transpiration fluctuations depend on the type of vegetation, season, and weather.[73] Hot, windy days produce maximum drawdowns, whereas cool, cloudy days show only small variations. Fluctuations begin with the appearance of foliage and cease after killing frosts. Cutting of plants eliminates or materially reduces amplitudes. Transpiration discharge does not occur in nonvegetated areas, such as plowed fields, or in areas where the water table is far below ground surface.* After rain on high water table vegetated land, the water table rises sharply as the increased soil moisture meets the transpiration demand and reduces the groundwater discharge; but on cleared land or when vegetation is dormant, little or no rise is evident.

An extensive study of the transpiration of groundwater by phreatophytes was undertaken by the U.S. Geological Survey in Lower Safford Valley, Arizona.[20] Six methods for determining water use by vegetation were employed:

1. Tank method—based on measurement of the quantity of water used by native vegetation growing in tanks in which a water table was maintained.

2. Transpiration-well method—based on measurement of diurnal water table fluctuations (see following section).

3. Seepage-run method—based on measurements of river discharge to determine seepage inflow to the river; water use was the difference in inflows between growing and nongrowing seasons.

4. Inflow-outflow method—based on a hydrologic inventory with water use calculated as the difference from all other factors.

5. Chloride-increase method—based on the increase in chloride concentration of groundwater as it flowed from the bottomland area to the river.

6. Slope-seepage method—based on difference between rates of groundwater inflow to the bottomland area and to river.

The computed transpiration value for each method was within 20 percent of the mean for all six methods. Results for a 12-month period showed a transpiration of 0.16 m from precipitation and 0.75 m from groundwater over the 3765-ha area.

Evapotranspiration Effects. From a practical standpoint it is often difficult to segregate evaporation and transpiration losses from groundwater; therefore, the combined loss, referred to as *evapo-*

---

*In an interesting field study, Lewis and Burgy[37] demonstrated that 6–12-m oak trees in California extracted water from a water table in fractured and jointed rock at a depth of 21 m. Tritium injected in a nearby well was found subsequently in water transpired by the trees (collected from plastic bags tied over ends of branches).

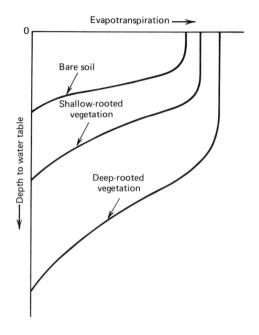

Fig. 6.14 Generalized variation of evapotranspiration from groundwater with water table depth for three ground-cover conditions (after Bouwer[3]).

transpiration (or *consumptive use*) is typically the quantity normally measured or calculated. The variation of evapotranspiration with water table depth is sketched in Fig. 6.14 for three groundcover conditions. It is apparent that the deeper the roots, the greater the depth at which water losses occur. Even with relatively deep water tables, evapotranspiration does not necessarily become zero because upward transport can still occur, albeit minimally, in the vapor phase.

The pattern of diurnal fluctuation resulting from discharge of groundwater is nearly identical for evaporation and transpiration. The maximum water table level occurs in midmorning (see Fig. 6.15) and represents a temporary equilibrium between discharge and recharge from surrounding groundwater. From midmorning until early evening, losses exceed recharge and the level falls. The steep slope near midday indicates maximum discharge associated with highest temperatures. The evening minimum again represents an equilibrium point, while the rise during the night hours is recharge in excess of discharge.

White[81] suggested a method for computing the total quantity of groundwater withdrawn by evapotranspiration during a day. If it is assumed that evapotranspiration is negligible from midnight to 4 A.M. and, further, that the water table level during this interval approximates the mean for the day, then the hourly recharge from midnight to 4 A.M. may be taken as the average rate for the day. Letting h equal the hourly rate of rise of the water table from midnight to 4 A.M., as shown by the upper curve in Fig. 6.15, and s the net fall or rise of the water table during the 24-hour period, then as a good approximation the diurnal volume of groundwater discharge per unit area

$$V_{ET} = S_y(24h \pm s) \tag{6.5}$$

where S_y is the specific yield near the water table. Actually, as pointed out by Troxell,[68] the rate of groundwater recharge to the vegetated area varies inversely with the water table level. Thus, the difference between the recharge rate and the slope of the groundwater level curve gives the evapotranspiration rate. The lower portion of Fig. 6.15 illustrates this; the area between the two curves is a measure of the daily volume of water released to the atmosphere.

Fig. 6.15 Interrelations of water table level, recharge, and evapotranspiration fluctuations (after Troxell[68]).

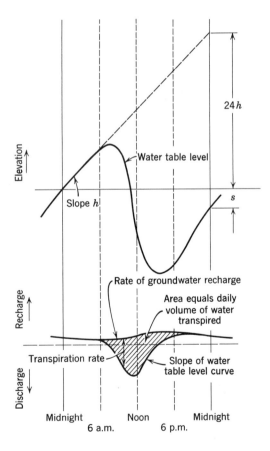

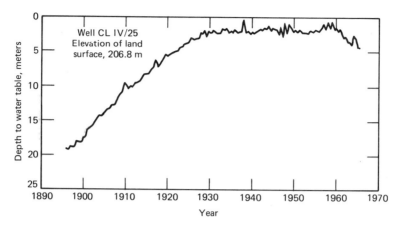

Fig. 6.16   Time variation of depth to water table for a well in the irrigated portion of the Indus River Plain, Pakistan. Leakage from unlined irrigation canals produced the rise until about 1930; thereafter, evapotranspiration losses near ground surface stabilized the level (after U.S. Geol. Survey Water Supply Paper 1608-0).

Evapotranspiration from groundwater can serve to stabilize the water table near ground surface. An example is shown in Fig. 6.16 for the Punjab region of Pakistan. Water tables began to rise about A.D. 1900 when a network of canals was constructed for irrigation. Leakage from the unlined canals into the permeable alluvium caused a steady rise in the water table until it was within 3–4 m of ground surface. The change in slope leading to a stabilized water table with well-defined annual fluctuations occurred as a result of an equilibrium between recharge and evapotranspiration.*

Fluctuations Due to Meteorological Phenomena

Atmospheric Pressure.   Changes in atmospheric pressure (barometric tides) produce sizable fluctuations in wells penetrating confined aquifers.[5,21] The relationship is inverse; that is, increases in atmospheric pressure produce decreases in water levels, and conversely. When atmospheric pressure changes are expressed in terms

---

*It is also important to point out that in areas where evapotranspiration serves to control the position of the water table, waterlogging and salination of soils in the root zone occur. Agricultural areas so afflicted can become saline wastelands unless adequate drainage systems are installed. Notable problem areas include portions of the Indus Valley in Pakistan, and of the San Joaquin Valley in California.

of a column of water, the ratio of water level change to pressure change expresses the *barometric efficiency* of an aquifer. Thus,

$$B = \frac{\gamma \Delta h}{\Delta p_a} \tag{6.6}$$

where B is barometric efficiency, γ is the specific weight of water, Δh is the change in piezometric level, and Δp_a is the change in atmospheric pressure. Most observations yield values in the range of 20 to 70 percent.

The effect is apparent in data shown in Fig. 6.17. The upper curve indicates observed water levels in a well at Iowa City, Iowa, penetrating a confined aquifer. The lower curve shows atmospheric pressure inverted, expressed in meters of water, and multiplied by 0.75. A close correspondence of major fluctuations exists in the two curves; the equality of amplitudes indicates that the barometric efficiency of the aquifer is about 75 percent.

The explanation of the phenomenon can be given by recognizing that aquifers are elastic bodies.[57,58] If Δp_a is the change in atmospheric pressure and Δp_w is the resulting change in hydrostatic pressure at the top of a confined aquifer, then

$$\Delta p_a = \Delta p_w + \Delta s_c \tag{6.7}$$

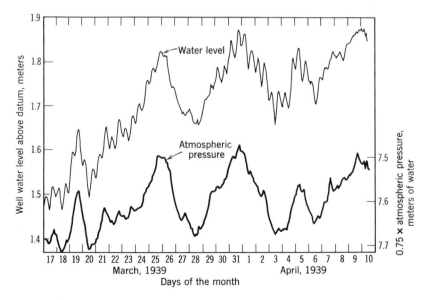

Fig. 6.17 Response of water level in a well penetrating a confined aquifer to atmospheric pressure changes, showing a barometric efficiency of 75 percent (after Robinson[58]).

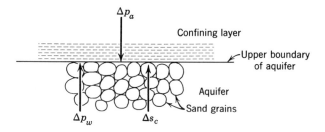

Fig. 6.18 Idealized distribution of forces at the upper boundary of a confined aquifer resulting from a change in atmospheric pressure.

where Δs_c is the increased compressive stress on the aquifer (Fig. 6.18). At a well penetrating the confined aquifer, the relation

$$p_w = p_a + \gamma h \tag{6.8}$$

exists as shown in Fig. 6.19a, where γ is the specific weight of water. Let the atmospheric pressure increase by Δp_a, then

$$p_w + \Delta p_w = p_a + \Delta p_a + \gamma h' \tag{6.9}$$

as shown in Fig. 6.19b. Substituting for p_w from Eq. 6.8 yields

$$\Delta p_w = \Delta p_a + \gamma(h' - h) \tag{6.10}$$

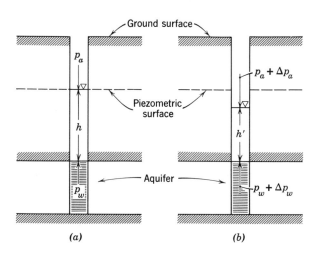

Fig. 6.19 Effect of an increase in atmospheric pressure on the water level of a well penetrating a confined aquifer.

But from Eq. 6.7 it is apparent that $\Delta p_w < \Delta p_a$, indicating that $h' < h$. Generally, therefore, the water level in a well falls with an increase in atmospheric pressure. It follows that the converse is also true.*

Jacob[30] developed expressions relating barometric efficiency of a confined aquifer to aquifer and water properties, including the storage coefficient. Gilliland[21] showed that changes in soil moisture from infiltrating precipitation can affect the magnitude of barometric efficiency.

For an unconfined aquifer, atmospheric pressure changes are transmitted directly to the water table, both in the aquifer and in a well; hence, no pressure difference occurs. Air entrapped in pores below the water table is affected by pressure changes, however, causing fluctuations similar to but smaller than that observed in confined aquifers.[44,48] Temperature fluctuations in the capillary zone will also induce water table fluctuations where entrapped air is present.[65,70]

Atmospheric pressure fluctuations do affect water tables substantially on small, permeable oceanic islands. The response of sea-level changes to atmospheric pressure is essentially isostatic; that is, sea level adjusts to a constant mass of the ocean-atmosphere column. This causes the ocean to act as an inverted barometer with sea level rising about 1 cm to compensate for a drop in atmospheric pressure of 1 mb. These fluctuations amount to about 20 cm in the open ocean and are transmitted as long-term tides to the water table (see the subsequent section on ocean tides). Data from Bermuda by Vacher[71] shown in Fig. 6.20 illustrate the fluctuations.

Rainfall. As described previously, rainfall is not an accurate indicator of groundwater recharge because of surface and subsurface losses as well as travel time for vertical percolation. The travel time may vary from a few minutes for shallow water tables in permeable formations to several months or years for deep water tables underlying sediments with low vertical permeabilities. Furthermore, in arid and semiarid regions, recharge from rainfall may be essentially zero. Shallow water tables show definite responses to rainfall, as Fig. 6.21 indicates. Water levels shown are the average for 25 observation wells; greatest fluctuations occurred in the upper portions of the basin and smallest near the basin outlet.

Groundwater levels may show seasonal variations due to rainfall, but often these include natural discharge and pumping effects as

---

*Atmospheric pressure waves created by nuclear explosions in the Soviet Union have caused fluctuations of the piezometric surface in limestone aquifers in England.[27] One fluctuation displayed an amplitude of 0.46 cm in response to a pressure wave of 900–1000 microbars.

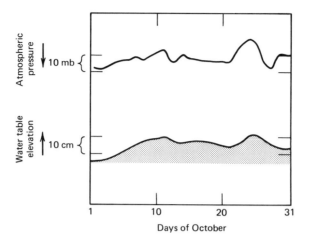

Fig. 6.20 Variations in atmospheric pressure and water table elevation during October 1973 at Devonshire Post Office, Bermuda. Note that the variation is inverse with a rise in pressure being associated with a decline in water level (after Vacher[71]).

well. Droughts extending over a period of several years contribute to declining water levels.

Where the unsaturated zone above a water table has a moisture content less than that of specific retention (see Chapter 2), the water table will not respond to recharge from rainfall until this deficiency has been satisfied. Thereafter, the rise Δh will amount to

$$\Delta h = P_i/S_y \tag{6.11}$$

where P_i is that portion of precipitation that percolates to the water table and S_y is specific yield.

An interesting phenomenon occasionally noted in observation wells is a nearly instantaneous response of shallow water tables to rainfall. This may be explained by the pressure increase of air trapped in the zone of aeration when rainfall seals surface pores and infiltrating water compresses the underlying air. If the zone containing interconnected air-filled pores (H in Fig. 6.22) is compressed to a thickness $H - m$, then the pressure above the water table is increased by $m/(H - m)$ of an atmosphere, causing the water level in an observation well to rise

$$\Delta h = \frac{m}{H - m}(10)\text{m} \tag{6.12}$$

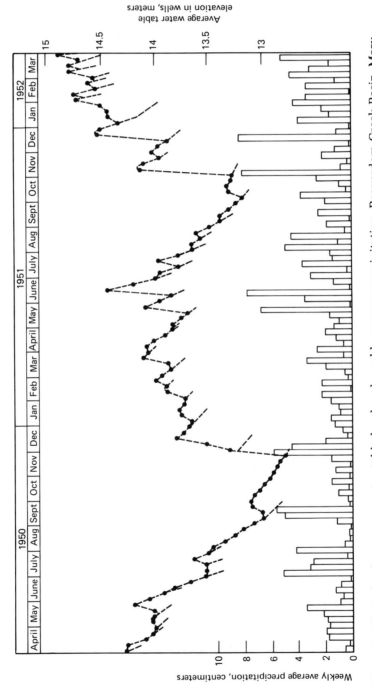

Fig. 6.21 Variation in average water table level and weekly average precipitation, Beaverdam Creek Basin, Maryland (after Rasmussen and Andreasen[53]).

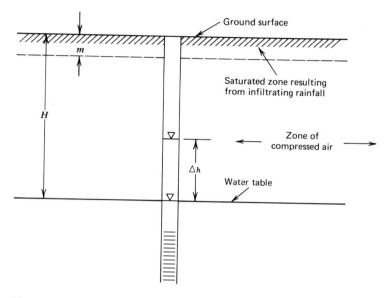

Fig. 6.22   Water table rise in an observation well resulting from infiltrating rainfall sealing the ground surface and compressing air above the water table.

For shallow water tables the rise (which occurs only in the well) can be an order of magnitude larger than the depth of infiltrating rainfall; however, escaping air soon dissipates the effect.*

Wind. Minor fluctuations of water levels are caused by wind blowing over the tops of wells. The effect is identical to the action of a vacuum pump. As a gust of wind blows across the top of a casing, the air pressure within the well is suddenly lowered and, as a consequence, the water level quickly rises. After the gust passes, the air pressure in the well rises and the water level falls. The effect is illustrated by Fig. 6.23, which shows a well record at Miami during the passage of a hurricane. The storm center passed north of Miami during the night of Oct. 18, 1944. Wind velocities reached 54 mph on October 18 and 65 mph on October 19; rapid fluctuations accompanying these winds are apparent on the well record.

Frost. In regions of heavy frost it has been observed that shallow water tables decline gradually during the winter and rise sharply in early spring before recharge from ground surface could occur (see

---

*Similarly, when water is applied uniformly to the top of dry column of sand in the laboratory, the air is compressed until released by a spontaneous upward eruption.

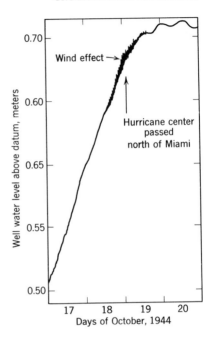

Fig. 6.23 Wind-induced water level fluctuations in a well at Miami, Florida, during passage of a hurricane (after Parker and Stringfield[47]).

Fig. 6.24).[61] This fluctuation can be attributed to the presence of a frost layer above the water table. During winter water moves upward from the water table by capillary movement and by vapor transfer to the frost layer where it freezes.* In early spring, approximately when the mean air temperature reaches 0°C, the frost layer begins thawing from the bottom; consequently, meltwater percolates downward to rejoin the water table.

Fluctuations Due to Tides

Ocean Tides. In coastal aquifers in contact with the ocean, sinusoidal fluctuations of groundwater levels occur in response to tides. If the sea level varies with a simple harmonic motion, a train of sinusoidal waves is propagated inland from the submarine outcrop of the aquifer. With distance, inland amplitudes of the waves decrease and the time lag of a given maximum increases. The problem has been solved by analogy to heat conduction in a semiinfinite solid subject to periodic temperature variations normal to the infinite dimension.[12,22,79]

---

*Vapor migration occurs in response to the thermal gradient and to the fact that vapor pressure over ice is less than that over liquid water at 0°C.

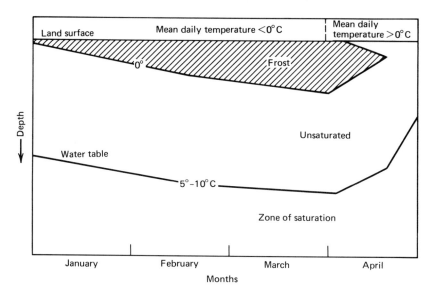

Fig. 6.24 Sketch illustrating the variation in depth to water table in response to winter frost conditions (after Schneider[61]).

For simplicity consider the one-directional flow in a confined aquifer as shown in Fig. 6.25a. From Eq. 3.76 the applicable differential equation governing the flow is

$$\frac{\partial^2 h}{\partial x^2} = \frac{S}{T}\frac{\partial h}{\partial t} \qquad (6.13)$$

where h is the net rise or fall of the piezometric surface with reference to the mean level, x is the distance inland from the outcrop, S is the storage coefficient of the aquifer, T is transmissivity, and t is time. Letting the amplitude, or half range, of the tide be h_0 (see Fig. 6.25a), the applicable boundary conditions include $h = h_0 \sin \omega t$ at $x = 0$ and $h = 0$ at $x = \infty$. The angular velocity is ω; for a tidal period t_0,

$$\omega = \frac{2\pi}{t_0} \qquad (6.14)$$

The solution of Eq. 6.13 with these boundary conditions is

$$h = h_0 e^{-x\sqrt{\pi S/t_0 T}} \sin\left(\frac{2\pi t}{t_0} - x\sqrt{\pi S/t_0 T}\right) \qquad (6.15)$$

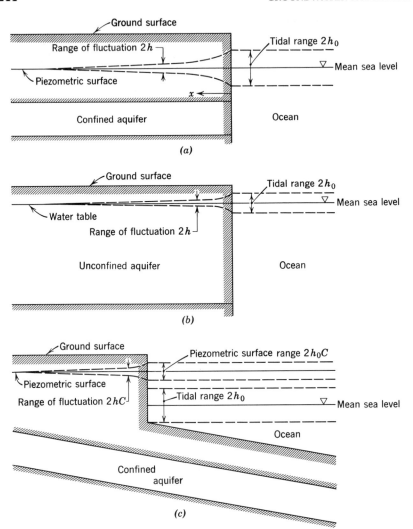

Fig. 6.25 Groundwater level fluctuations produced by ocean tides. (a) Confined aquifer. (b) Unconfined aquifer. (c) Loading of a confined aquifer.

From this it follows that amplitude h_x of groundwater fluctuations at a distance x from the shore equals

$$h_x = h_0 e^{-x\sqrt{\pi S/t_0 T}} \qquad\qquad (6.16)$$

The time lag t_L of a given maximum or minimum after it occurs in the ocean can be obtained by solving the quantity within the parentheses of Eq. 6.15 for t, so that

$$t_L = x\sqrt{t_0 S/4\pi T} \tag{6.17}$$

The waves travel with a velocity

$$v_w = \frac{x}{t_L} = \sqrt{4\pi T/t_0 S} \tag{6.18}$$

and the wavelength is given by

$$L_w = v_w t_0 = \sqrt{4\pi t_0 T/S} \tag{6.19}$$

Substituting the wavelength for x in Eq. 6.16 shows that the amplitude decreases by a factor $e^{-2\pi}$, or $1/535$, for each wavelength. Water flows into the aquifer during half of each cycle and out during the other half. By Darcy's law the quantity of flow V per half-cycle per foot of coast is

$$V = \int_{-t_0/8}^{3t_0/8} q\,dt = T\int_{-t_0/8}^{3t_0/8} \left(\frac{\partial h}{\partial x}\right)_{x=0} dt \tag{6.20}$$

where q is the flow per foot of coast. Differentiating Eq. 6.15 to obtain $\partial h/\partial x$ and integrating yields

$$V = h_0\sqrt{2t_0 ST/\pi} \tag{6.21}$$

The above analysis is also applicable as a good approximation to water table fluctuations of an unconfined aquifer if the range of fluctuation is small in comparison to the saturated thickness (Fig. 6.25b). In Fig. 6.26 are shown fluctuations in wells penetrating an unconfined aquifer at various distances from a surface water level varying approximately with a sinusoidal pattern.

Just as atmospheric pressure changes produce variations of piezometric levels, so do tidal fluctuations vary the load on confined aquifers extending under the ocean floor (Fig. 6.25c). Contrary to the atmospheric pressure effect, tidal fluctuations are direct; that is, as the sea level increases, the groundwater level does also. Figure 6.27 illustrates the effect for a well only 30 m from shore. The ratio of piezometric level amplitude to tidal amplitude is known as the *tidal efficiency* of the aquifer. Jacob[30] showed that tidal efficiency C is related to barometric efficiency B by

$$C = 1 - B \tag{6.22}$$

Thus, tidal efficiency is a measure of the incompetence of overlying confining beds to resist pressure changes. Aquifer response to loading rather than head change at the outcrop requires that the amplitude given by Eq. 6.15 be multiplied by C.

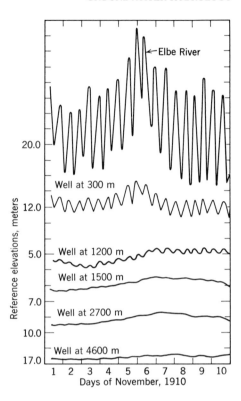

Fig. 6.26 Fluctuations of the Elbe River and water table levels in wells at various distances from the river (after Werner and Noren[79]).

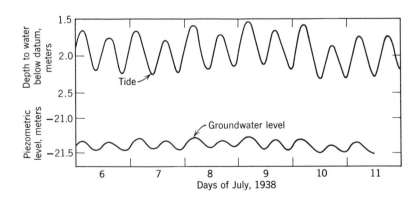

Fig. 6.27 Tidal fluctuations and induced piezometric surface fluctuations observed in a well 30 m from shore at Mattawoman Creek, Maryland (after Meinzer[40]).

Earth Tides. Regular semidiurnal fluctuations of small magnitude have been observed in piezometric surfaces of confined aquifers located at great distances from the ocean.[57] After correcting well levels for atmospheric pressure changes, these fluctuations appear quite distinctly in certain wells where the phenomenon has been investigated. Figure 6.28 shows fluctuations over a lunar cycle from a 250-m well tapping a confined aquifer in Iowa City.

These fluctuations result from earth tides, produced by the attraction exerted on the earth's crust by the moon and, to a lesser extent, the sun. Robinson's observations,[58] based on analyses of well records, make convincing evidence: (1) two daily cycles of fluctuations occur about 50 min later each day, as does the moon; (2) the average daily retardation of cycles agrees closely with that of the moon's transit; (3) the daily troughs of the water level coincide with the transits of the moon at upper and lower culmination; and (4) periods of large regular fluctuations coincide with periods of new and full moon, whereas periods of small irregular fluctuations coincide with periods of first and third quarters of the moon. All of these facts may be noted in the data of Fig. 6.28. Bredehoeft[4] has pointed out that wells serve as sensitive indicators of this dilatation of the earth's crust.

At times of new and full moon, the tide-producing forces of the moon and sun act in the same direction, then ocean tides display a greater than average range. But when the moon is in the first and third quarters, tide-producing forces of the sun and moon act perpendicular to each other, causing ocean tides of smaller than average range. The coincidence of the time of low water with that of the moon's transit can be explained by reasoning that at this time tidal attraction is maximum; therefore, the overburden load on the aquifer is reduced, allowing the aquifer to expand slightly.

Urbanization

The process of urbanization often causes changes in groundwater levels as a result of decreased recharge and increased withdrawal. In rural areas water supplies are usually obtained from shallow wells, while most of the domestic wastewater is returned to the ground through cesspools or septic tanks. Thus, a quantitative balance in the hydrologic system remains. As population increases, many individual wells are abandoned in favor of deeper public wells. Later, with the introduction of sewer systems, storm water and wastewater typically discharge to a nearby surface water body

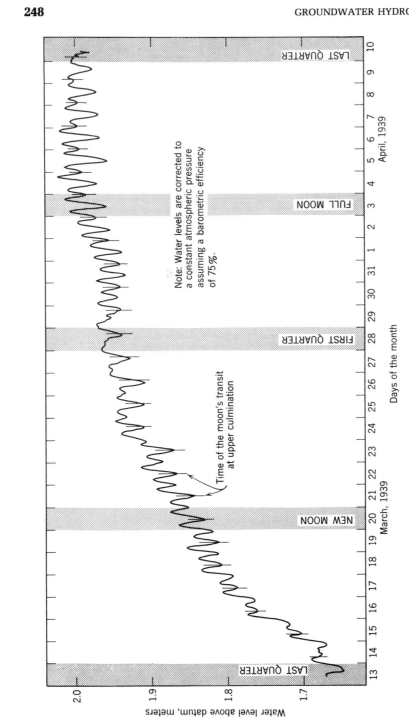

Fig. 6.28 Water level fluctuations in a confined aquifer produced by earth tides (after Robinson[58]).

(see Fig. 6.29). Thus, three conditions disrupt the subsurface hydrologic balance and produce declines in groundwater levels:

1. Reduced groundwater recharge due to paved surface areas and storm sewers.
2. Increased groundwater discharge by pumping wells.
3. Decreased groundwater recharge due to export of wastewater collected by sanitary sewers.

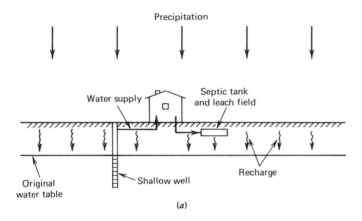

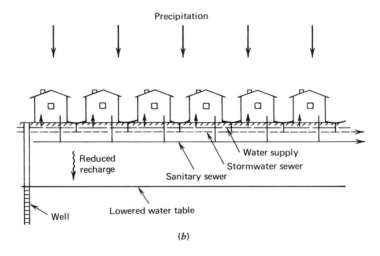

Fig. 6.29 Schematic diagram illustrating how urbanization can cause lowering of water table elevation. (*a*) Rural situation. (*b*) Urban development.

Effects of the urbanization trend are well illustrated on Long Island, New York.[6,15] Here the above conditions have all been present, leading not only to a decline in water tables but also to groundwater pollution (see Chapter 8), seawater intrusion (see Chapter 14), and reduced streamflow.* Artificial recharge efforts (see Chapter 13) are underway to counteract these undesirable results of urbanization.

Earthquakes

Observations reveal that earthquakes have a variety of effects on groundwater.[76,77] Most spectacular are sudden rises or falls of water levels in wells, changes in discharge of springs, appearance of new springs, and eruptions of water and mud out of the ground. More commonly, however, earthquake shocks produce small fluctuations (*hydroseisms*) in wells penetrating confined aquifers. A good example is furnished by the water level record on an expanded time scale shown in Fig. 6.30. This earthquake was centered on the Argentina–Chile border, nearly 8000 km from the recording well in Milwaukee. Although little is known of the quantitative effects of earthquakes on groundwater, these fluctuations result from com-

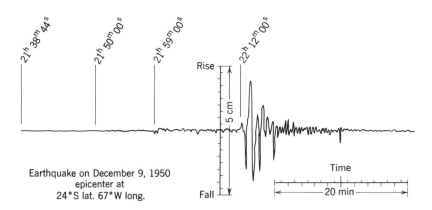

Fig. 6.30 Water level fluctuations in a well at Milwaukee, Wisconsin, resulting from an earthquake centered on the Argentina-Chile border (after Vorhis[77]).

*More than 95 percent of Long Island's streamflow is derived from groundwater. Thus, during the drought period 1961–1965 when groundwater levels in Nassau and Suffolk Counties declined 1.4 m, the average flow of 19 streams in the area decreased by 42 percent.

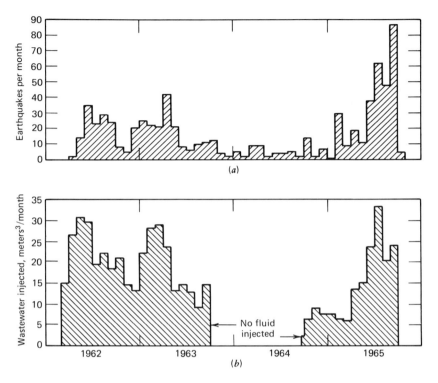

Fig. 6.31 (*a*) Earthquake frequency. (*b*) Volume of injected wastewater. Near Denver, Colorado, during the period 1962–1965 (after Healy, et al.[24]; copyright © 1968 by American Association for the Advancement of Science).

pression and expansion (dilatation) of elastic confined aquifers by the passage of earthquake (Rayleigh) waves.[8] These waves travel at speeds of approximately 200 km/min so that fluctuations appear after little more than one hour even from the most distant earthquake centers.

Looking at the converse situation, recent field studies have revealed that injection of wastewater into a deep well can trigger earthquakes.[24] Evidence stems from injection of chemical-manufacturing waste fluids near Denver, Colorado, into a well 3671 m deep and penetrating sedimentary rocks into Precambrian crystalline rocks. Figure 6.31 shows the time variation of fluid injection and earthquake frequency. Most earthquake magnitudes were small—within the range of 1.5 to 4.4 on the Richter scale. It is believed that the mechanism by which fluid injection triggered the earthquakes stems from a reduction of frictional resistance to faulting, a reduction that occurs with increase in pore pressure. Knowledge of this phenomenon has stimulated research into the possibility of injecting water into potentially dangerous fault zones. This might trigger minor earthquakes, ease stresses along a fault, and hence prevent the sudden release of accumulated energy that results in disastrous earthquakes.

The Hegben Lake, Montana, earthquake of August 1959 caused fluctuations in wells throughout the United States and even in Hawaii and Puerto Rico.[9] The main shock was recorded by water level fluctuations that ranged to more than 3.0 m; values exceeded 0.3 m in nine states. Even more dramatic were the hydroseismic responses to the Anchorage, Alaska, earthquake of March 27, 1964. This large-magnitude earthquake (8.4 to 8.8 on the Richter scale) affected groundwater levels throughout the United States, with the largest fluctuation exceeding 7.0 m at a well in South Dakota.[78] Hydroseisms were also recorded at such distant locations as Denmark, Egypt, South Africa, the Philippines, and Australia.

External Loads

The elastic property of confined aquifers results in changes in hydrostatic pressure when changes in loading occur. Some of the best examples are exhibited by wells located near railroads where passing trains produce measurable fluctuations of the piezometric surface. Figure 6.32 illustrates changes in water level produced by a train stopping and starting near a well at Smithtown, N.Y.

The application of a load compresses the aquifer and increases the hydrostatic pressure. Thereafter the pressure decreases and approaches its original value asymptotically as water flows radially away from the point where the load is applied. Thus, initially the load is shared by the confined water and the solid material of the aquifer; however, as the water flows radially outward, an increasing proportion of the load is borne by the structure of the aquifer. The schematic diagram after Jacob[29] in Fig. 6.33 shows this effect. Here a point load instantaneously applied is represented. The lower surface of the aquifer is assumed fixed; lengths of arrows indicate the relative magnitudes of flow velocities at various distances from the load. During the interval from A to B the hydrostatic pressure decreases and the deflection of the upper surface of the aquifer increases. Subsequently, when the load is removed, the pressure drops to a minimum and then recovers toward its initial value as shown by times C and D.

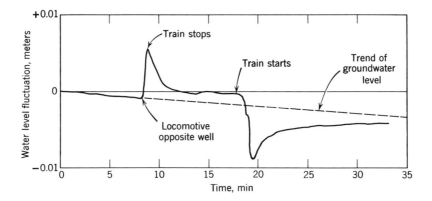

Fig. 6.32   Water level fluctuations in a confined aquifer produced by a train stopping and starting near an observation well (after Jacob[29]).

Land Subsidence and Groundwater

Changes in groundwater levels or subsurface moisture conditions may be responsible for subsidence of the land surface. This can severely damage wells and can create special problems in the design and operation of structures for drainage, flood protection, and water conveyance. At least four distinct phenomena have been identified.[1]

Lowering of Piezometric Surface.   Land subsidence has been observed to accompany extensive lowering of the piezometric surface in regions of heavy pumping from confined aquifers. Figure 6.34 shows the effect at San Jose, California. Here over a 15-year period (1920–1935), the average subsidence ratio equaled 1/13, indicating that the land surface subsided 1 m for every 13 m of lowering of the piezometric surface.

The explanation for this subsidence is based on fundamentals of soil mechanics.[10,38] Consider the pressure diagram for a confined aquifer overlain by an unconfined aquifer shown in Fig. 6.35. Initially, the total (geostatic) pressure p_t at any depth (see Fig. 6.35a) is

$$p_t = p_h + p_i \tag{6.23}$$

where p_h is the hydraulic pressure and p_i is the intergranular pressure. If pumping in the confined aquifer lowers the piezometric surface while the water table remains unchanged due to an imper-

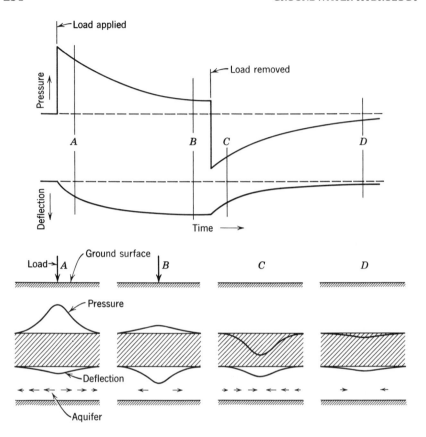

Fig. 6.33 Hydrostatic pressure variations and aquifer deflections resulting from a point load applied and later removed from the ground surface above a confined aquifer (after Jacob[29]).

meable clay layer separating the aquifers (Fig. 6.35b), then Eq. 6.23 becomes

$$p_t = p_h' + p_i' \tag{6.24}$$

Note that $p_h' < p_h$ and $p_i' > p_i$ for both the confined aquifer and the clay layer. Adjustments to these new pressure distributions will take place essentially instantaneously in the permeable, coarse-grained aquifer. But in the relatively impermeable, fine-grained clay, this adjustment may take months to years. Because clayey materials are highly compressible, the increased intergranular pressure $(p_i' - p_i)$ causes the clay layer to be compacted. This reduces its

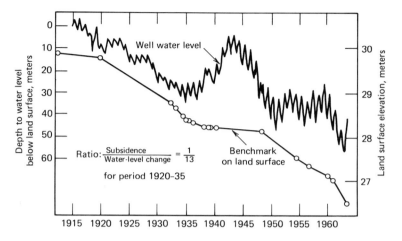

Fig. 6.34 Land subsidence and decline of the piezometric surface, 1912–1963, San Jose, California (after Poland and Davis[51]; courtesy The Geological Society of America, 1969).

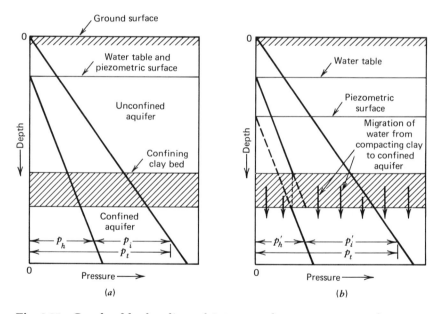

Fig. 6.35 Graph of hydraulic and intergranular pressures as a function of depth for an unconfined aquifer overlying a confined aquifer. (a) Initial condition with water table and piezometric surface at same elevation. (b) Subsequent condition with piezometric surface lowered (after Poland, et al.[52]).

porosity, while water contained in the clay pores is squeezed downward into the confined aquifer.

The volume of water displaced from the clay equals the reduction in clay volume and also the volumetric land surface subsidence. Similarly, the reduction in thickness of the clay layer equals the vertical land subsidence. The amount of compaction is a function of the thickness and vertical permeability of the clay, of the time and magnitude of piezometric surface decline, and of the microstructure of the clay.[51] Because sand and gravel deposits are relatively incompressible, the increased intergranular pressure has a negligible effect on the aquifer itself.

Land subsidence resulting from compaction of fine-grained sediments occurs at several locations in the United States[17] and throughout the world.[14,18] The problem has been extensively studied by Poland and others.[28,50,51,52] Table 6.1 lists areas of major land subsidence due to groundwater pumping.* Compaction of clay is largely inelastic and permanent; consequently, the only effective control measure for this type of subsidence is increasing piezometric levels by reducing pumping and by recharge of water through injection wells (see Chapter 13). Withdrawal of oil and gas produces the same problem of land subsidence; major areas affected exist in the United States, Italy, Japan, and Venezuela.[50]

Hydrocompaction. Collapse of the ground surface has been observed to occur when water is applied to certain types of soils. Particularly susceptible are (1) loose, moisture-deficient alluvial deposits, including mud flows, and (2) moisture-deficient loess deposits.[39,64] An example of this type of shallow subsidence is found on the arid west side of the San Joaquin Valley, California. Soils characteristically are desiccated with a high void content and low density (1.1 to 1.4 g/cm^3). Most of these soils have never been saturated since deposition, but when irrigation water, for example, is applied, their internal high void structure collapses, resulting in an erratic subsidence of the land surface.

To define the magnitude of this subsidence, a test pond 30 m by 30 m was constructed on flat land, and bench marks anchored at various depths were installed. Water to a depth of 0.6 m was ad-

---

*The periodic flooding of Piazza San Marco in Venice from the Grand Canal provides dramatic evidence that subsidence from groundwater pumpage threatens this beautiful city. A decline of 20 cm over the last 50 years has been reported. Recent restrictions on pumpage in the industrial suburb of Porto Marghera are expected to halt this subsidence rate.[18]

TABLE 6.1 Areas of Major Land Subsidence Due to Groundwater Overdraft (after Poland[50])

Location	Depositional Environment and Age	Depth Range of Compacting Beds, m	Maximum Subsidence, m	Area of Subsidence, sq km	Time of Principal Occurrence
Japan					
Osaka	Alluvial and shallow marine; Quaternary	10–400	3	190	1928–1968
Tokyo	As above	10–400	4	190	1920–1970+
Mexico					
Mexico City	Alluvial and lacustrine; late Cenozoic	10–50	9	130	1938–1970+
Taiwan					
Taipei basin	Alluvial and lacustrine; Quaternary	10–240	1.3	130	1961–1969+
United States					
Arizona, central	Alluvial and lacustrine; late Cenozoic	100–550	2.3	650	1948–1967
California Santa Clara Valley	Alluvial and shallow marine; late Cenozoic	55–300	4	650	1920–1970
San Joaquin Valley (three subareas)	Alluvial and lacustrine; late Cenozoic	60–1000	2.9–9	11,000 (>0.3 m)	1935–1970+
Lancaster area	Alluvial and lacustrine; late Cenozoic	60–300(?)	1	400	1955–1967+
Nevada Las Vegas	Alluvial; late Cenozoic	60–300	1	500	1935–1963

TABLE 6.1 (continued)

Location	Depositional Environment and Age	Depth Range of Compacting Beds, m	Maximum Subsidence, m	Area of Subsidence, sq km	Time of Principal Occurrence
Texas Houston-Galveston area	Fluvial and shallow marine; late Cenozoic	60–600(?)	1–1.5	6,860 (>0.15 m)	1943–1964+
Louisiana Baton Rouge	Fluvial and shallow marine; Miocene to Holocene	50–600(?)	0.3	650	1934–1965+

mitted in early October 1956. Subsidence of the various bench marks appears in Fig. 6.36. As the wetting front moved downward, the bench marks progressively subsided. At ground surface the change in level amounted to more than 3 m, while at the 45 m depth no effect was observed until after 16 months.

Shallow subsidence can influence irrigation, drainage, sewerage, and transportation systems.* Sprinkler irrigation and pipelines for water conveyance are best suited in these terrains.

Dewatering of Organic Soils. In flat peat or muck land with a shallow water table, lowering of the water table, such as by drainage, produces land subsidence. Causes include (1) shrinkage due to desiccation, (2) consolidation by loss of the buoyant force of groundwater, (3) compaction with tillage, (4) wind erosion, (5) burning, and (6) biochemical oxidation.[28] Investigations have shown that the rate of subsidence is proportional to the depth to the water table. To conserve the life of organic soils, the water table should be maintained as high as crop requirements and field conditions will permit.

Subsidence of organic soils has been noted in the Netherlands, the Soviet Union, and at various locations in the United States. During the last 50 years extensive drainage for agricultural purposes of islands in the Sacramento–San Joaquin Delta, California, has lowered the land surface 3 to 5 m below sea level over much of the area. This has necessitated construction of a vast network of perimeter levees to prevent inundation of the depressed islands by floods or high tides.

Sinkhole Formation. Catastrophic land subsidence leading to the formation of sinkholes can also be associated with declines in groundwater levels. Soluble rocks such as dolomite and limestone are slowly dissolved locally by groundwater. Eventually the ground surface sinks to form a cup-shaped depression. Over large areas of this type, a karstic sinkhole plain is formed with most of the drainage occurring in the subsurface.

New sinkholes often develop in regions where water tables have been lowered by pumping. For example, groundwater pumping from a dolomite aquifer for mine dewatering in the Far West Rand, South Africa, led to the formation of eight sinkholes larger than 50 m in

---

*More than 100 km along the California Aqueduct, a concrete-lined canal carrying some $25 \times 10^6 \, \text{m}^3/\text{day}$ of water to Southern California, crosses formations susceptible to shallow land subsidence. To avoid the danger of subsidence from canal leakage, large-scale spreading ponds were maintained along the alignment in order to preconsolidate the soils before construction of the aqueduct.

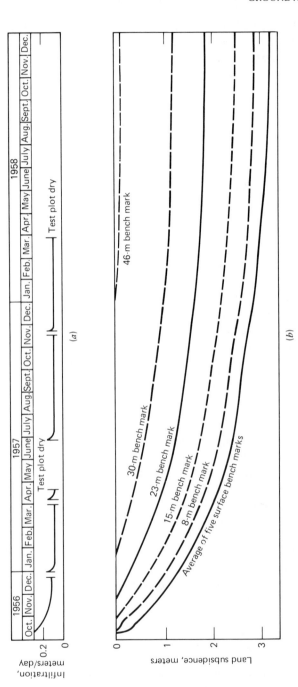

Fig. 6.36 Hydrocompaction and land subsidence resulting from water application on a test plot containing low density soil, San Joaquin Valley, California. (*a*) Infiltration from test plot. (*b*) Subsidence of bench marks anchored at ground surface and at various depth intervals (after Lofgren[39]; courtesy The Geological Society of America, 1969).

Fig. 6.37 Sinkhole formed suddenly in December 1973 with a diameter of 130 m and a depth of 45 m in Shelby County, Alabama. This is one of more than 1000 sinkholes that developed in Shelby County within a 15-year period. The concentration of sinkholes is attributed to a local lowering of the water table (courtesy U.S. Geological Survey).

diameter and deeper than 30 m within a period of 39 months.[13] Similarly, the massive Alabama sinkhole shown in Fig. 6.37 appeared suddenly after a local decline of the water table. As a water table is lowered, removal of the buoyant support from the subsurface clay above the cavern together with increased velocities of infiltrating water are believed to be responsible for the cave-ins.

Crustal Uplift. The opposite of land subsidence, crustal uplift can occur over large areas subject to heavy groundwater pumping. The tectonic uplift of land, involving an elastic expansion of the lithosphere, is caused by the removal of large masses of groundwater. The phenomenon has been noted in parts of Arizona, California, and Texas, where groundwater pumped from aquifers was removed by evapotranspiration by crops.[26] In the Santa Cruz River

Basin, Arizona, a crustal uplift of 6 cm was observed between 1948 and 1967, when 43.5 billion tons of groundwater were pumped from an 8070-km^2 area. This figure compares favorably with the 17-cm land surface depression, previously measured, resulting from the filling of Lake Mead on the Colorado River.

References

1. Amer. Soc. Civil Engrs., *Ground water management*, Manual Engrng. Practice 40, New York, 216 pp., 1972.
2. Barnes, B. S., The structure of discharge-recession curves, *Trans. Amer. Geophysical Union*, v. 20, pp. 721–725, 1939.
3. Bouwer, H., Predicting reduction in water losses from open channels by phreatophyte control, *Water Resources Research*, v. 11, pp. 96–101, 1975.
4. Bredehoeft, J. D., Response of well-aquifer systems to earth tides, *Jour. Geophysical Research*, v. 72, pp. 3075–3087, 1967.
5. Clark, W. E., Computing the barometric efficiency of a well, *Jour. Hydraulics Div.*, Amer. Soc. Civil Engrs., v. 93, no. HY4, pp. 93–98, 1967.
6. Cohen, P., et al., *An atlas of Long Island's water resources*, Bull. 62, New York Water Resources Comm., Albany, 117 pp., 1968.
7. Cooper, H. H., Jr., and M. I. Rorabaugh, Ground-water movements and bank storage due to flood stages in surface streams, *U.S. Geological Survey Water-Supply Paper* 1536-J, pp. 343–366, 1963.
8. Cooper, H. H., Jr., et al., The response of well-aquifer systems to seismic waves, *Jour. Geophysical Research*, v. 70, pp. 3915–3926, 1965.
9. Da Costa, J. A., Effect of Hegben Lake earthquake on water levels in wells in the United States, *U.S. Geological Survey Prof. Paper* 435, pp. 167–178, 1964.
10. Domenico, P. A., and M. D. Mifflin, Water from low-permeability sediments and land subsidence, *Water Resources Research*, v. 1, pp. 563–576, 1965.
11. Farvolden, R. N., Geologic controls on ground-water storage and base flow, *Jour. Hydrology*, v. 1, pp. 219–249, 1963.
12. Ferris, J. G., Cyclic fluctuations of water level as a basis for determining aquifer transmissibility, *Intl. Assoc. Sci. Hydrology Publ.* 33, pp. 148–155, 1951.
13. Foose, R. M., Sinkhole formation by groundwater withdrawal: Far West Rand, South Africa, *Science*, v. 157, pp. 1045–1048, 1967.
14. Fox, D. J., Man-water relationships in metropolitan Mexico, *Geogr. Review*, v. 55, pp. 523–545, 1965.
15. Franke, O. L., Double-mass-curve analysis of the effects of sewering on ground-water levels on Long Island, New York, *U.S. Geological Survey Prof. Paper* 600-B, pp. 205–209, 1968.
16. Freeze, R. A., Role of subsurface flow in generating surface runoff, *Water Resources Research*, v. 8, pp. 609–623, 1272–1283, 1972.

17. Gabrysch, R. K., and C. W. Bonnet, *Land-surface subsidence in the Houston-Galveston region, Texas,* Rept. 188, Texas Water Dev. Board, Austin, 19 pp., 1975.
18. Gambolati, G., and R. A. Freeze, Mathematical simulation of the subsidence of Venice, *Water Resources Research,* v. 9, pp. 721–733, 1973; v. 10, pp. 563–577, 1974.
19. Gardner, W. R., and M. Fireman, Laboratory studies of evaporation from soil columns in the presence of a water table, *Soil Sci.,* v. 85, pp. 244–249, 1958.
20. Gatewood, J. S., et al., Use of water by bottom-land vegetation in Lower Safford Valley, Arizona, *U.S. Geological Survey Water-Supply Paper* 1103, 210 pp., 1950.
21. Gilliland, J. A., A rigid plate model of the barometric effect, *Jour. Hydrology,* v. 7, pp. 233–245, 1969.
22. Gregg, D. O., An analysis of ground-water fluctuations caused by ocean tides in Glynn County, Georgia, *Ground Water,* v. 4, no. 3, pp. 24–32, 1966.
23. Hall, F. R., Base-flow recession—A review, *Water Resources Research,* v. 4, pp. 973–983, 1968.
24. Healy, J. H., et al., The Denver earthquakes, *Science,* v. 161, pp. 1301–1310, 1968.
25. Hellwig, D. H. R., Evaporation of water from sand, *Jour. Hydrology,* v. 18, pp. 93–118, 1973.
26. Holzer, T. J., Elastic expansion of the lithosphere caused by groundwater depletion, *Jour. Geophysical Research,* v. 84, pp. 4689–4698, 1979.
27. Ineson, J., Form of ground-water fluctuations due to nuclear explosions, *Nature,* v. 198, pp. 22–23, 1963.
28. Intl. Assoc. Sci. Hydrology, *Land subsidence,* Publ. nos. 88 and 89 (2 vols.), 661 pp., 1969.
29. Jacob, C. E., Fluctuations in artesian pressure produced by passing railroad-trains as shown in a well on Long Island, New York, *Trans. Amer. Geophysical Union,* v. 20, pp. 666–674, 1939.
30. Jacob, C. E., On the flow of water in an elastic artesian aquifer, *Trans. Amer. Geophysical Union,* v. 21, pp. 574–586, 1940.
31. Johnston, R. H., Base flow as an indicator of aquifer characteristics in the Coastal Plain of Delaware, *U.S. Geological Survey Prof. Paper* 750-D, pp. 212–215, 1971.
32. Keppel, R. V., and K. G. Renard, Transmission losses in ephemeral stream beds, *Jour. Hydraulics Div.,* Amer. Soc. Civil Engrs., v. 88, no. HY3, pp. 59–68, 1962.
33. Klein, M., and P. Kaser, *A statistical analysis of ground-water levels in twenty selected observation wells in Ohio,* Tech. Rept. 5, Ohio Dept. Natural Resources, Div. of Water, Columbus, 124 pp., 1963.
34. Knisel, W. G., Jr., Baseflow recession analysis for comparison of drainage basins and geology, *Jour. Geophysical Research,* v. 68, pp. 3649–3653, 1963.

35. Kunkle, G. R., The base flow-duration curve, a technique for the study of groundwater discharge from a drainage basin, *Jour. Geophysical Research*, v. 67, no. 4, pp. 1543–1554, 1962.

36. Kunkle, G. R., Computation of ground-water discharge to streams during floods, or to individual reaches during baseflow, by use of specific conductance, *U.S. Geological Survey Prof. Paper* 525-D, pp. 207–210, 1965.

37. Lewis, D. C., and R. H. Burgy, The relationship between oak tree roots and groundwater in fractured rock as determined by tritium tracing, *Jour. Geophysical Research*, v. 69, pp. 2579–2588, 1964.

38. Lofgren, B. E., Analysis of stresses causing land subsidence, *U.S. Geological Survey Prof. Paper* 600-B, pp. 219–225, 1968.

39. Lofgren, B. E., Land subsidence due to the application of water, *in* Varnes, D. J., and G. Kiersch (eds.), *Reviews in Engineering Geology*, v. 2, Geol. Soc. Amer., Boulder, Colo., pp. 271–303, 1969.

40. Meinzer, O. E., Ground water in the United States, *U.S. Geological Survey Water-Supply Paper* 836-D, pp. 157–232, 1939.

41. Meyboom, P., Estimating ground-water recharge from stream hydrographs, *Jour. Geophysical Research*, v. 66, pp. 1203–1214, 1961.

42. Minshall, N. E., Precipitation and base flow variability, *Intl. Assoc. Sci. Hydrology Publ.* 76, pp. 137–145, 1967.

43. Newcomb, R. C., and S. G. Brown, Evaluation of bank storage along the Columbia River between Richland and China Bar, Washington, *U.S. Geological Survey Water-Supply Paper* 1539-I, 13 pp., 1961.

44. Norris, S. E., and H. B. Eagon, Jr., Recharge characteristics of a watercourse aquifer system at Springfield, Ohio, *Ground Water*, v. 9, no. 1, pp. 30–41, 1971.

45. Norum, D. I., and J. N. Luthin, The effects of entrapped air and barometric fluctuations in the drainage of porous mediums, *Water Resources Research*, v. 4, pp. 417–424, 1968.

46. Olmsted, F. H., and A. G. Hely, Relation between ground water and surface water in Brandywine Creek Basin, Pennsylvania, *U.S. Geological Survey Prof. Paper* 417-A, 21 pp., 1962.

47. Parker, G. G., and V. T. Stringfield, Effects of earthquakes, trains, tides, winds, and atmospheric pressure changes on water in the geologic formations of Southern Florida, *Econ. Geol.*, v. 45, pp. 441–460, 1950.

48. Peck, A. J., The water table as affected by atmospheric pressure, *Jour. Geophysical Research*, v. 65, pp. 2383–2388, 1960.

49. Pinder, G. F., and J. F. Jones, Determination of the ground-water component of peak discharge from the chemistry of total runoff, *Water Resources Research*, v. 5, pp. 438–445, 1969.

50. Poland, J. F., Subsidence and its control, *in* Underground waste management and environmental implications, *Amer. Assoc. Petr. Geologists Memoir* 18, pp. 50–71, 1972.

51. Poland, J. F., and G. H. Davis, Land subsidence due to withdrawal of fluids, *in* Varnes, D. J., and G. Kiersch (eds.), *Reviews in Engineering Geology,* v. 2, Geol. Soc. Amer., Boulder, Colo., pp. 187–269, 1969.
52. Poland, J. F., et al., Studies in land subsidence, *U.S. Geological Survey Prof. Paper* 437-A to 437-H, vars. pp., 1964 to 1975.
53. Rasmussen, W. C., and G. E. Andreasen, Hydrologic budget of the Beaverdam Creek basin, Maryland, *U.S. Geological Survey Water-Supply Paper* 1472, 106 pp., 1959.
54. Remson, I., and J. R. Randolph, Application of statistical methods to the analysis of ground-water levels, *Trans. Amer. Geophysical Union,* v. 39, pp. 75–83, 1958.
55. Riggs, H. C., The base-flow recession curve as an indicator of ground water, *Intl. Assoc. Sci. Hydrology Publ.* 63, pp. 352–363, 1963.
56. Ripple, C. D., et al., Estimating steady-state evaporation rates from bare soils under conditions of high water table, *U.S. Geological Survey Water-Supply Paper* 2019-A, 39 pp., 1972.
57. Robinson, E. S., and R. T. Bell, Tides in confined well-aquifer systems, *Jour. Geophysical Research,* v. 76, pp. 1857–1869, 1971.
58. Robinson, T. W., Earth-tides shown by fluctuations of water-levels in wells in New Mexico and Iowa, *Trans. Amer. Geophysical Union,* v. 20, pp. 656–666, 1939.
59. Russell, R. R., *Ground-water levels in Illinois through 1961,* Rept. Inv. 45, Illinois State Water Survey, Urbana, 51 pp., 1963.
60. Schicht, R. J., and W. C. Walton, *Hydrologic budgets for three small watersheds in Illinois,* Rept. Inv. 40, Illinois State Water Survey, Urbana, 40 pp., 1961.
61. Schneider, R., Correlation of ground-water levels and air temperatures in the winter and spring in Minnesota, *U.S. Geological Survey Water-Supply Paper* 1539-D, 14 pp., 1961.
62. Singh, K. P., Some factors affecting baseflow, *Water Resources Research,* v. 4, pp. 985–999, 1968.
63. Singh, K. P., and J. B. Stall, Derivation of base flow recession curves and parameters, *Water Resources Research,* v. 7, pp. 292–303, 1971.
64. Shelton, M. J., and L. B. James, Engineer-geologist team investigates subsidence, *Jour. Pipeline Div.,* Amer. Soc. Civil Engrs., v. 85, no. PL2, 18 pp., 1959.
65. Smedema, L. B., and P. J. Zwerman, Fluctuations of the phreatic surface: 1. Role of entrapped air under a temperature gradient, *Soil Sci.,* v. 103, pp. 354–359, 1967.
66. Sokol, D., Position and fluctuations of water level in wells perforated in more than one aquifer, *Jour. Geophysical Research,* v. 68, pp. 1079–1080, 1963.
67. Todd, D. K., Ground-water flow in relation to a flooding stream, *Proc. Amer. Soc. Civil Engrs.,* v. 81, sep. 628, 20 pp., 1955.

68. Troxell, H. C., The diurnal fluctuation in the ground-water and flow of the Santa Ana River and its meaning, *Trans. Amer. Geophysical Union,* v. 17, pp. 496–504, 1936.

69. Tschinkel, H. M., Short-term fluctuation in streamflow as related to evaporation and transpiration, *Jour. Geophysical Research,* v. 68, pp. 6459–6469, 1963.

70. Turk, L. J., Diurnal fluctuations of water tables induced by atmospheric pressure changes, *Jour. Hydrology,* v. 26, pp. 1–16, 1975.

71. Vacher, H. L., Hydrology of small oceanic islands—influence of atmospheric pressure on the water table, *Ground Water,* v. 16, pp. 417–423, 1978.

72. van der Kamp, G. S. J. P., *Periodic flow of groundwater,* Rodopi, Amsterdam, 121 pp., 1973.

73. van Hylckama, T. E. A., Water use by saltcedar as measured by the water budget method, *U.S. Geological Survey Prof. Paper* 491-E, 30 pp., 1974.

74. Veihmeyer, F. J., and F. A. Brooks, Measurement of cumulative evaporation of bare soil, *Trans. Amer. Geophysical Union,* v. 35, pp. 601–607, 1954.

75. Visocky, A. P., Estimating the ground-water contribution to storm runoff by the electrical conductance method, *Ground Water,* v. 8, no. 2, pp. 5–10, 1970.

76. Vorhis, R. C., Interpretation of hydrologic data resulting from earthquakes, *Geologische Rundschau,* v. 43, pp. 47–52, 1955.

77. Vorhis, R. C., Earthquake-induced water-level fluctuations from a well in Dawson County, Georgia, *Seismological Soc. Amer. Bull.,* v. 54, pp. 1023–1133, 1964.

78. Vorhis, R. C., Hydrologic effects of the earthquake of March 27, 1964, outside Alaska, *U.S. Geological Survey Prof. Paper* 544-C, 54 pp., 1967.

79. Werner, P. W., and D. Noren, Progressive waves in non-artesian aquifers, *Trans. Amer. Geophysical Union,* v. 32, pp. 238–244, 1951.

80. Werner, P. W., and K. J. Sundquist, On the ground-water recession curve for large watersheds, *Intl. Assoc. Sci. Hydrology Publ.* 33, pp. 202–212, 1951.

81. White, W. N., A method of estimating ground-water supplies based on discharge by plants and evaporation from soil, *U.S. Geological Survey Water-Supply Paper* 659, pp. 1–105, 1932.

Quality of Groundwater

.... CHAPTER 7

It is now generally recognized that the quality of groundwater is just as important as its quantity. All groundwater contains salts in solution that are derived from the location and past movement of the water. The quality required of a groundwater supply depends on its purpose; thus, needs for drinking water, industrial water, and irrigation water vary widely. To establish quality criteria, measures of chemical, physical, biological, and radiological constituents must be specified, as well as standard methods for reporting and comparing results of water analyses. Dissolved gases in groundwater can pose hazards if their presence goes unrecognized. The uniformity of groundwater temperature is advantageous for water supply and industrial purposes, and underlying saline groundwaters are important because they offer potential benefits.

Sources of Salinity

All groundwater contains salts in solution; reported salt contents range from less than 25 mg/l in a quartzite spring to more than 300,000 mg/l in brines.[52] The type and concentration of salts depend on the environment, movement, and source of the groundwater. Ordinarily, higher concentrations of dissolved constituents are found

267

in groundwater than in surface water because of the greater exposure to soluble materials in geologic strata. Soluble salts in groundwater originate primarily from solution of rock materials. Bicarbonate, usually the primary anion in groundwater, is derived from carbon dioxide released by organic decomposition in the soil. Salinity varies with specific surface area of aquifer materials, solubility of minerals, and contact time; values tend to be highest where movement of groundwater is least; hence, salinity generally increases with depth. A common geochemical sequence in groundwater includes bicarbonate waters near ground surface varying to chloride waters in the deepest portions of formations.

Precipitation reaching the earth contains only small amounts of dissolved mineral matter. Once on earth the water reacts with the minerals of the soil and rocks in contact with it. The quantity and type of mineral matter dissolved depend on the chemical composition and physical structure of the rocks as well as the hydrogen-ion concentration (pH) and the redox potential (Eh) of the water.[8] Carbon dioxide in solution, derived from the atmosphere and from organic processes in the soil, assists the solvent action of water as it moves underground.[23] The geochemical cycle of surface water and groundwater shown in Fig. 7.1 illustrates the principal chemical changes involved in water as it travels through the hydrologic cycle from precipitation to groundwater.

In areas recharging large volumes of water underground, such as alluvial streams, channels, or artificial recharge areas, the quality of the infiltrating surface water can have a marked effect on that of the groundwater. Locally, absorbed gases of magmatic origin contribute dissolved mineral products to groundwater; mineralized thermal springs are an excellent example. Connate water is usually highly mineralized because it is derived from water originally entrapped in sedimentary strata since the time of deposition.*

Salts are added to groundwater passing through soils by soluble products of soil weathering and of erosion by rainfall and flowing water. Excess irrigation water percolating to the water table may contribute substantial quantities of salt. Water passing through the root zone of cultivated areas usually contains salt concentrations several times that of the applied irrigation water. Increases result primarily from the evapotranspiration process, which tends to con-

---

*It should be recognized that most connate water has been altered chemically by various chemical and physical processes and therefore does not necessarily represent the original water of deposition.

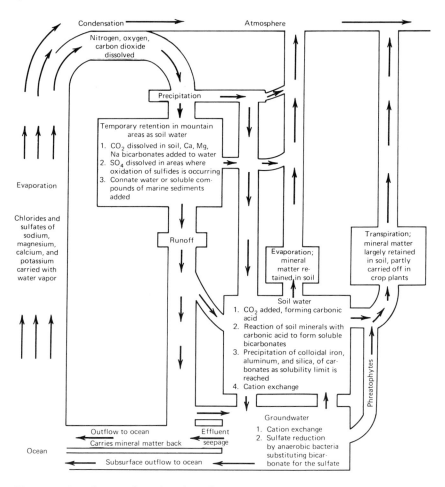

Fig. 7.1 Geochemical cycle of surface water and groundwater (after *USGS Water-Supply Paper* 1469).

centrate salts in drainage waters. In addition, soluble soil materials, fertilizers, and selective absorption of salts by plants will modify salt concentrations of percolating waters. Factors governing the increase include soil permeability, drainage facilities, amount of water applied, crops, and climate. Thus, high salinities may be found in soils and groundwater of arid climates where leaching by rainwater is not effective in diluting the salt solutions. Similarly, poorly drained areas, particularly basins having interior drainage, often contain high salt concentrations. Also, some regions contain remnants of sedimentary deposition under saline waters; the designation *badlands* implies in part the lack of productivity resulting from excess salt contents of the soil and water.

Groundwater passing through igneous rocks dissolves only very small quantities of mineral matter because of the relative insolubility of the rock composition. Percolating rainwater contains carbon dioxide derived from the atmosphere, which increases the solvent action of the water. The silicate minerals of igneous rocks result in silica being added to the groundwater.

Sedimentary rocks are more soluble than igneous rocks.* Because of their high solubility, combined with their great abundance in the earth's crust, they furnish a major portion of the soluble constituents to groundwater. Sodium and calcium are commonly added cations; bicarbonate and sulfate are corresponding anions. Chloride occurs to only a limited extent under normal conditions; important sources of chloride, however, are from sewage, connate water, and intruded seawater. Occasionally nitrate is an important natural constituent; high concentrations may indicate sources of past or present pollution. In limestone terrains calcium and bicarbonate ions are added to the groundwater by solution.

Illustrative of the variety of constituents dissolved in groundwater is the list in Table 7.1. Here the relative abundance of constituents is indicated by four broad and overlapping classes based on typical concentrations.

A summary of the natural sources and concentrations of the principal chemical constituents found in groundwater is presented in Table 7.2 together with their effects on the usability of the water. A comprehensive discussion of the chemical constituents of groundwater can be found in Hem.[23]

An important source of salinity in groundwater in coastal regions is airborne salts originating from the air-water interface over the sea. Detailed studies on a worldwide basis[16] and for Israel[28] suggest that salts are deposited on land both by precipitation and by dry fallout. Chloride deposition in coastal areas has been calculated to range from 4 to 20 kg Cl/ha. The deposition decreases inland, varying exponentially with distance from the sea. Thus, Israeli water measurements yielded the relation

$$N = 110 \ e^{-0.0133d} \tag{7.1}$$

---

*Important mineral sources in sedimentary rocks are feldspar, gypsum, and forms of calcium carbonate.

TABLE 7.1 Relative Abundance of Dissolved Solids in Potable Water (after Davis and DeWiest[14])

Major Constituents (1.0 to 1000 mg/l)	Secondary Constituents (0.01 to 10.0 mg/l)	Minor Constituents (0.0001 to 0.1 mg/l)	Trace Constituents (generally less than 0.001 mg/l)
Sodium	Iron	Antimony[a]	Beryllium
Calcium	Strontium	Aluminum	Bismuth
Magnesium	Potassium	Arsenic	Cerium[a]
Bicarbonate	Carbonate	Barium	Cesium
Sulfate	Nitrate	Bromide	Gallium
Chloride	Fluoride	Cadmium[a]	Gold
Silica	Boron	Chromium[a]	Indium
		Cobalt	Lanthanum
		Copper	Niobium[a]
		Germanium[a]	Platinum
		Iodide	Radium
		Lead	Ruthenium[a]
		Lithium	Scandium[a]
		Manganese	Silver
		Molybdenum	Thallium[a]
		Nickel	Thorium[a]
		Phosphate	Tin
		Rubidium[a]	Tungsten[a]
		Selenium	Ytterbium
		Titanium[a]	Yttrium[a]
		Uranium	Zirconium[a]
		Vanadium	
		Zinc	

[a]These elements occupy an uncertain position in the list.

TABLE 7.2 Principal Chemical Constituents in Groundwater—Their Sources, Concentrations, and Effect on Usability
(modified from C. N. Durfer, and E. Baker, USGS Water-Supply Paper 1812, 1964)

Constituent	Major Natural Sources	Concentration in Natural Water	Effect on Usability of Water
Silica (SiO_2)	Feldspars, ferromagnesium and clay minerals, amorphous silica, chert, opal	Ranges generally from 1.0 to 30 mg/l, although as much as 100 mg/l is fairly common; as much as 4000 mg/l is found in brines	In the presence of calcium and magnesium, silica forms a scale in boilers and on steam turbines that retards heat; the scale is difficult to remove. Silica may be added to soft water to inhibit corrosion of iron pipes
Iron (Fe)	Igneous rocks: amphiboles, ferromagnesian micas, ferrous sulfide (FeS), ferric sulfide or iron pyrite (FeS_2), magnetite (Fe_3O_4) Sandstone rocks: oxides, carbonates, and sulfides or iron clay minerals	Generally less than 0.50 mg/l in fully aerated water. Groundwater having a pH less than 8.0 may contain 10 mg/l; rarely as much as 50 mg/l may occur. Acid water from thermal springs, mine wastes, and industrial wastes may contain more than 6000 mg/l	More than 0.1 mg/l precipitates after exposure to air; causes turbidity, stains plumbing fixtures, laundry, and cooking utensils, and imparts objectionable tastes and colors to foods and drinks. More than 0.2 mg/l is objectionable for most industrial uses
Manganese (Mn)	Manganese in natural water probably comes most often from soils and sediments. Metamorphic and sedimentary rocks and mica biotite and amphibole hornblende minerals contain large amounts of manganese	Generally 0.20 mg/l or less. Groundwater and acid mine water may contain more than 10 mg/l	More than 0.2 mg/l precipitates upon oxidation; causes undesirable tastes, deposits on foods during cooking, stains plumbing fixtures and laundry, and fosters growths in reservoirs, filters, and distribution systems. Most industrial users

object to water containing more than 0.2 mg/l

Constituent	Source	Concentration	Remarks
Calcium (Ca)	Amphiboles, feldspars, gypsum, pyroxenes, aragonite, calcite, dolomite, clay minerals	Generally less than 100 mg/l; brines may contain as much as 75,000 mg/l	Calcium and magnesium combine with bicarbonate, carbonate, sulfate and silica to form heat-retarding, pipe-clogging scale in boilers and in other heat-exchange equipment. Calcium and magnesium combine with ions of fatty acid in soaps to form soap-suds; the more calcium and magnesium, the more soap required to form suds. A high concentration of magnesium has a laxative effect, especially on new users of the supply
Magnesium (Mg)	Amphiboles, olivine, pyroxenes, dolomite, magnesite, clay minerals	Generally less than 50 mg/l; ocean water contains more than 1000 mg/l, and brines may contain as much as 57,000 mg/l	
Sodium (Na)	Feldspars (albite); clay minerals; evaporites, such as halite (NaCl) and mirabilite ($Na_2SO_4 \cdot 10H_2O$); industrial wastes	Generally less than 200 mg/l; about 10,000 mg/l in seawater; about 25,000 mg/l in brines	More than 50 mg/l sodium and potassium in the presence of suspended matter causes foaming, which accelerates scale formation and corrosion in boilers. Sodium and potassium carbonate in recirculating cooling water can cause deterioration of wood in cooling towers. More than 65 mg/l of sodium can cause problems in ice manufacture
Potassium (K)	Feldspars (orthoclase and microcline), feldspathoids; some micas, clay minerals	Generally less than about 10 mg/l; as much as 100 mg/l in hot springs; as much as 25,000 mg/l in brines	

TABLE 7.2 Continued

Constituent	Major Natural Sources	Concentration in Natural Water	Effect on Usability of Water
Carbonate (CO_3)	Limestone, dolomite	Commonly less than 10 mg/l in groundwater. Water high in sodium may contain as much as 50 mg/l of carbonate	Upon heating, bicarbonate is changed into steam, carbon dioxide, and carbonate. The carbonate combines with alkaline earths—principally calcium and magnesium—to form a crustlike scale of calcium carbonate that retards flow of heat through pipe walls and restricts flow of fluids in pipes. Water containing large amounts of bicarbonate and alkalinity is undesirable in many industries
Bicarbonate (HCO_3)		Commonly less than 500 mg/l; may exceed 1000 mg/l in water highly charged with carbon dioxide	
Sulfate (SO_4)	Oxidation of sulfide ores; gypsum; anhydrite	Commonly less than 300 mg/l except in wells influenced by acid mine drainage. As much as 200,000 mg/l in some brines	Sulfate combines with calcium to form an adherent, heat-retarding scale. More than 250 mg/l is objectionable in water in some industries. Water containing about 500 mg/l of sulfate tastes bitter; water containing about 1000 mg/l may be cathartic
Chloride (Cl)	Chief source is sedimentary rock (evaporites); minor sources are igneous rocks	Commonly less than 10 mg/l in humid regions but up to 1000 mg/l in more arid regions.	Chloride in excess of 100 mg/l imparts a salty taste. Concentrations greatly in excess of

Constituent	Source	Concentration	Significance
		About 19,300 mg/l in seawater; and as much as 200,000 mg/l in brines	100 mg/l may cause physiological damage. Food processing industries usually require less than 250 mg/l. Some industries—textile processing, paper manufacturing, and synthetic rubber manufacturing—desire less than 100 mg/l
Fluoride (F)	Amphiboles (hornblende), apatite, fluorite, mica	Concentrations generally do not exceed 10 mg/l. Concentrations may be as much as 1600 mg/l in brines	Fluoride concentration between 0.6 and 1.7 mg/l in drinking water has a beneficial effect on the structure and resistance to decay of children's teeth. Fluoride in excess of 1.5 mg/l in some areas causes "mottled enamel" in children's teeth. Fluoride in excess of 6.0 mg/l causes pronounced mottling and disfiguration of teeth
Nitrate (NO_3)	Atmosphere; legumes, plant debris, animal excrement	Commonly less than 10 mg/l	Water containing large amounts of nitrate (more than 100 mg/l) is bitter tasting and may cause physiological distress. Water from shallow wells containing more than 45 mg/l has been reported to cause methemoglobinemia in

TABLE 7.2 Continued

Constituent	Major Natural Sources	Concentration in Natural Water	Effect on Usability of Water
			infants. Small amounts of nitrate help reduce cracking of high-pressure boiler steel
Dissolved solids	The mineral constituents dissolved in water constitute the dissolved solids	Commonly contains less than 5000 mg/l; some brines contain as much as 300,000 mg/l	More than 500 mg/l is undesirable for drinking and many industrial uses. Less than 300 mg/l is desirable for dyeing of textiles and the manufacture of plastics, pulp paper, rayon. Dissolved solids cause foaming in steam boilers; the maximum permissible content decreases with increases in operating pressure

where N is the annual amount of chloride precipitation in kg/ha and d is the distance from the sea in kilometers. In arid regions, where surface runoff is small and evapotranspiration is large, airborne salt deposition becomes intensified several fold in groundwater.

Measures of Water Quality

In specifying the quality characteristics of groundwater, chemical, physical, and biological analyses are normally required. A complete chemical analysis of a groundwater sample includes the determination of the concentrations of the inorganic constituents present; organic and radiological parameters are normally of concern only where human-induced pollution affects quality (see Chapter 8). Dissolved salts in groundwater of normal salinity occur as dissociated ions; in addition, other minor constituents are present and reported in elemental form. The analysis also includes measurement of pH and specific electrical conductance. Depending on the purpose of a water quality investigation, partial analyses of only particular constituents will sometimes suffice. Illustrative chemical analyses of groundwaters from a variety of geologic formations are shown in Table 7.3.

Properties of groundwater evaluated in a physical analysis include temperature, color, turbidity, odor, and taste. Biological analysis includes tests to detect the presence of coliform bacteria, which indicate the sanitary quality of water for human consumption. Because certain coliform organisms are normally found in intestines of humans and animals, the presence of these in groundwater is tantamount to its contact with sewage sources.

Standard methods of water analysis are specified by the American Public Health Association and others;[1,3] most laboratories conducting water analyses follow these procedures.

Chemical Analysis

Once a sample of groundwater has been analyzed in a laboratory, methods for reporting water analyses must be considered. From an understanding of expressions and units for describing water quality, standards can be established so that analyses can be interpreted in terms of the ultimate purpose of the water supply. In a chemical analysis of groundwater, concentrations of different ions are expressed by weight or by chemical equivalence. Total dissolved solids can be measured in terms of electrical conductance. These and other measures of chemical quality are described in the following sections.

TABLE 7.3 Chemical Analyses of Groundwater from Various Geologic Formations (after White, et al.[52])

Rock type Geologic age Location	Granite Carboniferous(?) McCormick County, South Carolina	Basalt Tertiary Moses Lake, Washington	Andesite tuff Paleozoic Randolph County, North Carolina	Sandstone Mississipian Crawford County Pennsylvania	Shale Mississipian Cuyahoga County, Ohio
Well depth, m	77	64	49	37	22
Chemical concentration, mg/l					
SiO_2	35	55	31	14	19
Al	0.1	–	0.2	0	–
Fe	0.18	0.03	0.16	1.3	1.3
Mn	0.13	–	0.03	0	0
Cu	0	–	0.01	0	–
Zn	0.09	–	–	0	–
Ca	13	29	14	44	123
Mg	4.3	19	5.6	11	70
Na	8.4	12	9.6	60	61
K	3.5	3.5	0.4	4.1	2.2
HCO_3	72	177	74	327	539
CO_3	0	0	0	0	0
SO_4	6.9	15	0.1	22	283
Cl	3.8	6.9	8.8	4.4	3.5
F	0.2	0.4	0	0.2	0.4
NO_3	0.4	9.7	6.8	2.0	0.1
PO_4	0.1	–	0	0	–
Specific Conductance	150	340	163	533	1180
pH	7.0	7.9	7.2	7.4	7.3

TABLE 7.3 Continued

Rock type	Limestone	Dolomite	Quartzite	Schist	Alluvium
Geologic age	Cretaceous	Silurian	Cambrian	Cambrian	Pleistocene
Location	Uvalde County, Texas	Milwaukee County, Wisconsin	Bucks County, Pennsylvania	Gwinnett County, Georgia	Franklin County, Ohio
Well depth, m	107	152	154	183	36
Chemical concentration, mg/l					
SiO_2	11	18	17	21	20
Al	–	0.2	–	0	–
Fe	0.08	0.39	1.6	0.11	2.3
Mn	–	0.03	–	0.02	0
Cu	–	0	–	0	–
Zn	–	0	–	0.02	–
Ca	74	35	25	27	126
Mg	9.5	33	5.1	5.7	43
Na	24	28	4.5	16	13
K	7.0	1.3	3.8	0.7	2.1
HCO_3	277	241	80	138	440
CO_3	0	0	0	0	0
SO_4	19	88	13	9.6	319
Cl	24	1.0	8.0	2.5	8.0
F	0.4	0.9	0.4	0.5	0.7
NO_3	4.1	1.2	0.3	0	0.2
PO_4	–	0	–	0	–
Specific Conductance	570	511	206	237	885
pH	7.0	8.2	7.1	8.0	7.6

Concentrations by Weight. Concentrations of the common ions found in groundwater are reported by weight-per-volume units of milligrams per liter (mg/l).* The total ionic concentration (or total dissolved solids) is also reported in this manner.

Chemical Equivalence. Positively charged cations and negative anions combine and dissociate in definite weight ratios. By expressing ion concentrations in equivalent weights, these ratios are readily determined because one equivalent weight of a cation will exactly combine with one equivalent weight of an anion. The combining weight of an ion is equal to its formula weight divided by its charge. When the concentration in milligrams per liter is divided by the combining weight, an equivalent concentration expressed in milliequivalents per liter (meq/l) results. Table 7.4 lists the reciprocals of combining weights of cations and anions; concentrations in milligrams per liter can be converted to milliequivalents per liter by multiplying by the appropriate conversion factor. For undissociated species with zero charge, of which silica is an example in groundwater quality, an equivalent weight cannot be computed.

In application, therefore, it may be expected that of the total dissolved solids in a groundwater sample, the sum of the cations and the sum of the anions when expressed in milliequivalents per liter will be equal. If the chemical analysis of the various ionic constituents indicates a difference from this balance, it may be concluded either that there are other undetermined constituents present or that errors exist in the analysis.

Total Dissolved Solids by Electrical Conductance. A rapid determination of total dissolved solids can be made by measuring the electrical conductance of a groundwater sample. Conductance is preferred rather than its reciprocal, resistance, because it increases with salt content. Specific electrical conductance defines the conductance of a cubic centimeter of water at a standard temperature of 25°C; an increase of 1°C increases conductance by about 2 percent.

Specific conductance is measured in microsiemens/cm (μS/cm).† Because natural water contains a variety of ionic and undissociated species, conductance cannot be simply related to total dissolved

---

*The units milligrams per liter have replaced parts per million; however, they are numerically equivalent up to a concentration of dissolved solids of about 7000 mg/l.

†The unit microsiemens/cm is equivalent to micromhos/cm. Because the definition of specific conductance already specifies the dimensions to which the measurement applies, the length in the units is often omitted in practice.

TABLE 7.4 Conversion Factors for Chemical Equivalence (after Hem[23]) (concentrations in mg/l times the conversion factor yields concentration in meq/l)

Chemical Constituent	Conversion Factor
Aluminum (Al^{+3})	0.11119
Ammonium (NH_4^+)	0.05544
Barium (Ba^{+2})	0.01456
Beryllium (Be^{+3})	0.33288
Bicarbonate (HCO_3^-)	0.01639
Bromide (Br^-)	0.01251
Cadmium (Cd^{+2})	0.01779
Calcium (Ca^{+2})	0.04990
Carbonate (CO_3^{-2})	0.03333
Chloride (Cl^-)	0.02821
Cobalt (Co^{+2})	0.03394
Copper (Cu^{+2})	0.03148
Fluoride (F^-)	0.05264
Hydrogen (H^+)	0.99209
Hydroxide (OH^-)	0.05880
Iodide (I^-)	0.00788
Iron (Fe^{+2})	0.03581
Iron (Fe^{+3})	0.05372
Lithium (Li^+)	0.14411
Magnesium (Mg^{+2})	0.08226
Manganese (Mn^{+2})	0.03640
Nitrate (NO_3^-)	0.01613
Nitrite (NO_2^-)	0.02174
Phosphate (PO_4^{-3})	0.03159
Phosphate (HPO_4^{-2})	0.02084
Phosphate ($H_2PO_4^-$)	0.01031
Potassium (K^+)	0.02557
Rubidium (Rb^+)	0.01170
Sodium (Na^+)	0.04350
Strontium (Sr^{+2})	0.02283
Sulfate (SO_4^{-2})	0.02082
Sulfide (S^{-2})	0.06238
Zinc (Zn^{+2})	0.03060

solids. However, conductance is easily measured and gives results that are convenient as a general indication of dissolved solids. An approximate relation for most natural water[29,38] in the range of 100 to 5000 μS/cm leads to the equivalencies 1 meq/l of cations = 100 μS/cm and 1 mg/l = 1.56 μS/cm.

Hardness. Hardness results from the presence of divalent metallic cations, of which calcium and magnesium are the most abundant in groundwater.* These ions react with soap to form precipitates and with certain anions present in the water to form scale. Because of their adverse action with soap, hard waters are unsatisfactory for household cleansing purposes; hence, water-softening processes for removal of hardness are needed.

The hardness in water is derived from the solution of carbon dioxide, released by bacterial action in the soil, in percolating rainwater.[39] Low pH conditions develop and lead to the solution of insoluble carbonates in the soil and in limestone formations to convert them into soluble bicarbonates. Impurities in limestone, such as sulfates, chlorides, and silicates, become exposed to the solvent action of water as the carbonates are dissolved so that they also pass into solution. Thus, hard water tends to originate in areas where thick topsoils overlie limestone formations. A map of hardness of groundwater in the United States is shown in Fig. 7.2.

Hardness H_T is customarily expressed as the equivalent of calcium carbonate. Thus,

$$H_T = Ca \times \frac{CaCO_3}{Ca} + Mg \times \frac{CaCO_3}{Mg} \qquad (7.2)$$

where H_T, Ca, and Mg are measured in milligrams per liter and the ratios in equivalent weights. Eq. 7.2 reduces to

$$H_T = 2.5 \, Ca + 4.1 \, Mg \qquad (7.3)$$

The degree of hardness in water is commonly based on the classification listed in Table 7.5.

TABLE 7.5 **Hardness Classification of Water
(after Sawyer and McCarty[39])**

Hardness, mg/l as $CaCO_3$	Water Class
0–75	Soft
75–150	Moderately hard
150–300	Hard
Over 300	Very hard

*The terms *hard* and *soft* as applied to water date from Hippocrates (460–354 B.C.), the father of medicine, in his treatise on public hygiene, *Air, Water and Places:* "consider the waters which the inhabitants use, whether they be marshy and soft, or hard and running from elevated and rocky situations, and then if saltish and unfit for cooking . . . for water contributes much to health."[10]

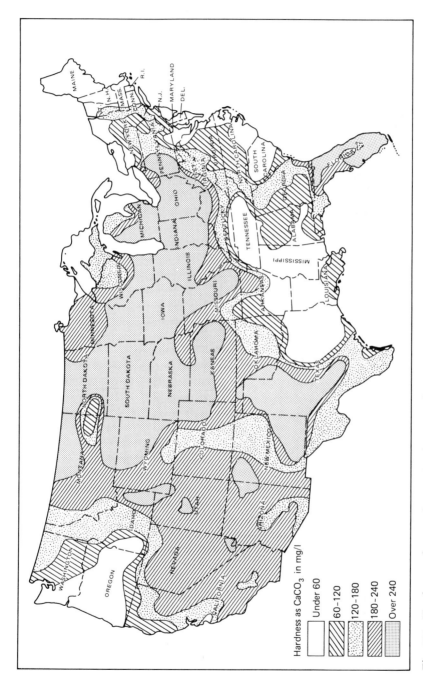

Fig. 7.2 Hardness of groundwater in the United States. Areas delineated represent average conditions on a generalized basis (after Ackerman and Löf, *Technology in American Water Development, Resources for the Future;* copyright © 1959 by The Johns Hopkins University Press).

Graphic Representations

Tables showing results of analyses of chemical quality of ground-water may be difficult to interpret, particularly where more than a few analyses are involved. To overcome this, graphic representations are useful for display purposes, for comparing analyses, and for emphasizing similarities and differences. Graphs can also aid in detecting the mixing of water of different compositions and in iden-tifying chemical processes occurring as groundwater moves. A va-riety of graphic techniques have been developed for showing the major chemical constituents; some of the more useful graphs are described and illustrated in the following paragraphs.

Figure 7.3 illustrates vertical bar graphs, widely used in the United States for portraying chemical quality. Each analysis appears as a

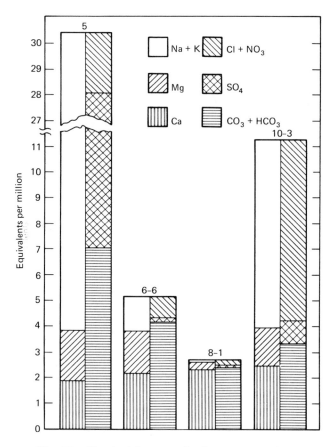

Fig. 7.3 Vertical bar graphs for representing analyses of groundwater quality (after Hem[23]).

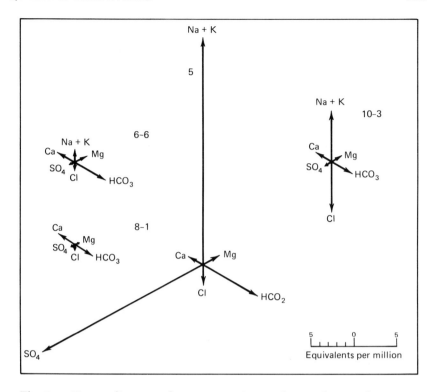

Fig. 7.4 Vector diagrams for representing analyses of groundwater quality (after Hem[23]).

vertical bar having a height proportional to the total concentration of anions or cations, expressed in milliequivalents per liter. The left half of a bar represents cations, and the right half anions. These segments are divided horizontally to show the concentrations of major ions or groups of closely related ions and identified by distinctive shading patterns.* The reference number of the analysis is shown at the top of the bar. This standard bar graph can be modified to include hardness and silica by additional segments.[23]

Another method for plotting chemical quality with radiating vectors is shown in Fig. 7.4. The lengths of the six vectors represent ionic concentrations in milliequivalents per liter.

Pattern diagrams, first suggested by Stiff,[45] for representing chemical analyses by four parallel axes are illustrated by Fig. 7.5. Concentrations of cations are plotted to the left of a vertical zero axis

---

*It should be noted that the water quality diagrams in Figs. 7.3 to 7.8 all depict the same four water analyses so as to facilitate comparisons. Also, equivalents per million is identical to meg/l.

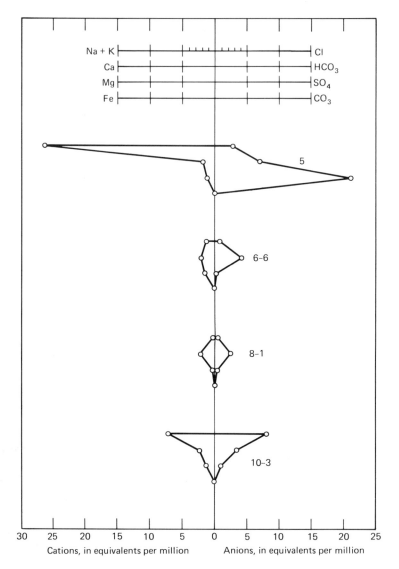

Fig. 7.5 Pattern diagrams for representing analyses of groundwater quality (after Hem[23]).

and anions to the right; all values are in milliequivalents per liter. The resulting points, when connected, form an irregular polygonal pattern; waters of a similar quality define a distinctive shape.

Figure 7.6 indicates circular diagrams of water quality with a special scale for the radii so that the area of a circle is proportional to

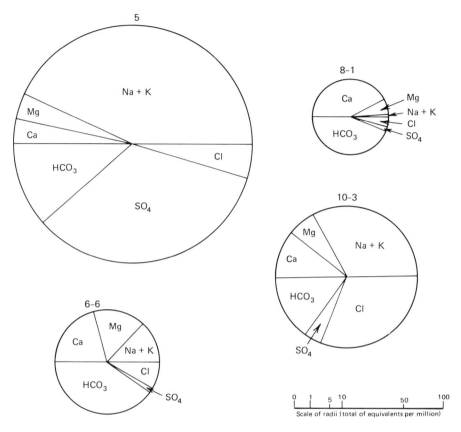

Fig. 7.6 Circular diagrams for representing analyses of groundwater quality (after Hem[23]).

the total ionic concentration of the analysis. Sectors within a circle show the fractions of the different ions expressed in milliequivalents per liter.

One of the most useful graphs for representing and comparing water quality analyses is the trilinear diagram by Piper[35] shown in Fig. 7.7. Here cations, expressed as percentages of total cations in milliequivalents per liter, plot as a single point on the left triangle; while anions, similarly expressed as percentages of total anions, appear as a point in the right triangle. These two points are then projected into the central diamond-shaped area parallel to the upper edges of the central area. This single point is thus uniquely related to the total ionic distribution; a circle can be drawn at this point with its area proportional to the total dissolved solids. The trilinear

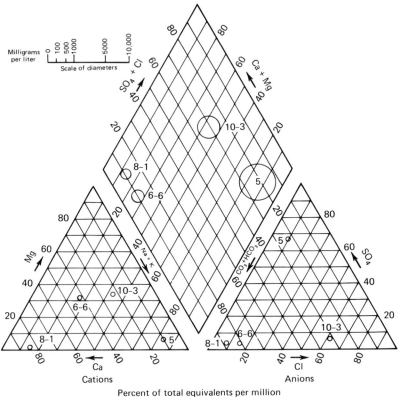

Fig. 7.7 Trilinear diagram for representing analyses of
groundwater quality (after Hem[23]).

diagram conveniently reveals similarities and differences among
groundwater samples because those with similar qualities will tend
to plot together as groups. Further, simple mixtures of two source
waters can be identified; for example, an analysis of any mixture of
two waters will plot on the straight line *AB* on the diagram, where
A and *B* are the positions of the analyses of the two component
waters.

   In Europe the semilogarithmic diagram developed by Schoeller[41]
is widely employed for comparing groundwater analyses. Here the
principal ionic concentrations, expressed in milliequivalents per
liter, are plotted on six equally spaced logarithmic scales in the arrangement shown by Fig. 7.8. The points thus plotted are joined by
straight lines. This type of graph shows not only the absolute value

of each ion but also the concentration differences among various groundwater analyses. Because of the logarithmic scale, if a straight line joining the points A and B of two ions in one water sample is parallel to another straight line joining the points A' and B' of the same two ions in another water sample, the ratio of the ions in both analyses is equal.*

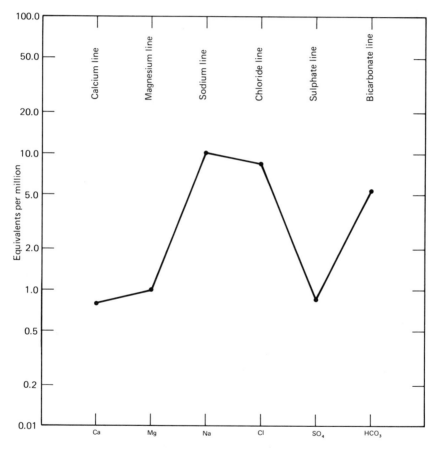

Fig. 7.8 Schoeller semilogarithmic diagram for representing analyses of groundwater quality (after Schoeller[41]).

*It should also be mentioned that the Schoeller diagram can be adapted to determine the degree of saturation of $CaCO_3$ and $CaSO_4$ in groundwater.[41]

Physical Analysis

In a physical analysis of groundwater, temperature is reported in degrees Celsius and necessarily must be measured immediately after collecting the sample. Color in groundwater may be due to mineral or organic matter in solution and is reported in mg/l by comparison with standard solutions. Turbidity is a measure of the suspended and colloidal matter in water, such as clay, silt, organic matter, and microscopic organisms. Measurements are often based on the length of a light path through the water which just causes the image of a flame of a standard candle to disappear. The natural filtration produced by unconsolidated aquifers largely eliminates turbidity, but other types of aquifers can produce turbid groundwater. Tastes and odors may be derived from bacteria, dissolved gases, mineral matter, or phenols. These characteristics are subjective sensations that can be defined only in terms of the experience of a human being. Quantitative determinations of taste and odor have been developed based on the maximum degree of dilution that can be distinguished from taste-free and odorfree water.[1]

Biological Analysis

As mentioned before, bacteriological analysis is important for detecting biological pollution of groundwater. Most pathogenic bacteria found in water are indigenous to the intestinal tract of animals and humans, but isolating them from natural water is difficult in the laboratory. Because bacteria of the coliform group are relatively easy to isolate and identify, standard tests to determine their presence or absence in a water sample are taken as a direct indication of the safety of the water for drinking purposes. Coliform test results are reported as the most probable number (MPN) of coliform group organisms in a given volume of water. By analysis of a number of separate portions of a water sample, the MPN is computed from probability tables for this purpose.

Groundwater Samples

In sampling groundwater for quality analysis, Pyrex glass or polyethylene bottles are generally satisfactory.[37] Volumes of one or two liters are usually sufficient for a normal routine chemical analysis. After rinsing the bottle with the water being sampled, the sample is then collected and securely sealed. The water should be stored in a cool place and transferred promptly to a laboratory for analysis. Samples should be taken from a well only after it has been pumped for some time, otherwise nonrepresentative samples of stagnant or polluted water may be obtained. With each sample a record should be made of well location, depth of sample, size of casing, date, water temperature, odor, color, turbidity, and operating conditions of the well immediately prior to the sampling. To analyze for organic and radiological constituents, special sampling and storage techniques are required.

The shorter the time that elapses between collection of a sample and its analysis, generally the more reliable will be the analytic results. For certain constituents and physical values, immediate analysis in the field is required to obtain dependable results; thus, determinations of temperature, pH, alkalinity, and dissolved gases should always be carried out in the field because changes are inevitable by the time samples reach a laboratory. Storage of samples prior to analysis can also affect results. Cations such as Fe, Cu, Al, Mn, Cr, and Zn are subject to loss by adsorption or ion exchange on the walls of glass containers.[37]

Finally, it should be noted that samples taken from a well penetrating stratified aquifers can yield solute concentrations that differ significantly from those occurring in the individual layers. Under these conditions it is possible to obtain water meeting specified quality criteria, whereas in individual strata concentrations could be entirely unacceptable.*

Water Quality Criteria

Whether a groundwater of a given quality is suitable for a particular purpose depends on the criteria or standards of acceptable quality for that use.[4,33,34,47] Quality limits of water supplies for drinking water, industrial purposes, and irrigation apply to groundwater because of its extensive development for these purposes.

Drinking Water Standards. Most drinking water supplies in the United States conform to standards established by the U.S. Environmental Protection Agency. A summary of the principal provisions relating to quantitative limits is given in Table 7.6.

Industrial Water Criteria. It should be apparent that the quality requirements of waters used in different industrial processes vary widely. Thus, makeup water for high-pressure boilers must meet extremely exacting criteria whereas water of as low a quality as sea-

---

*A multilevel well sampling device to overcome this problem has been described by Pickens, et al. See *Ground Water*, v. 16, pp. 322–327, 1978.

TABLE 7.6 Drinking Water Standards in the United States[a]

Physical Characteristics

Criterion	Recommended Limit[b]	Tolerance Limit[c]
Color, units	15	—
Odor, threshold number	3, inoffensive	—
Residue:		
Filtrable, mg/l	500	—
Taste	Inoffensive	—
Turbidity, units	5	—

Inorganic Chemicals, mg/l

Substance	Recommended Limit[b]	Tolerance Limit[c]
Alkyl benzene sulfonate (ABS)	0.5	—
Arsenic (As)	0.01	0.05
Barium (Ba)	—	1.0
Cadmium (Cd)	—	0.01
Carbon chloroform extract (CCE)	0.2	—
Chloride (Cl)	250	—
Chromium, hexavalent (Cr^{+6})	—	0.05
Copper (Cu)	1.0	—
Cyanide (CN)	0.01	0.2
Fluoride (F)	0.8–1.7[d,e]	1.4–2.4[e]
Iron (Fe)	0.3	—
Lead (Pb)	—	0.05
Manganese (Mn)	0.05	—
Mercury (Hg)	—	0.002
Nitrate (as N)	10	—
Phenolic compounds (as phenol)	0.001	—
Selenium (Se)	—	0.01
Silver (Ag)	—	0.05
Sulfate (SO_4)	250	—
Zinc (Zn)	5	—

Organic Chemicals, mg/l

Substance	Tolerance Limit
(A) Chlorinated hydrocarbons	
Endrin	0.0002
Lindane	0.004
Methoxychlor	0.1
Toxaphene	0.005
(B) Chlorophenoxys	
2, 4-D	0.1
2, 4, 5-TP Silvex	0.01

TABLE 7.6   (continued)

Biological Standards

Substance Examined	Maximum Permissible Limit
Standard 10-ml portions	Not more than 10 percent in one month shall show coliforms[f]
Standard 100-ml portions	Not more than 60 percent in one month shall show coliforms[f]

Radioactivity, $\mu\mu c/l$

Substance	Recommended Limit
Radium 226 (Ra^{226})	3
Strontium 90 (Sr^{90})	10
Gross beta activity	1000[g]

[a]Based on maximum contaminant levels of the U.S. Environmental Protection Agency (*Federal Register,* v. 40, no. 248, pp. 59566–59588, December 24, 1975).

[b]Concentrations that should not be exceeded where more suitable water supplies are available.

[c]Concentrations above this constitute grounds for rejection of the supply.

[d]Dependent on annual maximum daily air temperature.

[e]Where fluoridation is practiced, minimum recommended limits are also specified.

[f]Subject to further specified restrictions.

[g]In absence of strontium 90 and alpha emitters.

water can be satisfactorily employed for cooling of condensers. Even within each industry, criteria cannot be established; instead, only recommended limiting values or ranges can be stated.[2,30] Salinity, hardness, and silica are three parameters that usually are important for industrial water. Recommended limiting concentrations for process waters of selected industries are presented in Table 7.7.

Of almost equal importance for industrial purposes as quality of a water supply is the relative constancy of the various constituents. It is often possible to treat a poor quality water or adapt to it so that it is suitable for a given process, but if the quality fluctuates widely, continued attention and expense may be involved. Fluctuations of water temperature can be equally troublesome. From this standpoint, groundwater supplies are preferred to surface water supplies, which commonly display seasonal variations in chemical and physical quality. As a result, an adequate groundwater supply of suitable quality often becomes a primary consideration in selecting new industrial plant locations.

TABLE 7.7 Ranges in Recommended Limiting Concentrations for Industrial Process Waters (units are mg/l, except as noted) (after Amer. Water Works Assoc.[5])

Use	Turbidity, units	Color, units	Taste and odor threshold	Dissolved solids	Hardness, as $CaCO_3$	Alkalinity, as $CaCO_3$	pH, units	Chlorides, as Cl	Sulfates, as SO_4	Iron, as Fe	Manganese, as Mn	Iron plus manganese	Hydrogen sulfide	Fluorides, as F	Other requirements
Air-conditioning	–	–	Low	–	–	–	–	–	–	0.5	0.5	0.5	–	–	Not corrosive or slime-promoting
Baking	10	10	None-low	–	[a]	–	–	–	–	0.2	0.2	0.2	0.2	–	Potable
Boiler feed	–	–	–	–	–	–	–	–	–	–	–	–	–	–	Potable if steam is used for food preparation
Brewing	0–10	0–10	None-low	500–1,500[b]	[c]	75–80[d]	6.5–7.0[e]	60–100	–	0.1	0.1	0.1	0.2	1.0	Potable, numerous other requirements
Carbonated beverages	1–2	5–10	None-low	850	200–250	50–130	–	250	250	0.1–0.2	0.2	0.1–0.4	0–0.2	0.2–1.0	Potable; COD, 1.5; organic matter, infinitesimal; algae and protozoa, none
Confectionery	–	–	Low	50–100	Soft	–	>7.0	–	–	0.2	0.2	0.2	0.2	–	Potable
Dairy	–	None	None	500[f]	180	–	–	30	60	0.1–0.3	0.03–0.1	–	–	–	Potable: NO_3-N, 5.5; NO_2-N, 0; NH_3-N, trace only; COD as $KMnO_4$, 12
Drinking	5	15	3, inoffensive	500	–	–	–	250	250	0.3	0.05	–	–	1.4–2.4[e]	Potable.
Food canning and freezing	1–10	–	None-low	850	[h]	30–250	>7.5	–	–	0.2	0.2	0.2–0.3	1.0	1.0	Potable; free from saprophytic organisms; NaCl, 1,000–1,500; NO_3-N, 2.8; NH_3-N, 0.4
Food equipment, washing	1	5–20	None	850	10	–	–	250	–	–	–	0.1	–	1.0	Potable; organic matter, infinitesimal
Food processing, general	1–10	5–10	Low	850	10–250	30–250	–	–	–	0.2	0.2	0.2–0.3	–	1.0	Potable; SiO_2, 10
Ice manufacture	5	5	Low	170–1,300	–	–	–	–	–	0.2	0.2	0.2	–	[i]	Potable
Laundering	–	–	–	–	0–50	60	6.0–6.8[e]	–	–	0.2–1.0	0.2	0.2–1.0	–	–	–
Paper and pulp, fine	10	5	–	200	100[j]	75	–	–	–	0.1	0.05	–	–	–	Soluble SiO_2, 20; free CO_2, 10; residual Cl_2, 2
Paper, groundwood	50[k]	30	–	500	200	150	–	75	–	0.3	0.1	–	–	–	Soluble SiO_2, 50; free CO_2, 10

													Remarks
Paper, kraft, bleached	40	25	–	300	100	75	–	200	–	0.2	0.1	–	–	Soluble SiO_2, 50; free CO_2, 10
Paper, kraft, unbleached	100	100	–	500	200	150	–	200	–	1.0	0.5	–	–	Soluble SiO_2, 100; free CO_2, 10
Paper, soda, and sulfate pulps	25[k]	5	–	250	100[f]	75	–	75	–	0.1	0.05	–	–	Soluble SiO_2, 20; free CO_2, 10
Rayon and acetate fiber pulp production	5	5	–	100[f]	8	50–75	–	–	–	0.05	0.03	0.05	–	Al_2O_3, 8; Si, 25; Cu, 5
Rayon manufacture	0.3	–	–	–	55	–	7.8–8.3	–	20	0.0	0.0	0.0	–	–
Sugar	–	–	–	Low	Low	–	–	20	20	0.1	–	–	–	Ca, 20; Mg, 10; bicarbonate, as $CaCO_3$, 100; sterile, no saprophytic organisms
Tanning	20	10–100	–	–	50–500	130	6.0–8.0	–	–	0.1–0.2	0.1–0.2	0.2	–	Bicarbonate hardness, low; COD, 8; heavy metals, none; Ca, 10; Mg, 5; bicarbonate, as $CaCO_3$, 200
Textile	0.3–25	0–70	–	–	0–50	–	–	100	100	0.1–1.0	0.05–1.0	0.2–1.0	–	–

[a]Some calcium is necessary for yeast action. Too much hardness retards fermentation, but too little softens the gluten to produce soggy bread. Water of zero hardness is required for some cakes and crackers.

[b]Not more than 300 mg/l of any one substance.

[c]$CaSO_4$ less than 100 to 500 mg/l; $MgSO_4$ less than 50 to 200 mg/l.

[d]For dark beer alkalinity as $CaCO_3$ may be 80 to 150 mg/l.

[e]Range, lower to upper limits.

[f]Total solids.

[g]Tolerance limit depends on annual average of maximum daily air temperatures for a minimum of 5 years.

[h]For legumes, 25 to 75; for fruits and vegetables, 100 to 200; for peas, 200 to 400.

[i]1.5 mg/l of fluoride has been reported to cause embrittlement and cracking of ice.

[j]Calcium hardness, 50.

[k]No gritty material.

[l]Calcium hardness, 50; magnesium hardness, 50.

Irrigation Water Criteria. The suitability of groundwater for irrigation is contingent on the effects of the mineral constituents of the water on both the plant and the soil.[38,53] Salts may harm plant growth physically by limiting the uptake of water through modification of osmotic processes, or chemically by metabolic reactions such as those caused by toxic constituents. Effects of salts on soils, causing changes in soil structure, permeability, and aeration, indirectly affect plant growth.* Specific limits of permissible salt concentrations for irrigation water cannot be stated because of the wide variations in salinity tolerance among different plants; however, field-plot studies of crops grown on soils that are artificially adjusted to various salinity levels provide valuable information relating to salt tolerance.

In Table 7.8 relative tolerances of crops to soil-water salt concentrations are listed for major crop divisions. The criterion applied was the relative yield of the crop on a saline soil as compared to its yield on a nonsaline soil under similar growing conditions. Within each group, the crops are listed in order of increasing salt tolerance; electrical conductance values at the top and bottom of each column represent the range of salinity level at which a 50 percent decrease in yield may be expected. It should be noted that these concentrations refer to soil water, which may contain concentrations from five to ten times that of applied irrigation water. Soil type, climatic conditions, and irrigation practices may influence the reactions of a given crop to the salt constituents; therefore, the position of each crop in Table 7.8 reflects its relative salt tolerance under customary irrigation conditions.

An important factor allied to the relation of crop growth to water quality is drainage. If a soil is open and well drained, crops may be grown on it with the application of generous amounts of saline water; but, on the other hand, a poorly drained area combined with application of good-quality water may fail to produce as satisfactory a crop. Poor drainage permits salt concentrations in the root zone to build up to toxic proportions. Today, the necessity of adequate drainage is clearly recognized in order to maintain a favorable salt balance—where the total dissolved solids brought to the land annually by irrigation water is less than the total solids carried away annually by drainage water. It is believed that this factor accounted for the failure of many of the elaborate irrigation systems of historical times.

---

*An excellent analysis of irrigation water quality can be found in R. S. Ayers, Quality of water for irrigation, *Jour. Irrig. and Drainage Div.,* Amer. Soc. Civil Engrs., v. 103, no. IR2, pp. 135–154, 1977.

TABLE 7.8 Relative Tolerances of Crops to Salt Concentrations (after Richards[38])

Crop Division	Low Salt Tolerance	Medium Salt Tolerance	High Salt Tolerance
Fruit crops	Avocado	Cantaloupe	Date palm
	Lemon	Date	
	Strawberry	Olive	
	Peach	Fig	
	Apricot	Pomegranate	
	Almond		
	Plum		
	Prune		
	Grapefruit		
	Orange		
	Apple		
	Pear		
Vegetable crops	3000 μS/cm	4000 μS/cm	10,000 μS/cm
	Green bean	Cucumber	Spinach
	Celery	Squash	Asparagus
	Radish	Peas	Kale
	4000 μS/cm	Onion	Garden beet
		Carrot	12,000 μS/cm
		Potato	
		Sweet corn	
		Lettuce	
		Cauliflower	
		Bell pepper	

TABLE 7.8　(continued)

Cabbage
Broccoli
Tomato
10,000 μS/cm

Forage crops	2000 μS/cm	4,000 μS/cm	12,000 μS/cm
	Burnet	Sickle milkvetch	Bird's-foot trefoil
	Ladino clover	Sour clover	Barley (hay)
	Red clover	Cicer milkvetch	Western wheat grass
	Alsike clover	Tall meadow oat grass	Canada wild rye
	Meadow foxtail	Smooth brome	Rescue grass
	White Dutch clover	Big trefoil	Rhodes grass
	4000 μS/cm	Reed canary	Bermuda grass
		Meadow fescue	Nattall alkali grass
		Blue grame	Salt grass
		Orchard grass	Alkali sacaton
		Oats (hay)	18,000 μS/cm
		Wheat (hay)	
		Rye (hay)	
		Tall fescue	
		Alfalfa	
		Hubam clover	
		Sudan grass	
		Dallis grass	
		Strawberry clover	
		Mountain brome	

Field crops

4000 µS/cm	6000 µS/cm	10,000 µS/cm	12,000 µS/cm
Field bean	Castor bean	Cotton	Perennial rye grass
	Sunflower	Rape	Yellow sweet clover
	Flax	Sugar beet	White sweet clover
	Corn (field)	Barley (grain)	
	Sorghum (grain)	16,000 µS/cm	
	Rice		
	Oat (grain)		
	Wheat (grain)		
	Rye (grain)		
	10,000 µS/cm		

NOTE: Electrical conductance values represent salinity levels of the saturation extract at which a 50 percent decrease in yield may be expected as compared to yields on nonsaline soil under comparable growing conditions. The saturation extract is the solution extracted from a soil at its saturation percentage.

In place of rigid limits of salinity for irrigation water, quality is commonly expressed by classes of relative suitability. Most classification systems include limits on specific conductance (expressing total dissolved solids), sodium content, and boron concentration.

Sodium concentration is important in classifying an irrigation water because sodium reacts with soil to reduce its permeability (see following section). Soils containing a large proportion of sodium with carbonate as the predominant anion are termed alkali soils; those with chloride or sulfate as the predominant anion are saline soils. Ordinarily, either type of sodium-saturated soil will support little or no plant growth. Sodium content is usually expressed in terms of percent sodium (also known as sodium percentage and soluble-sodium percentage), defined by

$$\% \ Na = \frac{(Na + K)100}{Ca + Mg + Na + K} \tag{7.4}$$

where all ionic concentrations are expressed in milliequivalents per liter. The Salinity Laboratory of the U.S. Department of Agriculture[38] recommends the sodium adsorption ratio (SAR) because of its direct relation to the adsorption of sodium by soil. It is defined by

$$SAR = \frac{Na}{\sqrt{(Ca + Mg)/2}} \tag{7.5}$$

where the concentrations of the constituents are expressed in milliequivalents per liter.

Boron is necessary in very small quantities for normal growth of all plants, but in larger concentrations it becomes toxic. Quantities needed vary with the crop type; sensitive crops require minimum amounts whereas tolerant crops will make maximum growth on several times these concentrations. Relative boron tolerances of a number of crops are summarized in Table 7.9.

An example of an irrigation water classification system is shown in Table 7.10. Applications of the classification to groundwaters from alluvium in California are listed in Table 7.11. A graphic classification by the U.S. Salinity Laboratory, shown in Fig. 7.9, is based on SAR and conductance.

In investigations relating to quality of water for irrigation and the salinity of the soil solution, particularly regarding sampling programs, it is important to take cognizance of the salt distribution within the soil. As an illustration, the salt distribution under irrigated cotton plants is shown in Fig. 7.10. It is apparent that the leaching effect of the irrigation water in the furrow, together with the move-

TABLE 7.9 Relative Tolerance of Plants to Boron
(after Richards[38])
(listed in order of increasing tolerance)

Sensitive	Semitolerant	Tolerant
Lemon	Lima bean	Carrot
Grapefruit	Sweet potato	Lettuce
Avocado	Bell pepper	Cabbage
Orange	Pumpkin	Turnip
Thornless blackberry	Zinnia	Onion
Apricot	Oat	Broadbean
Peach	Milo	Gladiolus
Cherry	Corn	Alfalfa
Persimmon	Wheat	Garden beet
Kadota fig	Barley	Mangel
Grape	Olive	Sugar beet
Apple	Ragged robin rose	Date palm
Pear	Field pea	Palm
Plum	Radish	Asparagus
American elm	Sweetpea	Athel
Navy bean	Tomato	
Jerusalem artichoke	Cotton	
English walnut	Potato	
Black walnut	Sunflower	
Pecan		

ment toward the plants on the ridges, creates wide variations of salt concentration within short distances.

Changes in Chemical Composition

As groundwater moves underground it tends to develop a chemical equilibrium by chemical reactions with its environment. Effects of some equilibria in groundwater have important applications —for example, artificial recharge, movement of pollutants, and clogging of wells.

Chemical precipitation may remove ions in solution by forming insoluble compounds. Precipitation of calcium carbonate and release of dissolved carbon dioxide may result from a decrease in pressure and/or an increase in temperature. Ferrous iron in solution oxidizes on exposure to air and is deposited as ferric hydroxide.[24]

Ion exchange involves the replacement of ions adsorbed on the surface of fine-grained materials in aquifers by ions in solution. Because the exchange involves principally cations (sodium, calcium,

TABLE 7.10 Quality Classification of Water for Irrigation (after Wilcox[53])

Water Class	Percent Sodium	Specific Conductance, $\mu S/cm$	Boron, mg/l		
			Sensitive Crops	Semitolerant Crops	Tolerant Crops
Excellent	<20	<250	<0.33	<0.67	<1.00
Good	20–40	250–750	0.33–0.67	0.67–1.33	1.00–2.00
Permissible	40–60	750–2000	0.67–1.00	1.33–2.00	2.00–3.00
Doubtful	60–80	2000–3000	1.00–1.25	2.00–2.50	3.00–3.75
Unsuitable	>80	>3000	>1.25	>2.50	>3.75

TABLE 7.11 Chemical Analyses and Classifications of Selected Groundwaters in California (after Doneen, L. D., *Calif. Agric.*, v. 4, no. 11, 1950)

Number	Specific Conductance, μS/cm	B, mg/l	Major Constituents, meq/l						Percent Na	Water Class[a]
			Ca	Mg	Na	CO$_3$ +HCO$_3$	Cl	SO$_4$		
1	260[b]	0.13	1.41	0.44	0.89	1.88	0.34	0.33	32	Good
2	270	0.10	0.21	0.05	2.42	1.20	0.68	0.67	90	Unsuitable
3	790	6.90	0.24	0.02	7.28	2.39	2.47	2.48	96	Unsuitable
4	900	0.51	2.49	5.81	2.83	8.87	1.13	1.02	25	Permissible
5	1090	. . .	1.20	2.00	8.10	8.10	1.00	2.60	72	Doubtful
6	1370	0.25	8.30	0.75	3.96	2.46	2.73	4.47	30	Permissible
7	1740	0.71	2.14	0.08	12.67	1.02	12.04	1.80	85	Unsuitable
8	2550	0.50	11.40	5.70	12.90	2.80	2.80	23.00	45	Doubtful
9	4330	1.63	12.37	16.71	27.39	2.75	8.55	41.74	49	Unsuitable

[a]Based on classification in Table 7.10.
[b]Underlined values determine water class.

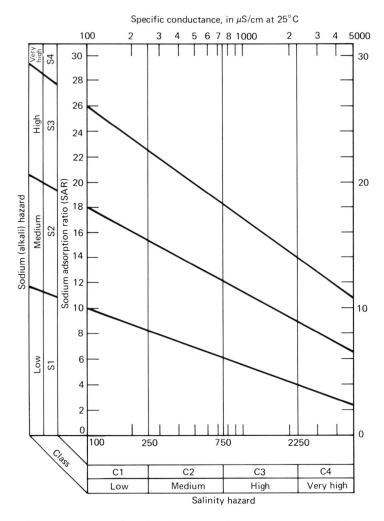

Fig. 7.9 Diagram for classification of irrigation waters (after Richards[38]).

and magnesium), the process is known as base, or cation, exchange. The direction of the exchange is toward an equilibrium of bases present in the water and on the finer materials of the aquifer. Base exchange is known to soften groundwater naturally and to produce, in coastal regions where seawater has entered an aquifer, groundwater having a quality other than a simple mixture of the two source waters (see Chapter 14).

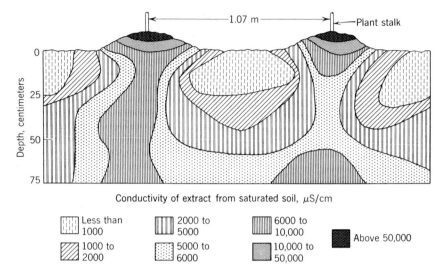

Fig. 7.10 Salt distribution under furrow-irrigated cotton for soil initially salinized to 0.2 percent salt (3100 μS/cm) and irrigated with water of medium salinity (after Wadleigh and Fireman[49]; reproduced from *Soil Science Society of America Proceedings*, Vol. 13, p. 529, 1948 by permission of the Soil Science Society of America).

Base exchange causes changes in the physical properties of soils. When high-sodium water is applied to a soil, the number of sodium ions combined with the soil increases, while an equivalent quantity of calcium, or other, ions is displaced. These reactions cause deflocculation and reduction of permeability. In the opposite case where calcium is the dominant cation, the exchange occurs in the reverse direction, creating a flocculated and more permeable soil. The advantage of adding gypsum ($CaSO_4$) to a soil is that by base exchange the soil texture and drainability can be improved.

Chemical reduction of oxidized sulfur ions to sulfate ions or to the sulfide state occurs frequently in groundwater. The reaction is believed to take place in the presence of certain bacteria.[23] Typically, waters experiencing sulfate reduction have high bicarbonate and carbon dioxide contents and contain hydrogen sulfide.

The equilibrium achieved by the various chemical reactions described above tends to produce a quality that remains stable with time because of the slow movement of groundwater and its long residence time within a given geologic formation. In general, quality variations are more noticeable in shallow aquifers where seasonal

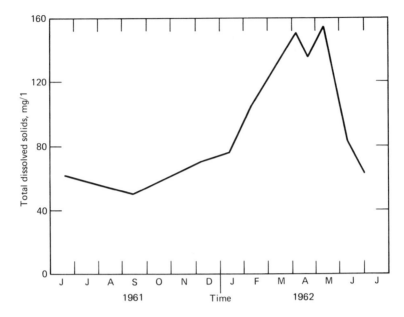

Fig. 7.11   Variation in chemical quality of shallow groundwater at a site in northwestern Alaska due to seasonal effects of freezing (after Fuelner and Schupp[20]).

variations in recharge and discharge create corresponding fluctuations in salinity. Interestingly, freezing of shallow groundwater in arctic regions can also cause seasonal changes in chemical quality, as shown in Fig. 7.11. Here a marked increase in salinity during winter months results from freezing, which not only reduces the diluting effect of recharge from precipitation, but also selectively concentrates the mineral content in the unfrozen groundwater.

Finally, it should be noted that human acts can markedly change the natural quality of groundwater, usually for the worse. The resulting degradation of groundwater has become an important international concern; Chapter 8 treats the subject of groundwater pollution.

Dissolved Gases

Although it is not generally recognized, most groundwater contains dissolved gases derived from natural sources. Those involved in the normal geochemical cycle of groundwater include the atmospheric gases: carbon dioxide (CO_2), oxygen (O_2), and nitrogen (N_2).

Others derived from underground biochemical processes include the flammable gases methane (CH_4) and hydrogen sulfide (H_2S). With its distinctive rotten-egg odor, hydrogen sulfide is readily detected at concentrations of less than 1 mg/l and rarely accumulates to dangerous proportions.

Methane, which is colorless, tasteless, and odorless, occurs frequently in groundwater and can be a much more serious problem. The gas is a decomposition product of buried plant and animal matter in unconsolidated and geologically young deposits.* The minimum concentration of methane in water sufficient to produce an explosive methane-air mixture above the water from which it bubbles out of solution depends on the temperature, pressure, quantity of water pumped, and volume of air into which the gas evolves. Theoretically, water containing as little as 1 to 2 mg/l of methane can produce an explosion in a poorly ventilated air space. Fires and explosions in well pits, basements, and water tanks have occurred from methane emitted by groundwater; safety measures include analyses to detect the presence of the gas, aeration of water before use, and adequate ventilation where the water is being used. The danger exists of people suffocating in dug wells and pump pits where high methane concentrations form.

Temperature

Variations in solar energy received at the earth's surface create periodicities, both diurnal and annual, in temperature below ground surface. The insulating qualities of the earth's crust rapidly damp the large temperature range found at ground surface so that only shallow groundwater displays any appreciable fluctuation in temperature.[22] In marked contrast to the large seasonal variation of surface water temperatures (except in tropical regions), groundwater temperatures tend to remain relatively constant—an important advantage for drinking water and industrial uses.

An example of the variability of groundwater temperature in relation to that of a surface water body is shown by Fig. 7.12 for the Schenectady, N.Y., well field adjoining the Mohawk River. The annual variation in groundwater temperature is only a small fraction of that observed in the river water. Furthermore, the tongue of larger temperature variation pointing toward the well field indicates the

---

*For example, a buried oak log, only 1 ft in diameter and 5 ft long, can generate enough methane when it decomposes underground to form an explosive mixture in 400 to 1200 m³ of air.[6]

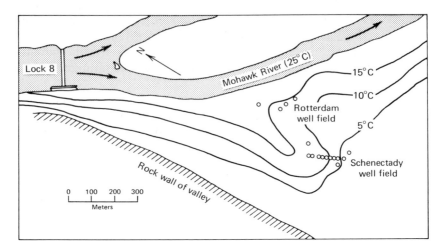

Fig. 7.12 Map of annual variation in groundwater temperature near Mohawk River, New York. Numbered lines show locations of equal annual groundwater temperature change in °C; annual river water change was 25°C (after Winslow[54]).

principal flow path of groundwater from the river to the Schenectady well field through a permeable zone of the shallow aquifer.*

Assuming that ground surface temperature is a sinusoidal function of time, the temperature T_z at a depth z can be given by

$$T_z = T_o + Ae^{-z\sqrt{\pi/a\tau}} \sin\left(\frac{2\pi t}{\tau} - z\sqrt{\pi/a\tau}\right) \qquad (7.6)$$

where T_o is the mean ground surface temperature, A is the amplitude of the surface temperature variation, τ is the oscillation period (one day or one year), a is the thermal diffusivity of the subsurface material (approximating 0.005 cm²/sec), and t is time. The depth of nearly uniform temperature occurs at about 10 m in the tropics and increases to about 20 m in polar regions,[52] although influences such as rock type, elevation, precipitation, cloudiness, and wind can produce significant local deviations. Illustrative of annual fluctuations and the effect of depth are the temperatures in two nearby shallow wells shown in Fig. 7.13.

---

*It should be noted generally that the velocity of groundwater can have an appreciable influence on the distribution of underground temperatures. Where sufficient field data are available, it may be possible to calculate flow velocities from temperature measurements and even to determine aquifer permeability from a combination of water level and temperature measurements.[40,44]

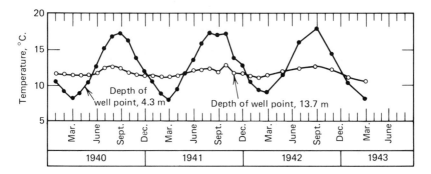

Fig. 7.13 Monthly groundwater temperatures in two nearby wells on Long Island, New York (after *New York State Water Resources Comm. Bull. 55*).

Shallow groundwater temperatures can also be influenced by the type of overlying surface environment. Thus, a study on Long Island, New York[36] showed that the mean annual temperature and the annual range in temperature of groundwater were larger beneath cleared areas than beneath wooded areas. The difference can be attributed to the absence of shade and the lack of an insulating layer of organic material on the ground in the cleared areas.

Below the zone of surface influence groundwater temperatures increase approximately 2.9° C for each 100 m of depth in accordance with the geothermal gradient of the earth's crust.* Thus, groundwater pumped from deep wells is appreciably warmer than shallow groundwater. For example, if shallow groundwater has a temperature of 15° C, that obtained from an aquifer at 500 m might be expected to have a temperature near 30° C.

In the United States it has been found that the temperature of groundwater occurring at a depth of 10 to 20 m will generally be about 1° to 2° C higher than the local mean annual air temperature.[13] On this basis a groundwater temperature map, as shown in Fig. 7.14, can be constructed from climatological data. As extremes, groundwater temperatures may vary from below freezing to above the boiling point in geothermal areas and in superheated water emerging from geysers.

---

*Most geothermal gradients measured in the United States fall within the range 1.8–3.6° C per 100 m. The heat flux from the interior of the earth is estimated to average 1.3×10^{-6} cal/cm²sec.

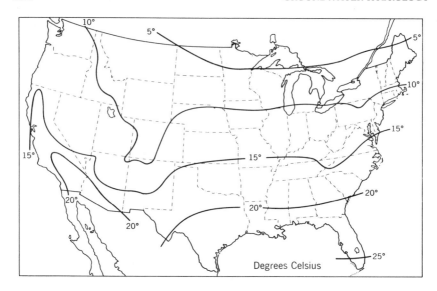

Fig. 7.14 Approximate temperature of groundwater in the United States at depths of 10 to 20 meters (after Collins[13]).

Saline Groundwater

Many groundwaters contain dissolved salts in such concentrations as to make them unusable for ordinary water supply purposes.[27] *Saline groundwater* is a general term referring to any groundwater containing more than 1000 mg/l total dissolved solids. Various classification schemes based on dissolved solids have been proposed; the simplicity of the one shown in Table 7.12 makes it particularly convenient.

A map of the United States showing the minimum depth to aquifers containing groundwater exceeding 1000 mg/l is given in Fig. 7.15.

TABLE 7.12   Classification of Saline Groundwater (after Carroll[12])

	Total Dissolved Solids, mg/l
Fresh water	0–1,000
Brackish water	1,000–10,000
Saline water	10,000–100,000
Brine	>100,000

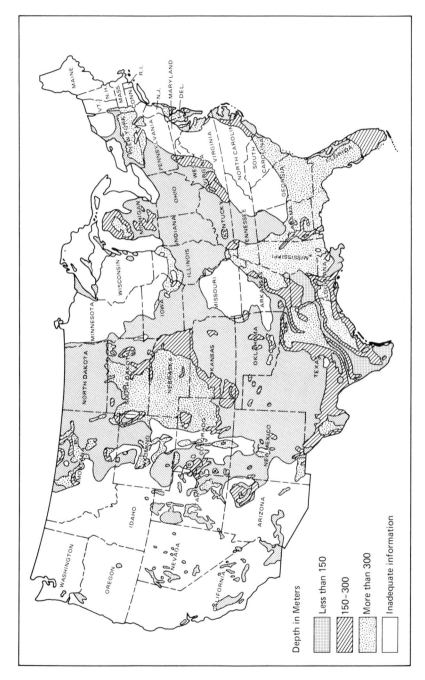

Fig. 7.15 Depth to saline groundwater in the United States (after Feth, et al.[18]).

Approximately two-thirds of the country is known to contain such waters; blank areas indicate areas where either well yields are less than 40 m³/day or no data on saline groundwater are available. Four types of occurrence are recognized: connate water, intruded seawater (see Chapter 14), water salinized by contact with soluble salts in the formation where it is situated, and water in regions with shallow water tables where evapotranspiration concentrates the salts in solution.

Although saline groundwater has traditionally been regarded as an undesirable resource, modern technological advances may reverse this role. Rapid advances in desalination techniques suggest that saline groundwater may be a potentially important water supply source where shortages are imminent. Industrial use of saline groundwater for cooling purposes has yet to be exploited fully; high costs, involving deep wells, corrosion control, and disposal, have hindered such developments. Saline aquifers also serve useful purposes for temporary storage of fresh water and energy in the form of hot water and for disposal of wastewater (see Chapter 13).

References

1. Amer. Public Health Assoc., Amer. Water Works Assoc., and Water Pollution Control Fed., *Standard methods for the examination of water and wastewater*, 14th ed., Amer. Public Health Assoc., Washington, D.C., 1200 pp., 1975.
2. Amer. Soc. Testing Matls., *Manual on industrial water and industrial waste water*, 2nd ed., Philadelphia, 992 pp., 1966.
3. American Soc. Testing Matls., *Manual on water*, ASTM Spl. Tech. Publ. no. 442, Philadelphia, 360 pp., 1969.
4. Amer. Soc. Testing Matls., *Water quality criteria*, ASTM Spl. Tech. Publ. 416, Philadelphia, 120 pp., 1967.
5. Amer. Water Works Assoc., *Water quality and treatment*, McGraw-Hill, New York, 654 pp., 1971.
6. Anon., Gas in ground water, *Jour. Amer. Water Works Assoc.*, v. 61, pp. 413–414, 1969.
7. Back, W., Hydrochemical facies and ground-water flow patterns in northern part of Atlantic Coastal Plain, *U.S. Geological Survey Prof. Paper 498-A*, 42 pp., 1966.
8. Back, W., and B. B. Hanshaw, Chemical geohydrology, in *Advances in Hydroscience* (V. T. Chow, ed.), v. 2, Academic Press, New York, pp. 49–109, 1965.
9. Back, W., and B. B. Hanshaw, Geochemical interpretations of groundwater flow systems, *Water Resources Bull.* v. 7, pp. 1008–1016, 1971.
10. Baker, M. N., *The quest for pure water*, Amer. Water Works Assoc., New York, 527 pp., 1948.

11. Brown, E., et al., Methods for collection and analysis of water samples for dissolved minerals and gases, *U.S. Geological Survey Techniques for Water-Resources Investigations,* Bk. 5, Chap. A1, 160 pp., 1970.
12. Carroll, D., Rainwater as a chemical agent of geologic processes—A review, *U.S. Geological Survey Water-Supply Paper* 1535-G, 18 pp., 1962.
13. Collins, W. D., Temperature of water available for industrial use in the United States, *U.S. Geological Survey Water-Supply Paper* 520-F, pp. 97–104, 1925.
14. Davis, S. N., and R. J. M. DeWiest, *Hydrogeology,* John Wiley & Sons, New York, 463 pp., 1966.
15. Durum, W. H., and J. Haffty, Occurrence of minor elements in water, *U.S. Geological Survey Circ.* 445, 11 pp., 1961.
16. Eriksson, E., Atmospheric transport of oceanic constituents in their circulation in nature, *Tellus,* v. 11. pp. 1–72, 1959.
17. Everett, L. G., et al., *Monitoring groundwater quality: methods and costs,* Rept. EPA-600/4-76-023, U.S. Environmental Protection Agency, Las Vegas, 140 pp., 1976.
18. Feth, J. H., et al., Preliminary map of the conterminous United States showing depth to and quality of shallowest ground water containing more than 1000 parts per million dissolved solids, *U.S. Geological Survey Hydrologic Inv. Atlas* HA-199, 31 pp., 1965.
19. Feth, J. H., et al., Sources of mineral constituents in water from granitic rocks, Sierra Nevada, California and Nevada, *U.S. Geologic Survey Water-Supply Paper* 1535-I, 70 pp., 1964.
20. Feulner, A. J., and R. G. Schupp, Seasonal changes in the chemical quality of shallow ground water in northwestern Alaska, *U.S. Geological Survey Prof. Paper* 475-B, pp. 189–191, 1963.
21. Harder, A. H., and W. R. Holden, Measurement of gas in groundwater, *Water Resources Research,* v. 1, pp. 75–82, 1965.
22. Heath, R. C., Seasonal temperature fluctuations in surficial sand near Albany, New York, *U.S. Geological Survey Prof. Paper* 475-D, pp. 204–208, 1964.
23. Hem, J. D., Study and interpretation of the chemical characteristics of natural water, 2nd ed., *U.S. Geological Survey Water-Supply Paper* 1473, 363 pp., 1970.
24. Hem, J. D., et al., Chemistry of iron in natural water, *U.S. Geological Survey Water-Supply Paper* 1459, 268 pp., 1962.
25. Kohout, F. A., and N. D. Hoy, Some aspects of sampling salty ground water in coastal aquifers, *Ground Water,* v. 1, pp. 28–32, 43, 1963.
26. Krieger, R. A., The chemistry of saline waters, *Ground Water,* v. 1, no. 4, pp. 7–12, 1963.
27. Krieger, R. A., et al., Preliminary survey of the saline-water resources of the United States, *U.S. Geological Survey Water-Supply Paper* 1374, 172 pp., 1957.
28. Loewengart, S., Airborne salts—The major source of the salinity of waters in Israel, *Bull. Research Council Israel,* v. 10G, pp. 183–206, 1961.

29. Logan, J., Estimation of electrical conductivity from chemical analysis of natural waters, *Jour. Geophysical Research*, v. 66, pp. 2479–2483, 1961.

30. McKee, J. E., and H. W. Wolf (eds.), *Water quality criteria*, Publ. no. 3-A, California State Water Resources Control Board, Sacramento, 548 pp., 1963.

31. McMillion, L. G., and J. W. Keeley, Sampling equipment for ground-water investigations, *Ground Water*, v. 6, no. 2, pp. 9–11, 1968.

32. Mink, J. F., Groundwater temperatures in a tropical island environment, *Jour. Geophysical Research*, v. 69, pp. 5225–5230, 1964.

33. Natl. Acad. of Sciences, Natl. Acad. of Engrng., *Water quality criteria 1972*, Washington, D.C., 594 pp., 1972.

34. Natl. Tech. Adv. Comm., *Water quality criteria*, Federal Water Pollution Control Admin., Washington, D.C., 234 pp., 1968.

35. Piper, A. M., A graphic procedure in the geochemical interpretation of water-analyses, *Trans. Amer. Geophysical Union*, v. 25, pp. 914–928, 1944.

36. Pluhowski, E. J., and I. H. Kantrowitz, Influence of land-surface conditions on ground-water temperatures in southwestern Suffolk County, Long Island, New York, *U.S. Geological Survey Prof. Paper 475-B*, pp. 186–188, 1963.

37. Rainwater, F. H., and L. L. Thatcher, Methods for collection and analysis of water samples, *U.S. Geological Survey Water-Supply Paper 1454*, 301 pp., 1960.

38. Richards, L. A. (ed.), *Diagnosis and improvement of saline and alkali soils*, Agric. Handbook 60, U.S. Dept. Agric., Washington, D.C., 160 pp., 1954.

39. Sawyer, C. N., and P. L. McCarty, *Chemistry for sanitary engineers*, 2nd ed., McGraw-Hill, New York, 518 pp., 1967.

40. Schneider, R., An application of thermometry to the study of ground water, *U.S. Geological Survey Water-Supply Paper 1544-B*, 16 pp., 1962.

41. Schoeller, H., *Les eaux souterraines*, Masson & Cie, Paris, 642 pp., 1962.

42. Scott, R. C., and F. B. Barker, Radium and uranium in ground water in the United States, *Proc. 2d United Nations Intl. Conf. on Peaceful Uses of Atomic Energy*, v. 2, pp. 153–157, 1958.

43. Silvey, W. D. Occurrence of selected minor elements in the waters of California, *U.S. Geological Survey Water-Supply Paper 1535-L*, 25 pp., 1967.

44. Stallman, R. W., Computation of ground-water velocity from temperature data, *U.S. Geological Survey Water-Supply Paper 1544-H*, pp. 36–46, 1963.

45. Stiff, H. A., Jr., The interpretation of chemical water analysis by means of patterns, *Jour. Petr. Technology*, v. 3, no. 10, pp. 15–17, 1951.

46. Summers, W. K., Factors affecting the validity of chemical analyses of natural water, *Ground Water*, v. 10, no. 2, pp. 12–17, 1972.

47. U.S. Environmental Protection Agency, *Quality criteria for water*, Washington, D.C., 501 pp., 1976.

48. U.S. Public Health Service, *Drinking water standards* 1962, Publ. 956, Washington, D.C., 61 pp., 1962.
49. Wadleigh, C. H., and M. Fireman, Salt distribution under furrow and basin irrigated cotton and its effect on water removal, *Soil Sci. Soc. Amer. Proc.*, v. 13, pp. 527–530, 1948.
50. White, D. E., Magmatic, connate, and metamorphic waters, *Bull. Geol. Soc. Amer.*, v. 68, pp. 1659–1682, 1957.
51. White, D. E., Thermal waters of volcanic origin, *Bull. Geol. Soc. Amer.*, v. 68, pp. 1637–1658, 1957.
52. White, D. E., et al., Data of geochemistry—chemical composition of subsurface waters, 6th ed., *U.S. Geological Survey Prof. Paper* 440-F, 67 pp., 1963.
53. Wilcox, L. V., *Classification and use of irrigation waters*, U.S. Dept. Agric. Circ. 969, Washington, D.C., 19 pp., 1955.
54. Winslow, J. D., Effect of stream infiltration on ground-water temperatures near Schenectady, N.Y., *U.S. Geological Survey Prof. Paper* 450-C, pp. 125–128, 1962.
55. Zaporozec, A., Graphical interpretation of water-quality data, *Ground Water*, v. 10, pp. 32–43, 1972.

Pollution
of Groundwater

CHAPTER 8 ··

Groundwater pollution may be defined as the artificially induced degradation of natural groundwater quality. Pollution can impair the use of water and can create hazards to public health through toxicity or the spread of disease.* Most pollution originates from the disposal of wastewater following the use of water for any of a wide variety of purposes. Thus, a large number of sources and causes can modify groundwater quality, ranging from septic tanks to irrigated agriculture.[4,7,8,57] In contrast with surface water pollution, subsurface pollution is difficult to detect, is even more difficult to control,

---

*One need not go far back in history to find instances of major diseases transmitted by groundwater. A classic example was the outbreak of cholera in London, England, in 1854. Dr. John Snow, while investigating the epidemic, noted with considerable astuteness that more than 500 persons died from cholera in ten days within 250 yards of a public water supply well; this led to his conclusion:

The result of the inquiry then was, that there had been no particular outbreak or increase of cholera, in this part of London, except among the persons who were in the habit of drinking the water of the above-mentioned pump-well. I had an interview with the Board of Guardians of St. James parish on the evening of Thursday, 7th September, and represented the above circumstances to them. In consequence of what I said, the handle of the pump was removed on the following day (*On the Mode of Communication of Cholera*, 1855).

and may persist for decades. With the growing recognition of the importance of underground water resources, efforts are increasing to prevent, reduce, and eliminate groundwater pollution.

Pollution in Relation to Water Use

The possible pollutants in groundwater are virtually limitless. A wide range of pollutants found in groundwater samples are listed in Table 8.1 according to organic and inorganic chemical, biological, physical, and radiological types. The sources and causes of groundwater pollution are closely associated with human use of water.[87,91] A complex and interrelated series of modifications to natural water quality is created by the diversity of human activities impinging on the hydrologic cycle.

The principal sources and causes of groundwater pollution are listed in Table 8.2 under four categories—municipal, industrial, agricultural, and miscellaneous. Most pollution stems from disposal of wastes on or into the ground.[58] Methods of disposal include placing wastes in percolation ponds, on the ground surface (spreading or irrigation), in seepage pits or trenches, in dry streambeds, in landfills, into disposal wells, and into injection wells. Common methods for disposal associated with each pollution source are identified in Table 8.2; a few nondisposal causes of groundwater pollution can also be recognized.

All sources and causes of pollution can be classified as to their geometry, also shown in Table 8.2. A point source originates from a singular location, a line source has a predominantly linear alignment, and a diffuse source occupies an extensive area that may or may not be clearly defined.

In the following sections the principal sources and causes of pollution* are briefly described with regard to their occurrence and their effects on groundwater quality.

Municipal Sources and Causes

Sewer Leakage. Sanitary sewers are intended to be watertight; however, in reality leakage of sewage into the ground is a common occurrence, especially from old sewers. Leakage may result from poor workmanship, defective sewer pipe, breakage by tree roots,

---

*Considerable confusion exists in the literature over the distinction between *pollution* and *contamination*. Here pollution shall signify any degradation of natural water quality, while contamination shall be reserved for pollution that constitutes a hazard to human health.

TABLE 8.1 Parameters and Constituents that May Be Involved in Analysis of Polluted Groundwater[a] (after Todd, et al.[89])

Chemical–Organic	Units	Chemical–Inorganic	Units
Biochemical oxygen demand (BOD)	mg/l	Nickel (Ni)	μg/l
Carbon chloroform extract (CCE)	μg/l	Nitrite (NO_2)	mg/l
		Nitrate (NO_3)	mg/l
Chemical oxygen demand (COD)	mg/l	Nitrogen (N)	mg/l
		Oil and grease	mg/l
		Oxygen (O_2)	mg/l
Chlorinated phenoxy acid herbicides	μg/l	pH	pH units
		Phosphate (PO_4)	mg/l
Detergents (surfactants)	mg/l	Potassium (K)	mg/l
Organic carbon (C)	mg/l	Selenium (Se)	μg/l
		Silver (Ag)	μg/l
Organophosphorus pesticides	μg/l	Silica (SiO_2)	mg/l
		Sodium (Na)	mg/l
Phenols	mg/l	Solids, dissolved	mg/l
Tannins and lignins	mg/l	Solids, suspended	mg/l
		Strontium (Sr)	μg/l
Chemical–Inorganic		Sulfate (SO_4)	mg/l
Acidity	mg/l	Sulfide (S)	mg/l
Alkalinity	mg/l	Sulfite (SO_3)	mg/l
Aluminum (Al)	μg/l	Tin (Sn)	μg/l
Ammonia (NH_4)	mg/l	Titanium (Ti)	μg/l
Antimony (Sb)	μg/l	Vanadium (V)	μg/l
Arsenic (As)	μg/l	Zinc (Zn)	μg/l
Barium (Ba)	μg/l		

Beryllium (Be)	μg/l
Bicarbonate (HCO$_3$)	mg/l
Boron (B)	μg/l
Bromide (Br)	μg/l
Cadmium (Cd)	μg/l
Calcium (Ca)	mg/l
Carbonate (CO$_3$)	mg/l
Chloride (Cl)	mg/l
Chromium (Cr)	μg/l
Cobalt (Co)	μg/l
Conductance, specific	μS/cm at 25°C
Copper (Cu)	μg/l
Cyanide (CN)	μg/l
Fluoride (F)	μg/l
Hardness	mg/l
Hydroxide (OH)	mg/l
Iodide (I)	μg/l
Iron (Fe)	μg/l
Lead (Pb)	μg/l
Lithium (Li)	μg/l
Magnesium (Mg)	mg/l
Manganese (Mn)	μg/l
Mercury (Hg)	μg/l
Molybdenum (Mo)	μg/l

Biological	
Coliform bacteria	Coliforms/100 ml
Fecal coliform bacteria	Fecal coliforms/100 ml
Fecal streptococci bacteria	Fecal streptococci/100 ml
Physical	
Color	PCU
Odor	TO
Temperature	°C
Turbidity	TU
Radiological	
Barium–140 (^{140}Ba)	pc/l
Cerium–141 and 144 (^{141}Ce, ^{144}Ce)	pc/l
Cesium–134 and 137 (^{134}Cs, ^{137}Cs)	pc/l
Gamma spectrometry	pc/l
Gross alpha	pc/l
Gross gamma	nc/l
Iodine–131 (^{131}I)	pc/l
Neptunium–239 (^{239}Np)	pc/l
Radium (Ra)	pc/l
Thorium (Th)	μg/l
Tritium (^{3}H)	pc/l
Uranium (U)	μg/l

[a]This list is intended to be illustrative rather than comprehensive.

TABLE 8.2 Principal Sources and Causes of Groundwater Pollution
(after Todd, et al.[88])

Source or Cause	Pollution Geometry			Disposal Method						
	Point	Line	Diffuse	Percolation Pond	Surface Spreading and Irrigation	Seepage Pits and Trenches	Dry Streambeds	Landfills	Disposal Wells	Injection Wells
Municipal										
Sewer leakage	○	○								
Liquid wastes	○	○	○	○	○	○	○		○	
Solid wastes	○				○			○		
Industrial										
Liquid wastes	○			○		○			○	○
Tank and pipeline leakage		○								
Mining activities	○	○	○	○		○	○	○	○	○
Oilfield brines	○	○	○	○	○	○	○			○
Agriculture										
Irrigation return flows	○				○					
Animal wastes			○	○	○	○		○		
Fertilizers and soil amendments			○		○					
Pesticides			○		○					

Miscellaneous

Spills and surface
discharges

Stockpiles

Septic tanks and
cesspools

Roadway deicing

Saline water
intrusion

Interchange through
wells

Surface water

321

TABLE 8.3 Normal Range of Increase in
Mineral Constituents Found in
Domestic Sewage (after Miller, et al.[58])

Mineral	Range of Increase, mg/l
Dissolved solids	100–300
Boron (B)	0.1–0.4
Sodium (Na)	40–70
Potassium (K)	7–15
Magnesium (Mg)	3–6
Calcium (Ca)	6–16
Total Nitrogen (NO_3)	20–40
Phosphate (PO_4)	20–40
Sulfate (SO_4)	15–30
Chloride (Cl)	20–50
Alkalinity (as $CaCO_3$)	100–150

ruptures from heavy loads or soil slippage, fractures from seismic activity, loss of foundation support, shearing due to differential settlement at manholes, and infiltration causing sewage flow into abandoned sewer laterals[48]. Because suspended solids in sewage tend to clog sewer cracks and because the surrounding soil tends to become clogged due to anaerobic conditions, leakage from minor sewer openings is often small.

Sewer leakage can introduce high concentrations of BOD, COD, nitrate, organic chemicals, and possibly bacteria into groundwater. The normal ranges of increases in inorganic constituents from water supply to domestic sewage are shown in Table 8.3. Where sewers serve industrial areas, heavy metals such as arsenic, cadmium, chromium, cobalt, copper, iron, lead, manganese, and mercury may enter the wastewater.

Liquid Wastes.   Wastewater in an urban area may originate from domestic uses (see Fig. 8.1), industries, or storm runoff. Most of this highly variable mix of waters receives some degree of treatment and is then discharged into surface waters. There is an increasing trend for treated wastewater to be recharged into the ground where it mingles with naturally occurring groundwater and subsequently becomes available for reuse.[15,31,55]*

*Reclamation of wastewater for reuse through groundwater recharge can be an economical source of supplemental water; consequently, interest is growing in such operations. In 1972 only 2 percent of the total wastewater produced in California was

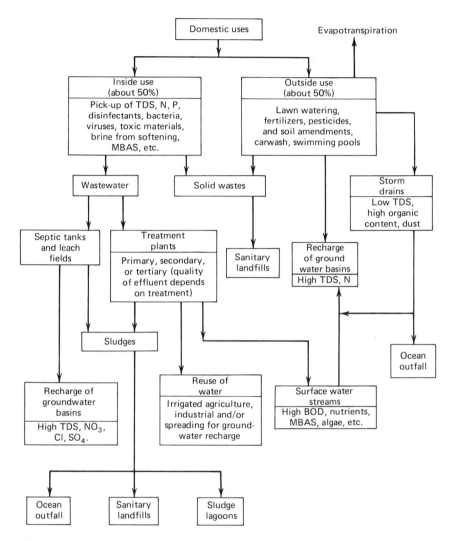

Fig. 8.1 Domestic uses of water and their effects on water quality (after Hassan[38]).

Land application of municipal effluent is accomplished by one of three methods (see Fig. 8.2): irrigation, infiltration-percolation, or overland flow. The selection of a method at a given site is primarily governed by the drainability of the soil, because this property deter-

recharged to groundwater in planned operations; in 1975 proposals were pending for recharging five times that amount[16].

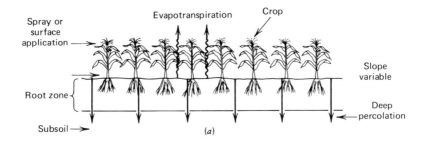

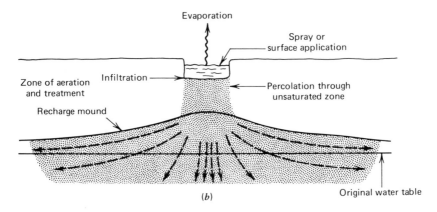

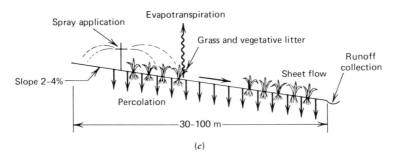

Fig. 8.2 Methods of land application of municipal wastewaters.
(a) Irrigation. (b) Infiltration-percolation. (c) Overland flow (after U.S.
Environmental Protection Agency[90]).

mines the allowable liquid loading rate. In irrigation systems wastewater is applied by spraying, ridge and furrow (see Chapter 13),
and flooding; some water is lost by evapotranspiration. For the
infiltration-percolation method effluent is applied by spreading in
basins or by spraying; almost all of the water so applied reaches the

groundwater. In the overland flow technique wastewater is sprayed over the upper reaches of sloped terraces and allowed to flow across a vegetated surface to runoff collection ditches; percolation to groundwater is minor here because surface runoff and evapotranspiration account for most of the applied water. Table 8.4 summarizes the characteristics of the three methods together with the expected removal of pollutants.

Municipal wastewaters can introduce bacteria, viruses, and inorganic and organic chemicals into groundwater. Where the recharged water is later extracted for potable use, concerns exist regarding health aspects of this reclaimed water, particularly involving viruses, trace elements and heavy metals, and stable organics.[16] Furthermore, chlorination of wastewater effluent can produce additional potential pollutants.

Shallow wells are widely employed to place surface runoff and sometimes treated municipal wastewater underground in freshwater aquifers (see Chapter 13). Such *disposal wells* have been criticized from a health standpoint because of the potential for pollutants to be released directly into an aquifer. The problem is most critical where disposal wells are near pumping wells (Fig. 8.3) and where the beneficial effects of water passing through fine-grained materials may be absent, such as in basalt and limestone aquifers.

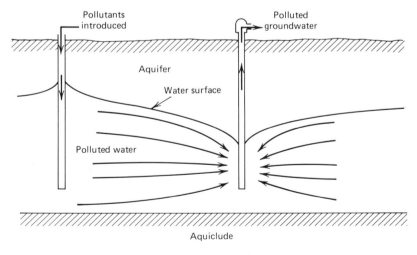

Fig. 8.3 Diagram showing movement of pollutants from a disposal well to a nearby pumping well (after Deutsch[21]).

TABLE 8.4 Characteristics of Irrigation, Infiltration-Percolation, and Overland Flow Systems for Land Application of Municipal Wastewater (after U.S. Environmental Protection Agency[90])

| Factor | Irrigation | | Infiltration–Percolation | Overland Flow |
	Low-Rate	High-Rate		
Liquid loading rate, cm/wk	1.3 to 3.8	3.8 to 10.0	10 to 300	5 to 14
Annual application, m/yr	0.6 to 1.2	1.2 to 5.5	5.5 to 150	2.5 to 7
Land required for 3000 m^3/day flowrate, ha[a]	90 to 180	20 to 180	1 to 20	15 to 45
Application techniques	Spray or surface		Usually surface	Usually spray
Crop production	Excellent	Fair	Poor	Fair
Soils	Moderately permeable soils with good productivity when irrigated		Rapidly permeable soils, such as sands, loamy sands, and sandy loams	Slowly permeable soils, such as clay loams and clays
Climatic constraints	Growing season only	Storage often needed	Reduce loadings in freezing weather	Storage often needed
Wastewater lost to:	Evaporation and percolation		Percolation	Surface runoff and evaporation with some percolation
Expected treatment performance				
BOD and SS removal, %	98+		85 to 99	92+
Nitrogen removal, %	85+[a]		0 to 50	70 to 90
Phosphorus removal, %	80 to 99		60 to 95	40 to 80

[a]Dependent on crop uptake.

Solid Wastes. The land disposal of solid wastes creates an important source of groundwater pollution. A landfill may be defined as any land area serving as a depository of urban, or municipal, solid waste. Most landfills are simply refuse dumps; only a fraction can be regarded as sanitary landfills, indicating that they were designed and constructed according to engineering specifications. Leachate from a landfill can pollute groundwater if water moves through the fill material.[6,14,39,100]. Possible sources of water include precipitation, surface water infiltration, percolating water from adjacent land, and groundwater in contact with the fill. Ordinary mixed refuse usually has a moisture content less than that of field capacity; therefore, leachate from a landfill can be minimized if water from the above sources can be kept from the fill material. In a properly constructed sanitary landfill, any leachate generated can be controlled and prevented from polluting groundwater.*

The problem of pollution from landfills is greatest where high rainfall and shallow water tables occur. Important pollutants frequently found in leachate include BOD, COD, iron, manganese, chloride, nitrate, hardness, and trace elements. Hardness, alkalinity, and total dissolved solids are often increased, while generation of gases, such as methane, carbon dioxide, ammonia, and hydrogen sulfide, are further by-products of landfills.

Industrial Sources and Causes

Liquid Wastes. The major uses of water in industrial plants are for cooling, sanitation, and manufacturing and processing. The quality of the wastewater varies with type of industry and type of use. A generalized flow diagram of industrial water use and its effects on water quality is shown in Fig. 8.4. Cooling water that is softened before use to inhibit scale formation produces wastewaters with salts and heat as important pollutants. Groundwater pollution can occur where industrial wastewaters are discharged into pits, ponds, or lagoons, thereby enabling the wastes to migrate down to the water table.

Cooling water is sometimes recharged underground through shallow disposal wells because its quality, except for the addition of heat, may be unimpaired;[80] industrial recharge wells on Long Island are described in Chapter 13.

---

*In addition, it is assumed that the landfill is properly located, operated, and monitored.

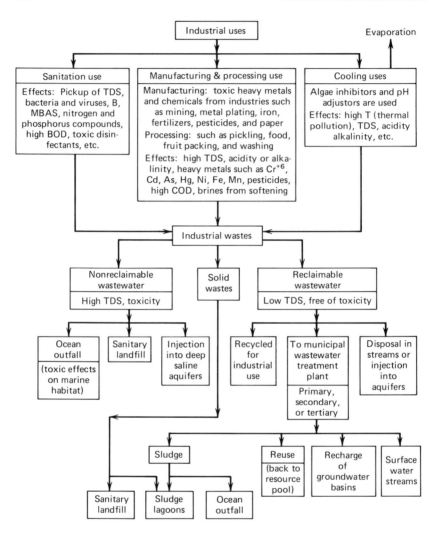

Fig. 8.4 Industrial uses of water and their effects on water quality (after Hassan[38]).

   The disposal of hazardous and toxic industrial wastes is some-times accomplished by means of deep injection wells that place the fluids into saline water formations far below developed freshwater aquifers.[42,95,98] The few hundred such wells existing in the United States have generally been constructed with the guidance of government regulatory agencies so that reported cases of groundwater pollution from these sources are rare.

Tank and Pipeline Leakage.   Underground storage and transmission of a wide variety of fuels and chemicals are common practices for industrial and commercial installations. These tanks and pipelines are subject to structural failures so that subsequent leakage becomes a source of groundwater pollution. Petroleum and petroleum products are responsible for much of the pollution. Leakage is particularly frequent from gasoline station and home fuel oil tanks.[51,63] An immiscible liquid, such as oil, leaking underground moves downward through permeable soils until it reaches the water table. Thereafter it spreads to form a layer on top of the water table and migrates laterally with groundwater flow (see Fig. 8.5a). Liquid radioactive wastes are sometimes stored in underground tanks; leakage from such installations, which has occurred, can cause serious pollution problems in local groundwater.

A documented example of gasoline pollution in groundwater occurred in Glendale, California, in 1968.[54,99] An estimated 400 m^3 of gasoline were discovered floating up to 0.75 m deep on the water table. The gasoline was initially removed by bailing and by skimmer pumps. Subsequently, pumping wells created drawdown cones to contain the leakage, as shown in Fig. 8.5b, and to remove a mixture of gasoline and water.

Mining Activities.   Mines can produce a variety of groundwater pollution problems.[9,26,59] Pollution depends on the material being extracted and the milling process: coal, phosphate, and uranium mines are major contributors; metallic ores for production of iron, copper, zinc, and lead are also important; stone, sand, and gravel quarries, although numerous, are chemically much less important. Both surface and underground mines invariably extend below the water table so that dewatering to expand mining is common. Water so pumped may be highly mineralized and is frequently referred to as *acid mine drainage*. Normal characteristics include low pH and high iron, aluminum, and sulfate.[58]

Coal deposits are often associated with pyrite (FeS_2). This is stable for conditions below the water table, but if the water table is lowered, oxidation occurs. Oxidation of pyrite followed by contact with water produces ferrous sulfate ($FeSO_4$) and sulfuric acid (H_2SO_4) in solution; groundwater intermingling with this water will have a reduced pH and an increase in iron and sulfate contents. Pollution of groundwater can also result from the leaching of old mine tailings and settling ponds; therefore, pollution problems can be associated with both active and abandoned mines.

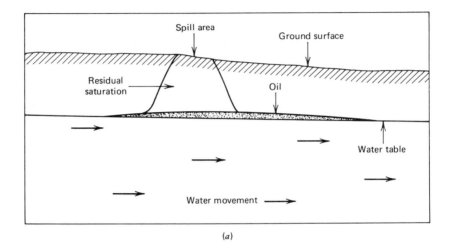

(a)

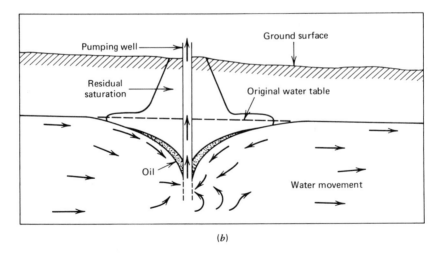

(b)

Fig. 8.5 Oil pollution of groundwater and control by a pumping well. (a) Fluid oil floating on water table and elongating in direction of groundwater flow. (b) Oil on water table contained in cone of depression created by pumping well (after Engrng. and Tech. Research Comm.[27]).

Oil-field Brines. The production of oil and gas is usually accompanied by substantial discharges of wastewater in the form of brine[41]. Constituents of brine include sodium, calcium, ammonia, boron, chloride, sulfate, trace metals, and high total dissolved solids. In the past oil-field brine disposal was handled by discharge to streams or "evaporation ponds." In both instances brine-polluted aquifers became commonplace in oil production areas as the infiltrating water reached the underlying groundwater. Today, such disposal methods are prohibited by most regulatory agencies; however, regulation is often ineffective so that many brine-affected areas remain and will persist for years into the future.[82] Figure 8.6 provides a graphic example of aquifer pollution from an oil-field brine disposal pit.

Oil and gas producers now inject most brines through wells into deep formations that are geologically isolated from overlying freshwater aquifers. Properly designed injection wells contain an injection tubing inside the casing to prevent ruptures and to facilitate the detection of leaks.[95] Even so, brine disposal can cause pollution because surrounding abandoned and unplugged oil and gas wells and test holes provide vertical pathways for injected brines to rise into overlying aquifers.

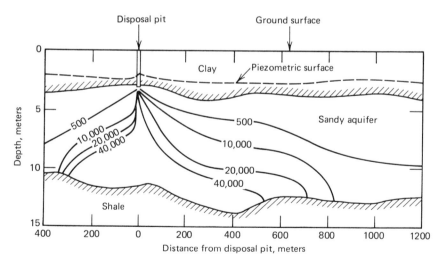

Fig. 8.6 Distribution of saline water in confined aquifer resulting from an oil-field brine disposal pit in southwestern Arkansas. Numbered lines are isochlors in mg/l (after Fryberger[33]).

Agricultural Sources and Causes

Irrigation Return Flows.   Approximately one-half to two-thirds of the water applied for irrigation of crops is consumed by evapotranspiration; the remainder, termed *irrigation return flow,* drains to surface channels or joins the underlying groundwater. Irrigation increases the salinity of irrigation return flow from three to ten times that of the applied water.[40] The degradation results from the addition of salts by dissolution during the irrigation process, from salts added as fertilizers or soil amendments, and from the concentration of salts by evapotranspiration (see Fig. 8.7). Principal cations include calcium, magnesium, and sodium; major anions include bicarbonate, sulfate, chloride and nitrate. Because irrigation is the primary use for water in arid and semiarid regions, irrigation return flow can be the major cause of groundwater pollution in such regions.[34]

Animal Wastes.   Where animals are confined within a limited area, as for beef or milk production, large amounts of wastes are deposited on the ground. Thus, for the 120 to 150 days that a beef animal remains in a feedlot, it will produce over a half-ton of manure on a dry-weight basis. With thousands of animals in a single feedlot, the natural assimilative capacity of the soil can become overtaxed. Storm runoff in contact with the manure carries highly concentrated pollutants to surface and subsurface waters. Animal wastes may transport salts, organic loads, and bacteria into the soil. Nitrate-nitrogen is the most important persistent pollutant that may reach the water table.[2,36,76]

Fertilizers and Soil Amendments.   When fertilizers are applied to agricultural land, a portion usually leaches through the soil and to the water table. The primary fertilizers are compounds of nitrogen, phosphorus, and potassium. Phosphate and potassium fertilizers are readily adsorbed on soil particles and seldom constitute a pollution problem. But nitrogen in solution is only partially used by plants or adsorbed by the soils, and it is the primary fertilizer pollutant. Fertilizers are extensively used and will undoubtedly increase in the future.

Soil amendments are applied to irrigated lands to alter the physical or chemical properties of the soil. Lime, gypsum, and sulfur are widely used for this purpose; substantial amounts of these soil amendments may eventually leach to the groundwater, thereby increasing its salinity.

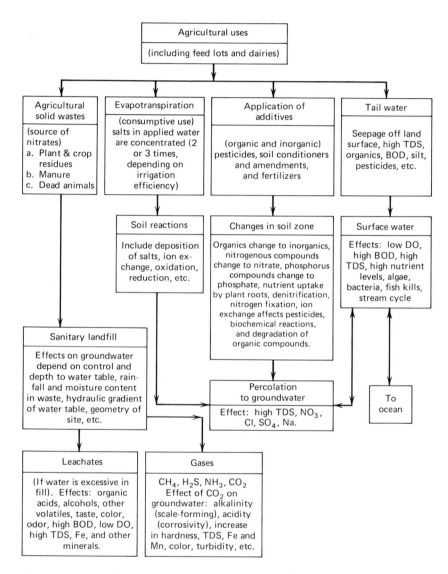

Fig. 8.7 Agricultural uses of water and their effects on water quality (after Hassan[38]).

Pesticides. Pesticides* can be significant in agricultural areas as a diffuse source of groundwater pollution. The presence of these materials in groundwater, even in minute concentrations, can have serious consequences in relation to the potability of the water. The impact of pesticides on groundwater quality depends on the properties of the pesticide residue, rainfall or irrigation rates, and soil characteristics.[13,87] Most pesticides are relatively insoluble in water, while others are readily adsorbed by soil particles or are subject to microbial degradation.[16] Analytic tests of water in California during 1979 revealed that some 100 water supply wells contained trace amounts of DBCP (dibromochloropropane), formerly a widely used pesticide and a suspected carcinogenic compound.

Miscellaneous Sources and Causes

Spills and Surface Discharges. Liquids discharged onto the ground surface in an uncontrolled manner can migrate downward to degrade groundwater quality. At industrial sites causal activities may include boilovers, losses during transfers of liquids, leaks from pipes and valves, and inadequate control of wastes and storm runoff. Washing aircraft with solvents and spills of fuel at airports can form a layer of hydrocarbons floating on the water table. Pollution can also occur from the intermittent dumping of fluids on the ground, especially near gasoline stations, small commercial establishments, and construction sites. It has been estimated that millions of gallons of automobile waste oil are discharged on the ground surface annually.[58] Finally, accidents involving aboveground pipelines, storage tanks, railroad cars, and trucks can release large quantities of a pollutant at a particular site. Hazardous and flammable liquids are often flushed by water from highways; this action may actually aid in transporting the pollutant to the water table.

Stockpiles. Solid materials are frequently stockpiled near industrial plants, construction sites, and large agricultural operations. These may be raw materials awaiting use, or they may be solid wastes placed for temporary or permanent storage. Precipitation falling on unsheltered stockpiles causes leaching to occur into the soil; this may transport heavy metals, salts, and other inorganic and organic constituents as pollutants to the groundwater.

---

*The term *pesticide* is here broadly interpreted to embrace any chemical applied to control, destroy, or mitigate pests.

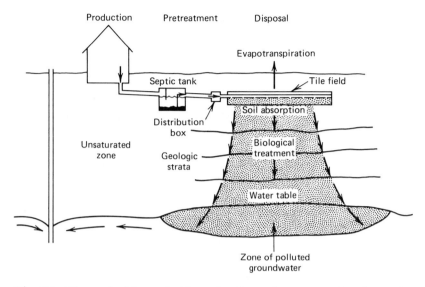

Fig. 8.8 Disposal of household wastes through a conventional septic tank system (after Miller[58]).

Septic Tanks and Cesspools. The most numerous and widely distributed potential sources of groundwater pollution are septic tanks and cesspools. Approximately 40 million persons in the United States are served by individual household wastewater treatment systems. It follows that some 2.5 billion gallons of partially treated sewage is discharged from residences directly into the ground every day.[89] In addition, commercial establishments, hospitals, industrial plants, and resorts employ septic tanks in areas where community sewer systems are not available.* A septic tank is a watertight basin intended to separate floating and settleable solids from the liquid fraction of domestic sewage and to discharge this liquid with its dissolved and suspended solids into the biologically active zone of the soil mantle through a subsurface percolation system such as a tile field, a seepage bed, or a sand-covered sand filter (see Fig. 8.8). A cesspool is a large buried chamber with porous walls designed to

*In the United States, the largest concentrations of septic systems, and hence the largest potential areas for groundwater pollution from this source, are located in suburban residential areas constructed shortly after World War II. Through inadequate knowledge and control at that time, septic tanks with badly constructed percolation systems were installed on small lots in subdivisions, sometimes containing thousands of houses.[57]

receive and percolate raw sewage. Domestic sewage adds minerals to groundwater, as indicated in Table 8.3. Bacteria and viruses are normally removed by the soil system; phosphorus is generally retained by the soil, but significant quantities of nitrogen can be added to groundwater.[74]

Roadway Deicing. Pollution of groundwater results from the application of deicing salts to streets and highways in winter.[30,85] Urban areas with winter temperatures below freezing are affected the most. Salt reaches the groundwater in solution after spreading on roadways and also from stockpiles. Salt application rates range from 2 to 11 m-tons per single-lane kilometer per winter season. Sodium chloride accounts for most of applications, with calcium chloride the remainder. A steady growth in roadway deicing has stemmed from the demand to maintain streets and highways in safe driving condition throughout the winter. The salts have produced widespread degradation of groundwater quality and have also hastened the corrosion of wells.

Illustrative of the effect is the dramatic increase in chloride measured in a well in Burlington, Massachusetts, shown in Fig. 8.9. When

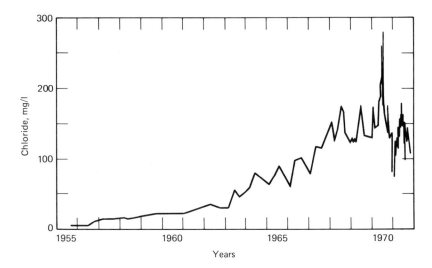

Fig. 8.9 Time variation of chloride in groundwater at Burlington, Massachusetts, due to local roadway deicing (after Terry[85]; reprinted with permission from *Road Salt, Drinking Water, and Safety,* copyright © 1974, Ballinger Publishing Company).

chloride exceeded the 250 mg/l drinking water standard in 1970, local use of deicing chemicals on city streets was banned.

Saline Water Intrusion.   Salt water may invade freshwater aquifers to create point or diffuse pollution sources. In coastal aquifers seawater is the pollutant, while in inland aquifers underlying saline water may be responsible. Because of the worldwide importance of saline water intrusion, its occurrence, mechanisms, and control methods are discussed more fully in Chapter 14.

Interchange through Wells.   Because wells form highly permeable vertical connections between aquifers, they can serve as avenues for groundwater pollution where inadequate attention is given to the proper construction, sealing, or abandonment of wells. Pollution occurs where there is incomplete hydraulic separation within a well and where a vertical difference in hydraulic head exists between two aquifers. Two flow conditions with a well connecting an unconfined aquifer with a confined aquifer are illustrated in Fig. 8.10. In addition, if the top of a well is not constructed so as to divert surface water away from a well, it can admit a wide variety of surface pollutants. Although governmental regulations in most locations require plugging of abandoned wells, many such wells remain unplugged and serve as ongoing interchanges between aquifers.

Surface Water.   Polluted surface water bodies that contribute to groundwater recharge become sources of groundwater pollution. The recharge may occur naturally from a losing stream, or it may be induced by a nearby pumping well, as indicated in Fig. 8.11. Because many municipal water supply wells are located adjacent to rivers to ensure adequate flows, they can serve as important groundwater pollution mechanisms.

Attenuation of Pollution

Pollutants in groundwater tend to be removed or reduced in concentration with time and with distance traveled. Mechanisms involved include filtration, sorption, chemical processes, microbiological decomposition, and dilution. The rate of pollution attenuation depends on the type of pollutant and on the local hydrogeologic situation.[16,64,89] Attenuation mechanisms tend to localize groundwater pollution near its source; they also are responsible for the interest in groundwater recharge as a water reclamation technique.

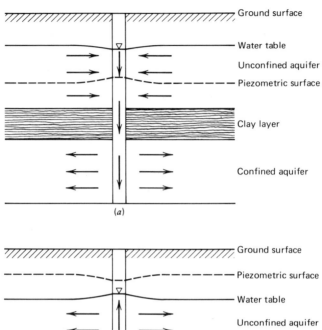

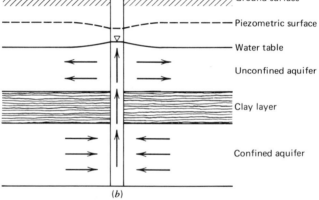

Fig. 8.10 Diagrams showing aquifer leakage by
vertical movement of water through a nonpumping
well. (*a*) Water table above the piezometric surface.
(*b*) Piezometric surface above the water table (after
Todd, et al.[89]).

Filtration. Filtration removes suspended materials; hence, this
action is most important at ground surface where polluted surface
water is infiltrating into the ground. In groundwater filtration can
remove particulate forms of iron and manganese as well as precipitates formed by chemical reactions.

Sorption. Sorption serves as a major mechanism for attenuating
groundwater pollution. Clays, metallic oxides and hydroxides, and
organic matter function as sorptive materials. Most pollutants can be

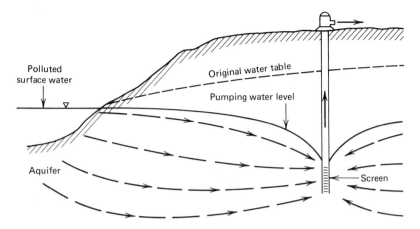

Fig. 8.11   Diagram showing how polluted water can be induced to flow from a surface stream to a pumping well (after Deutsch[21]).

sorbed under favorable conditions with the general exception of chloride and to a lesser extent, nitrate and sulfate. The sorption process depends on the type of pollutant and the physical and chemical properties of both the solution and the subsurface materials; a substantial clay content in the strata above the water table is a key factor. The sorptive capacity of geologic materials is finite for most inorganic substances; however, for biodegradable substances such as bacteria and ammonia, the sorptive capacity may be renewed indefinitely.

Chemical Processes. Precipitation in groundwater can occur where appropriate ions are in solution in sufficient quantities. The most important precipitation reactions for the major constituents involve calcium, magnesium, bicarbonate, and sulfate. Trace elements having important precipitation potential include arsenic, barium, cadmium, copper, cyanide, fluoride, iron, lead, mercury, molybdenum, radium, and zinc. In arid regions, where moisture in the near-surface zone may be minimal, chemical precipitation becomes a major attenuation mechanism.

In the zone above the water table, oxidation of organic matter acts as an important attenuation mechanism. A generalized concept of aerobic organic matter decomposition is shown in Fig. 8.12. Complex organic compounds are oxidized stepwise to more simple organic compounds until CO_2 and H_2O are formed along with numerous inorganic ions and compounds. Both oxidation and reduction reactions can occur underground in conjunction with other mech-

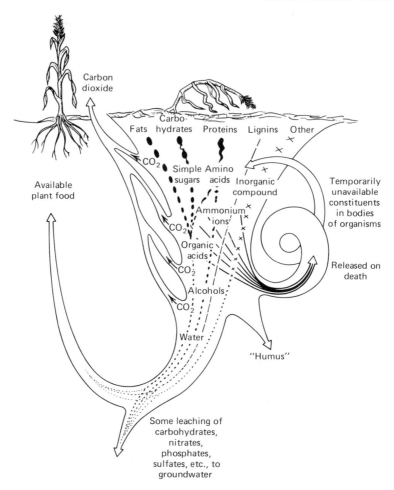

Fig. 8.12 Schematic diagram of organic matter decomposition in the soil (after Flack[31]; courtesy Burns Sabey, Colorado State University).

anisms, leading to precipitates, deposits of insoluble trace metals, and gases. Volatilization and loss as a gas apply particularly to reactions involving nitrate and sulfate. Radioactive decay, based on the half-life of a radioisotope, acts as an attenuation mechanism for radioactive pollutants.

Microbiological Decomposition. Most pathogenic microorganisms in the soil do not flourish in the soil and hence are subject to ultimate destruction, the timing of which depends on different species and environmental conditions.[23] Bacteria and viruses as particulate matter suspended in water tend to move slower through a porous media than water. Field studies indicate that these pathogens are largely removed by passage through as little as one meter of soil, provided reasonable amounts of silt and clay are present.[22,78]

Dilution. Pollutants in groundwater flowing through porous media tend to become diluted in concentration due to hydrodynamic dispersion occurring on both microscopic and macroscopic scales (see Chapter 3). These mixing mechanisms produce a longitudinal and lateral spreading of a pollutant within the groundwater so that the volume affected increases and the concentration decreases with distance traveled. Dilution is the most important attenuation mechanism for pollutants after they reach the water table.

In summary, attenuation above the water table is generally effective for most pollutants with the exception of the major inorganic constituents. Other exceptions include boron and tritium, as well as some trace elements and organic chemicals. Maximum attenuation requires an adequate distance to the water table and the presence of fine-grained geologic materials such as silt and clay. Without these conditions almost any pollutant can be introduced directly into the saturated zone. Pollutant attenuation below the water table occurs more slowly with dilution serving as the principal mechanism.

Distribution of Pollution Underground

In a large region or basin, bodies of polluted groundwater can be visualized as innumerable scattered dots on a map, while in some areas diffuse sources, such as from irrigation return flows, would create areas with appreciable horizontal extent. Entry of pollutants into shallow aquifers occurs by percolation from ground surface, through wells, from surface waters, and by saline water intrusion. The extent of pollution in groundwater from a point source decreases as pollutants move away from the source until a harmless or very low concentration level is reached. Because each constituent of a pollution source may have a different attenuation rate, the distance to which pollutants travel will vary with each quality component.

A hypothetical example of a waste-disposal site is shown in Fig. 8.13 with groundwater flowing toward a river. Zones *A, B, C, D,* and *E* represent essentially stable limits for different contaminants resulting from the steady release of wastes of unchanging composition. Pollutants, once entrained in the saturated groundwater flow, tend

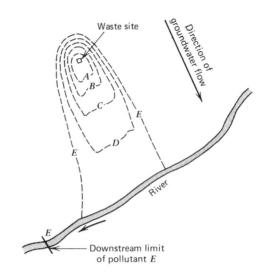

Fig. 8.13 Plan view of an unconfined aquifer showing areal extent to which various pollutants of mixed wastes at a disposal site disperse and move to insignificant levels (after LeGrand[49]).

to form *plumes* (analogous to smoke from a smokestack as it drifts downwind in the atmosphere) of polluted water extending downstream from the pollution source until they attenuate to a minimum quality level. Only zone E reaches the river in Fig. 8.13 and is subsequently diluted by surface water.

The shape and size of a plume depend on the local geology, the groundwater flow, the type and concentration of pollutant, the continuity of waste disposal, and any human modifications of the groundwater system, such as pumping wells.[49] Where groundwater is moving relatively rapidly, a plume from a point source tends to be long and thin, such as that shown in Fig. 8.14, but where the flow rate is low, the pollutant tends to spread more laterally to form a wider plume. Irregularly shaped plumes can be created by local influences such as pumping wells and nonuniformities in permeability.

Plumes tend to become stable areas where there is a constant input of waste into the ground. This occurs for two reasons: enlargement as pollutants continue to be added at a point source is counterbalanced by attenuation mechanisms, or the pollutant reaches a location of groundwater discharge, such as a stream, and emerges from the underground. When a waste is first released into groundwater, the plume expands until a quasi-equilibrium stage is reached. If sorption is important, a steady inflow of pollution will cause a slow expansion of the plume as the earth materials within it reach a sorption capability limit.

An approximately stable plume will expand or contract generally in response to changes in the rate of waste discharge. Figure 8.15

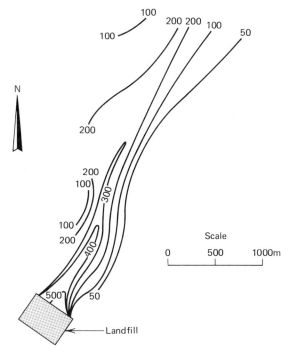

Fig. 8.14 Plume of groundwater pollution from a landfill near Munich, West Germany. Numbered lines are isochlors in mg/l (after Cole[18]).

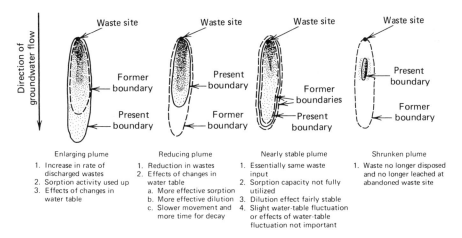

Fig. 8.15 Changes in groundwater pollution plumes and causal factors (after LeGrand[49]).

shows changes in plumes that can be anticipated from variations in waste inputs.

An important aspect of groundwater pollution is the fact that it may persist underground for years, decades, or even centuries. This is in marked contrast to surface water pollution. Reclaiming polluted groundwater is usually much more difficult, time consuming, and expensive than reclaiming polluted surface water. Underground pollution control is achieved primarily by regulating the pollution source, and secondarily by physically entrapping and, when feasible, removing the polluted water from the underground.

Evaluation of Pollution Potential

To provide guidance in evaluating the potential pollution from a given source, LeGrand[50] developed an empirical point-count system. The concept is applicable to waste disposal sites and to wells; water table aquifers are presumed. The physical factors considered to influence pollution include: depth to water table, sorption above the water table, aquifer permeability, water table gradient, and horizontal distance. The rating chart shown in Fig. 8.16 illustrates the evaluation procedure for these factors at sites in unconsolidated alluvial materials. A numerical value is read above the line for each of the five factors, based on the corresponding data below the line.

The measure of pollution potential is given by the sum of the numerical ratings of the five factors in Fig. 8.16. Total point values may be interpreted in terms of possibility of pollution as follows:

Total Points	Possibility of Pollution
0–4	Imminent
4–8	Probable or possible
8–12	Possible but not likely
12–25	Very improbable
25–35	Impossible

Although this procedure is imperfect, it has the advantages of providing quantitative evaluations and of permitting relative comparisons of potential pollution from alternative waste disposal sites.

Monitoring Groundwater Quality

To protect a groundwater resource against pollution, a water quality monitoring program—defined as a scientifically designed surveillance system of continuing measurements, observation, and

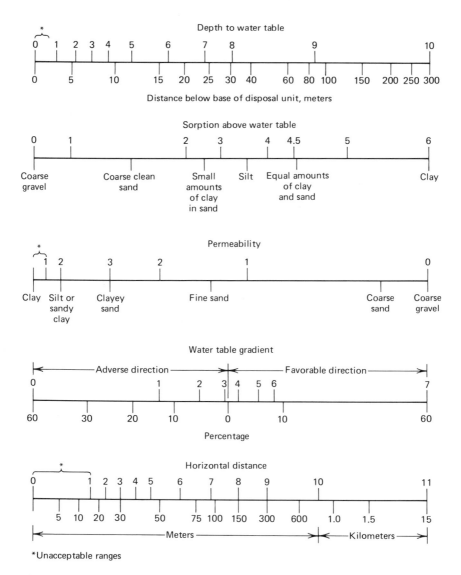

Fig. 8.16 Chart for evaluating the pollution potential of unconfined aquifers consisting of unconsolidated alluvium (after LeGrand[50]; reprinted from *Journal American Water Works Association*, Vol. 56, by permission of the Association; copyright © 1964 by American Water Works Association, 6666 West Quincy Avenue, Denver, Col. 80235).

evaluations—is necessary. Monitoring methods may include not only sampling and analyses of groundwater quality, but also determination of groundwater levels and flow directions, measurements of moisture in the unsaturated zone, geophysical surveys, evaluations of wastes and other materials contributing to subsurface pollution, testing of pipelines and tanks, and aerial surveillance.[28]

A generalized methodology for monitoring groundwater quality has been prepared.[86,88] Its purpose is to assist a local governmental agency in implementing a monitoring program. The procedure involves a series of action steps arranged in chronological order.

1. Select the area or basin to be monitored.
2. Identify all sources and causes of pollution.
3. Identify potential pollutants in the sources and causes.
4. Define groundwater usage in terms of location, type of use, and quantity.
5. Define the local hydrogeologic situation.
6. Evaluate the existing groundwater quality.
7. Evaluate the infiltration rate of pollutants at ground surface.
8. Evaluate the mobility of pollutants from ground surface to the water table.
9. Evaluate the attenuation of pollutants within the saturated zone.
10. Prioritize the sources and causes of pollution in terms of their importance or potential impact on groundwater quality.
11. Determine monitoring activities already in existence.
12. Determine methods, locations, and frequencies for monitoring.
13. Select and implement the monitoring program on a prioritized basis.
14. Review and interpret monitoring data.
15. Summarize and transmit monitoring information to appropriate public agencies and private organizations.

The above procedure is flexible, allowing for changes over time in land use and water use and also for variations in financial resources to implement and improve such a monitoring program. Specialized monitoring techniques for injection wells and mining activities have been described by Warner.[95,96]

References

1. Abegglen, D. E., et al., *The effects of drain wells on the ground-water quality of the Snake River Plain,* Pamphlet 148, Idaho Bur. Mines and Geol., Moscow, 51 pp., 1970.
2. Adriano, D. C., et al., Nitrate and salt in soils and ground waters from land disposal of dairy manure, *Soil Sci. Soc. Amer. Proc.,* v. 35, pp. 759–762, 1971.
3. Anderson, J. R., and J. N. Dornbush, Influence of sanitary landfill on ground water quality, *Jour. Amer. Water Works Assoc.,* v. 59, pp. 457–470, 1967.

4. Anon., Ground water pollution, *Water Well Jour.*, v. 24, no. 7, pp. 31–61, 1970.

5. Anon., *Ground water pollution from subsurface excavations,* Rept. EPA-430/9-73-012, U.S. Environmental Protection Agency, Washington, D.C., 224 pp., 1973.

6. Apgar, M. A., and D. Langmuir, Ground-water pollution potential of a landfill above the water table, *Ground Water,* v. 9, no. 6, pp. 76–96, 1971.

7. Bader, J. S., et al., *Selected references—Ground-water contamination, The United States of America and Puerto Rico,* U.S. Geological Survey, Washington, D.C., 103 pp., 1973.

8. Ballentine, R. K., et al., *Subsurface pollution problems in the United States,* Tech. Studies Rept. TS-00-72-02, U.S. Environmental Protection Agency, Washington, D.C., 29 pp., 1972.

9. Barnes, I., and F. E. Clarke, Geochemistry of ground water in mine drainage problems, *U.S. Geological Survey Prof. Paper* 473-A, 6 pp., 1964.

10. Born, S. M., and D. A. Stephenson, Hydrogeologic considerations in liquid waste disposal, Jour. Soil Water Conservation, v. 24, no. 2, pp. 52–55, 1969.

11. Brown, R. H., Hydrologic factors pertinent to ground-water contamination, *Ground Water,* v. 2, no. 1, pp. 5–12, 1964.

12. Burt, E. M., The use, abuse, and recovery of a glacial aquifer, *Ground Water,* v. 10, no. 1, pp. 65–72, 1972.

13. California Dept. Water Resources, *The fate of pesticides applied to irrigated agricultural land,* Bull. 174-1, Sacramento, 30 pp., 1968.

14. California Dept. Water Resources, *Sanitary landfill studies: Appendix A—Summary of selected previous investigations,* Bull. 147-5, Sacramento, 115 pp., 1969.

15. California Dept. Water Resources, *Feasibility of reclamation of water from wastes in the Los Angeles metropolitan area,* Bull. 80, Sacramento, 183 pp., 1961.

16. California State Water Resources Control Bd., Dept. Water Resources, and Dept. Health, *A "state-of-the-art" review of health aspects of wastewater reclamation for groundwater recharge,* Water Information Center, Huntington, N.Y., 240 pp., 1978.

17. Cameron, R. D., The effects of solid waste landfill leachates on receiving waters, *Jour. Amer. Water Works Assoc.,* vol. 70, pp. 173–176, 1978.

18. Cole, J. A. (ed.), *Groundwater pollution in Europe,* Water Information Center, Port Washington, N.Y., 347 pp., 1975.

19. Committee for Hydrological Research TNO, *Groundwater pollution,* Proc. Tech. Meeting 31, The Hague, Netherlands, 106 pp., 1976.

20. Dallaire, G., EPA's hazardous-waste program: will it save our groundwater?, *Civil Engrng.,* v. 48, no. 12, pp. 39–45, 1978.

21. Deutsch, M., Groundwater contamination and legal controls in Michigan, *U.S. Geological Survey Water-Supply Paper* 1691, 79 pp., 1963.

22. Drewry, W. A., and R. Eliassen, Virus movement in ground water, *Jour. Water Poll. Control Fed.,* v. 40, pp. 257–272, 1968.

23. Dunlap, W. J., and J. F. McNabb., *Subsurface biological activity in*

relation to ground water pollution, Rept. EPA-660/2-73-014, U.S. Environmental Protection Agency, Corvallis, Oreg., 60 pp., 1973.

24. Dunlap, W. J., et al., Probable impact of NTA on ground water, *Ground Water,* v. 10, no. 1, pp. 107–116, 1972.

25. Dunlap, W. J., et al., *Sampling for organic chemicals and microorganisms in the subsurface,* EPA-600/2-77-176, U.S. Environmental Protection Agency, Ada, Okla., 27 pp., 1977.

26. Emrich, G. H., and G. L. Merritt, Effects of mine drainage on ground water, *Ground Water,* v. 7, no. 3, pp. 27–32, 1969.

27. Engineering and Technical Research Committee, *The migration of petroleum products in soil and ground water,* Publ. no. 4149, Amer. Petroleum Inst., 36 pp., 1972.

28. Everett, L. G., et al., *Monitoring groundwater quality: methods and costs,* Rept. EPA-600/4-76-023, U.S. Environmental Protection Agency, Las Vegas, 140 pp., 1976.

29. Fenn, D., et al., *Procedures manual for ground water monitoring at solid waste disposal facilities,* Rept. EPA/530/SW-611, U.S. Environmental Protection Agency, Cincinnati, 269 pp., 1977.

30. Field, R., et al., *Water pollution and associated effects from street salting,* Rept. EPA-R2-73-257, U.S. Environmental Protection Agency, Cincinnati, 48 pp., 1973.

31. Flack, J. E. (ed.), *Proceedings of the symposium on land treatment of secondary effluent,* Info. Ser. no. 9, Environmental Resources Center, Colorado State Univ., Fort Collins, 257 pp., 1973.

32. Fried, J. J., *Groundwater pollution, theory, methodology, modelling and practical rules,* Elsevier, New York, 330 pp., 1975.

33. Fryberger, J. S., Investigation and rehabilitation of a brine-contaminated aquifer, *Ground Water,* v. 13, no. 2, pp. 155–160, 1975.

34. Fuhriman, D. K., and J. R. Barton, *Groundwater pollution in Arizona, California, Nevada, and Utah,* Water Poll. Cont. Research Ser. 16060 ERU, U.S. Environmental Protection Agency, Washington, D.C., 249 pp., 1971.

35. Geraghty, J. J., and D. W. Miller, Status of groundwater contamination in the U.S., *Jour. Amer. Water Works Assoc.,* v. 70, pp. 162–167, 1978.

36. Gillham, R. W., and L. R. Webber, Nitrogen contamination of groundwater by barnyard leachates, *Jour. Water Poll. Control Fed.,* v. 41, no. 10, pp. 1752–1762, 1969.

37. Hajek, B. F., Chemical interactions of wastewater in a soil environment, *Jour. Water Poll. Control Fed.,* v. 41, no. 10, pp. 1775–1786, 1969.

38. Hassan, A. A., Water quality cycle—Reflection of activities of nature and man, *Ground Water,* v. 12, no. 1, pp. 16–21, 1974.

39. Hughes, G., et al., *Pollution of groundwater due to municipal dumps,* Tech. Bull. no. 42, Canada Dept. Energy, Mines and Resources, Inland Waters Branch, Ottawa, 98 pp., 1971.

40. Jenke, A. L., *Evaluation of salinity created by irrigation return flows,* Rept. EPA 430/9-74-006, U.S. Environmental Protection Agency, Washington, D.C., 128 pp., 1974.

41. Karubian, J. F., *Polluted groundwater: estimating the effects of man's activities,* Rept. EPA 600/4-74-002, U.S. Environmental Protection Agency, Washington, D.C., 99 pp., 1974.
42. Kaufman, M. I., Subsurface wastewater injection, Florida, *Jour. Irrig. Drain. Div.,* Amer. Soc. Civil Engrs., v. 99, no. IR1, pp. 53-70, 1973.
43. Kaufman, W. J., Chemical pollution of ground waters, *Jour. Amer. Water Works Assoc.,* vol. 66, pp. 152-159, 1974.
44. Kimmel, G. E., and O. C. Braids, Leachate plumes in a highly permeable aquifer, *Ground Water,* v. 12, no. 6, pp. 388-393, 1974.
45. Kirkham, D., and S. B. Affleck, Solute travel times to wells, *Ground Water,* v. 15, pp. 231-242, 1977.
46. Klaer, F. H., Jr., Bacteriological and chemical factors in induced infiltration, *Ground Water,* v. 1, no. 1, pp. 38-43, 1963.
47. Knowles, D. B., Hydrologic aspects of the disposal of oil-field brines in Alabama, *Ground Water,* v. 3, no. 2, pp. 22-27, 1965.
48. LeGrand, H. E., Environmental framework of ground-water contamination, *Ground Water,* v. 3, no. 2, pp. 11-15, 1965.
49. LeGrand, H. E., Patterns of contaminated zones of water in the ground, *Water Resources Research,* v. 1, pp. 83-95, 1965.
50. LeGrand, H. E., System for evaluation of contamination potential of some waste disposal sites, *Jour. Amer. Water Works Assoc.,* v. 56, pp. 959-974, 1964.
51. Matis, J. R., Petroleum contamination of ground water in Maryland, *Ground Water,* v. 9, no. 6, pp. 57-61, 1971.
52. Maxey, G. B., and R. N. Farvolden, Hydrogeologic factors in problems of contamination in arid lands, *Ground Water,* v. 3, no. 4, pp. 29-32, 1965.
53. McGauhey, P. H., *Engineering management of water quality,* McGraw-Hill, New York, 295 pp., 1968.
54. McKee, J. E., et al., Gasoline in groundwater, *Jour. Water Poll. Control Fed.,* v. 44, no. 2, pp. 293-302, 1972.
55. McMichael, F. C., and J. E. McKee, *Wastewater reclamation at Whittier Narrows,* Publ. no. 33, California State Quality Control Board, Sacramento, 100 pp., 1966.
56. McNabb, J. F., et al., *Nutrient, bacterial, and virus control as related to ground-water contamination,* Rept. EPA-600/8-77-010, U.S. Environmental Protection Agency, Ada, Okla., 18 pp., 1977.
57. Meyer, C. F. (ed.), *Polluted groundwater: some causes, effects, controls, and monitoring,* Rept. EPA-600/4-73-0016, U.S. Environmental Protection Agency, Washington, D.C., 282 pp., 1973.
58. Miller, D. W. (ed.), *Waste disposal effects on ground water,* Premier Press, Berkeley, Calif., 512 pp., 1980.
59. Mink, L. L., et al., Effect of early day mining operations on present day water quality, *Ground Water,* v. 10, no. 1, pp. 17-26, 1972.
60. National Water Well Assoc., *Proceedings of the Third National Ground Water Quality Symposium,* EPA-600/9-77-014, U.S. Environmental Protection Agency, Ada, Okla., 232 pp., 1977.

61. Nelson, R. W., Evaluating the environmental consequences of ground-water contamination, *Water Resources Research,* v. 14, pp. 409–450, 1978.

62. New York State Dept. Health, *The Long Island ground water pollution study,* New York State Dept. Environmental Conservation, Albany, 396 pp., 1972.

63. Osgood, J. O., Hydrocarbon dispersion in ground water: significance and characteristics, *Ground Water,* v. 12, no. 6, pp. 427–438, 1974.

64. Palmquist, R., and L. V. A. Sendlein, The configuration of contamination enclaves from refuse disposal sites on floodplains, *Ground Water,* v. 13, pp. 167–181, 1975.

65. Page, H. G., et al., Behavior of detergents (ABS), bacteria, and dissolved solids in water-saturated soils, *U.S. Geological Survey Prof. Paper* 450-E, pp. 179–181, 1963.

66. Perlmutter, N. M., and A. A. Guerrera, Detergents and associated contaminants in ground water at three public-supply well fields in southwestern Suffolk County, Long Island, N.Y., *U.S. Geological Survey Water-Supply Paper* 2001-B, 22 pp., 1970.

67. Perlmutter, N. M., and M. Lieber, Dispersal of plating wastes and sewage contaminants in ground water and surface water, South Farmingdale–Massapequa area, Nassau County, N.Y., *U.S. Geological Survey Water-Supply Paper* 1879-G, 67 pp., 1970.

68. Perlmutter, N. M., and E. Koch, Preliminary hydrogeologic appraisal of nitrate in ground water and streams, Southern Nassau County, Long Island, New York, *U.S. Geological Survey Prof. Paper* 800-P, pp. 225–235, 1972.

69. Phillips, K. J., and L. W. Gelhar, Contaminant transport to deep wells, *Jour. Hydraulics Div.,* Amer. Soc. Civil Engrs., v. 104, no. HY6, pp. 807–819, 1978.

70. Piper, A. M., Disposal of liquid wastes by injection underground—neither myth nor millennium, *U.S. Geological Survey Circ.* 631, 15 pp., 1969.

71. Pojasek, R. B., (ed.), *Drinking water quality enhancement through source protection,* Ann Arbor Science, Ann Arbor, 614 pp., 1977.

72. Public Health Service, *Ground water contamination,* Proceedings of the 1961 symposium, Tech. Rept. W61-5, U.S. Dept. of Health, Education, and Welfare, Cincinnati, 218 pp., 1961.

73. Qasim, S. R., and J. C. Burchinal, Leaching of pollutants from refuse beds, *Jour. San. Engrng. Div.,* Amer. Soc. Civil Engrs., v. 96, no. SA1, pp. 49–58, 1970.

74. Quan, E. L., et al., Subsurface sewage disposal and contamination of ground water in East Portland, Oregon, *Ground Water,* v. 12, no. 6, pp. 356–368, 1974.

75. Rima, D. R., et al., Subsurface waste disposal by means of wells—selective annotated bibliography, *U.S. Geological Survey Water-Supply Paper* 2020, 305 pp., 1971.

76. Robbins, J. W. D., and G. J. Kriz, Relation of agriculture to groundwater pollution: a review, *Trans. Amer. Soc. Agric. Engrs.*, v. 12, pp. 397–403, 1969.

77. Robertson, J. M., et al., *Organic compounds entering ground water from a landfill,* Rept. EPA-660/2-74-077, U.S. Environmental Protection Agency, Washington, D.C., 47 pp., 1974.

78. Romero, J. C., The movement of bacteria and virus through porous media, *Ground Water,* v. 8, no. 2, pp. 37–48, 1970.

79. Runnells, D. D., Wastewaters in the vadose zone of arid regions: geochemical reactions, *Ground Water,* vol. 14, pp. 374–385, 1976.

80. Sasman, R. T., Thermal pollution of ground water by artificial recharge, *Water and Sewage Works,* v. 119, no. 12, pp. 52–55, 1972.

81. Scalf, M. R., et al., Movement of DDT and nitrates during ground-water recharge, *Water Resources Research,* v. 5, pp. 1041–1052, 1969.

82. Scalf, M. R., et al., *Ground water pollution in the South Central States,* Rept. EPA-R2-73-268, U.S. Environmental Protection Agency, Corvallis, Oreg., 181 pp., 1973.

83. Scalf, M. R., et al., *Environmental effects of septic tank systems,* Rept. EPA-600/3-77-096, U.S. Environmental Protection Agency, Ada, Okla., 35 pp., 1977.

84. Schmidt, K. D., Nitrate in ground water of the Fresno-Clovis metropolitan area, California, *Ground Water,* v. 10, no. 1, pp. 50–64, 1972.

85. Terry, R. C., Jr., *Road salt, drinking water, and safety,* Ballinger, Cambridge, Mass., 161 pp., 1974.

86. Tinlin, R. M. (ed.), *Monitoring groundwater quality: illustrative examples,* Rept. EPA-600/4-76-036, U.S. Environmental Protection Agency, Las Vegas, 81 pp., 1976.

87. Todd, D. K., and D. E. O. McNulty, *Polluted groundwater,* Water Information Center, Port Washington, N.Y., 179 pp., 1976.

88. Todd, D. K., et al., A groundwater quality monitoring methodology, *Jour. Amer. Water Works Assoc.,* v. 68, pp. 586–593, 1976.

89. Todd, D. K., et al., *Monitoring groundwater quality: monitoring methodology,* Rept. EPA-600/4-76-026, U.S. Environmental Protection Agency, Las Vegas, 154 pp., 1976.

90. U.S. Environmental Protection Agency, *Land application of wastewater,* Rept. EPA 903/9-75-017, Philadelphia, 94 pp., 1975.

91. U.S. Environmental Protection Agency, *Subsurface water pollution—a selective annotated bibliography,* Pt. I—Subsurface waste injection, Pt. II—Saline water intrusion, Pt. III—Percolation from surface sources, Washington, D.C., 479 pp., 1972.

92. Utah State Univ. Fndn., *Characteristics and pollution problems of irrigation return flow,* Federal Water Pollution Control Admin., Ada, Okla., 237 pp., 1969.

93. van der Leeden, F., et al., *Ground-water pollution problems in the Northwestern United States,* Rept. EPA-660/3-75-018, U.S. Environmental Protection Agency, Corvallis, Oreg., 361 pp., 1975.

94. van der Waarden, M., et al., Transport of mineral oil components to groundwater—1. Model experiments on the transfer of hydrocarbons from a residual oil zone to trickling water, *Water Research*, vol. 5, pp. 213–226, 1971.

95. Warner, D. L., *Deep-well injection of liquid waste*, Publ. no. 999-WP-21, U.S. Public Health Service, 55 pp., 1965.

96. Warner, D. L., *Rationale and methodology for monitoring groundwater polluted by mining activities*, Rept. EPA-600/4-74-003, U.S. Environmental Protection Agency, Washington, D.C., 76 pp., 1974.

97. Warner, D. L., *Monitoring disposal-well systems*, Rept. EPA-680/4-74-008, U.S. Environmental Protection Agency, Las Vegas, 99 pp., 1975.

98. Warner, D. L., and J. H. Lehr, *An introduction to the technology of subsurface wastewater injection*, Rept. EPA-600/2-77-240, U.S. Environmental Protection Agency, Ada, Okla., 344 pp., 1977.

99. Williams, D. E., and D. G. Wilder, Gasoline pollution of a groundwater reservoir—a case history, *Ground Water*, v. 9, no. 6, pp. 50–56, 1971.

100. Zanoni, A. E., Ground water pollution and sanitary landfills—a critical review, *Ground Water*, v. 10, no. 1, pp. 3–16, 1972.

Management
of Groundwater

Maximum development of groundwater resources for beneficial use involves planning in terms of an entire groundwater basin. Recognizing that a basin is a large natural underground reservoir, it follows that utilization of groundwater by one landowner affects the water supply of all other landowners. Management objectives must be selected in order to develop and operate the basin. These involve not only geologic and hydrologic considerations but also economic, legal, political, and financial aspects. Typically, optimum economic development of water resources in an area requires an integrated approach that coordinates the use of both surface water and groundwater resources. After evaluation of total water resources and preparation of alternative management plans, action decisions can then be made by appropriate public bodies or agencies.

Concepts of Basin Management

The management of a groundwater basin implies a program of development and utilization of subsurface water for some stated purpose, usually of a social or economic nature.[9] In general, the desired goal is to obtain the maximum quantity of water to meet

353

predetermined quality requirements at least cost.* Because a groundwater basin can be visualized as a large natural underground reservoir, it follows that extraction of water by wells at one location
influences the quantity of water available at other locations within
the basin.

Groundwater is extracted from the ground just as are other minerals such as oil, gas, or gold. Water typically carries a special con-
straint: it is regarded as a renewable natural resource. Thus, when a
water well is drilled, people presume that production of water will
continue indefinitely with time. In effect, this can only occur if there
exists a balance between water recharged to the basin from surface
sources and water pumped from within the basin by wells.

Development of water supplies from groundwater begins typically
with a few pumping wells scattered over a basin. With time more
wells are drilled and the rate of extraction increases. As wells become more numerous, development of the basin reaches and exceeds
its natural recharge capability. Continued development thereafter
without a management plan could eventually deplete the groundwater resource.

By regulating inflow to and outflow from the basin, an underground
reservoir can be made to function beneficially and indefinitely just
as a surface water reservoir.[3,14] The increasing demand for water
in the United States and throughout the world has produced the
realization that the vast underground reservoirs formed by aquifers
constitute invaluable water storage facilities; proper management of
them, therefore, has become a matter of considerable interest.[1,8,50]
Some of the pros and cons of subsurface and surface reservoirs are
summarized in Table 9.1.

Forecasts of future water demand suggest that mismanagement—or
lack of management—of major groundwater basins cannot be per-
mitted if adequate ongoing water supplies are to be provided. The
management objective consists of providing an economic and continuous water supply to meet a usually growing demand from an
underground water resource of which only a small portion is perennially renewable.

Equation of Hydrologic Equilibrium

To manage a groundwater basin, knowledge of the quantity of
water that can be developed is a prerequisite. Determination of the
available water within a basin requires evaluation of the elements

---

*As Bear and Levin[4] succinctly stated: "The basic idea is to regard the aquifer as a
system which has to be operated in an optimal manner."

TABLE 9.1 Advantages and Disadvantages of Subsurface and Surface Reservoirs (after U.S. Bureau of Reclamation[51])

Subsurface Reservoirs	Surface Reservoirs
Advantages	Disadvantages
1. Many large-capacity sites available	1. Few new sites available
2. Slight to no evaporation loss	2. High evaporation loss even in humid climate
3. Require little land area	3. Require large land area
4. Slight to no danger of catastrophic structural failure	4. Ever-present danger of catastrophic failure
5. Uniform water temperature	5. Fluctuating water temperature
6. High biological purity	6. Easily contaminated
7. Safe from immediate radioactive fallout	7. Easily contaminated by radioactive material
8. Serve as conveyance systems— canals or pipeline across lands of others unnecessary	8. Water must be conveyed
Disadvantages	Advantages
1. Water must be pumped	1. Water may be available by gravity flow
2. Storage and conveyance use only	2. Multiple use
3. Water may be mineralized	3. Water generally of relatively low mineral content
4. Minor flood control value	4. Maximum flood control value
5. Limited flow at any point	5. Large flows
6. Power head usually not available	6. Power head available
7. Difficult and costly to investigate, evaluate, and manage	7. Relatively easy to evaluate, investigate, and manage
8. Recharge opportunity usually dependent on surplus surface flows	8. Recharge dependent on annual precipitation
9. Recharge water may require expensive treatment	9. No treatment required of recharge water
10. Continuous expensive maintenance of recharge areas or wells	10. Little maintenance required of facilities

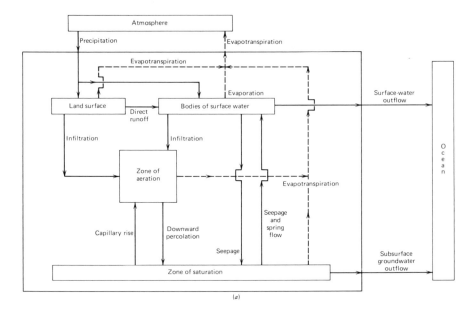

(a)

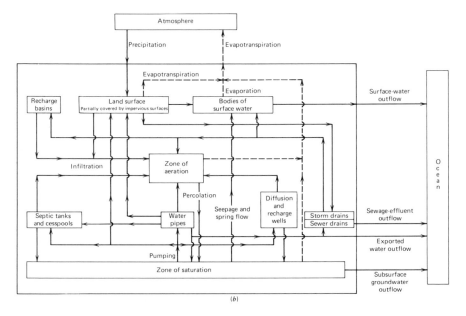

(b)

Fig. 9.1 Flow diagram of a hydrologic system. (*a*) Natural conditions.
(*b*) Urban and suburban development. Solid lines represent flow of liquid
water; dashed lines represent movement of water vapor (after Franke and
McClymonds[20]).

constituting the hydrologic cycle. A flow diagram of the hydrologic system for a basin under natural conditions is shown in Fig. 9.1*a*, while the more complex system for a basin containing urban and suburban development is depicted in Fig. 9.1*b*.

In terms of the hydrologic cycle for a particular groundwater basin, a balance must exist between the quantity of water supplied to the basin and the amount leaving the basin. The equation of hydrologic equilibrium provides a quantitative statement of this balance. In its most general form it may be expressed as in Eq. 9.1.

$$
\begin{bmatrix}
\text{surface inflow} + \text{subsurface inflow} + \text{precipitation} \\
+ \text{imported water} + \text{decrease in surface storage} \\
+ \text{decrease in groundwater storage}
\end{bmatrix} \tag{9.1}
$$

$$
= \begin{bmatrix}
\text{surface outflow} + \text{subsurface outflow} + \text{consumptive use} \\
+ \text{exported water} + \text{increase in surface storage} \\
+ \text{increase in groundwater storage}
\end{bmatrix}
$$

In this form the equation includes all waters—surface and subsurface—entering and leaving a basin. There are many situations in which it is possible to eliminate certain items from the equation because they are negligible or because they do not affect the solution. For example, a confined aquifer may have a hydrologic equilibrium independent of overlying surface waters; therefore, items of surface flow, precipitation, consumptive use, imported and exported water, and changes in surface storage can be omitted from the equation.

Each item of the equation represents a discharge, a volume of water per unit of time. Any consistent units of volume and time can be adopted. The water year, extending from October 1 to September 30, is preferable to the calendar year. The equation can be applied to areas of any size, although for meaningful results a hydrologic entity, such as an aquifer, a groundwater basin, or a river valley, is best.

The equation of hydrologic equilibrium in theory must balance. In practice, if all items can be evaluated, it will rarely balance exactly. This may be attributed to inaccuracies of measurements, lack of adequate basic data, or incorrect approximations. The amount of unbalance should not exceed the limits of accuracy of the basic data. In order to achieve a balance, adjustments should be made in items subject to large error. If the unbalance exceeds the limits of accuracy of the basic data, further investigation is necessary. Application of the equation requires good judgment, adequate hydrologic data, and careful analysis of the geology and hydrology of the particular area.

With the equation the quantity of water available from a groundwater basin can be determined under existing conditions as well as under any specified future conditions. Also, any one unknown item can be determined if all others are known. This last application can be misleading, however, for inaccuracies in one or more of the known quantities may exceed the magnitude of the unknown quantity.

Groundwater Basin Investigations

Ideally, before groundwater is developed in a basin, an investigation of the underground water resources should be made. In practice this rarely occurs; instead, a study is usually initiated either after extensive development with a view toward further development or after overdevelopment when a problem threatening the water supply appears imminent. Investigations are seldom concerned with simply locating groundwater supplies. More commonly the concerns involve evaluating the quantity and quality of groundwater resources already known to exist or determining the impact of human plans or activities on the quantity and quality of groundwater. Figure 9.2 illustrates the sequence of activities preceding the start of a groundwater management investigation.

Groundwater management studies are usually undertaken by local government agencies. Four levels of study are generally recognized, although not all are required.[2] In brief these include:

1. *Preliminary Examination*—Based largely on judgment by experienced personnel, this study identifies the management possibilities of meeting a defined need for a specified area.

2. *Reconnaissance*—This study considers possible alternatives in the formulation of a water management plan to meet a defined need for an area, including estimates of benefits and costs. The investigation draws on available data and generally necessitates a minimum of new data collection.

3. *Feasibility*—This study requires detailed engineering, hydrogeologic, and economic analyses together with cost and benefit estimates to ensure that the selected project is an optimum development. The sequence of activities normally involved in a feasibility investigation is outlined in Fig. 9.3. Typically, the investigation concludes with a report recommending approval and funding for the project.

4. *Definite Project*—This investigation involves planning studies necessary for defining specific features of the selected project. The completed report forms the basis for starting final design and preparation of plans and specifications.

The following section briefly outlines the types of data and the tasks involved in the physical portion of a reconnaissance or feasibility study for groundwater management.

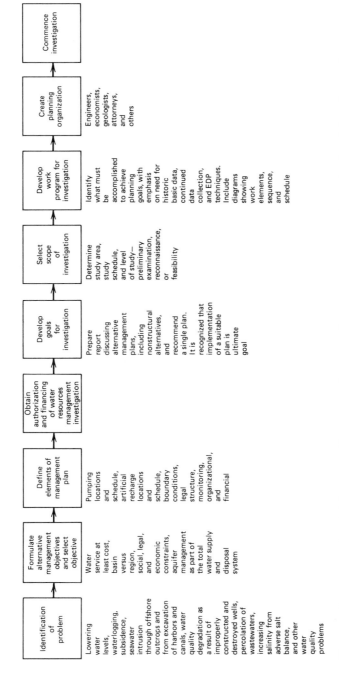

Fig. 9.2 Sequence of activities preceding start of a groundwater management investigation (after Amer. Soc. Civil Engrs.[2]).

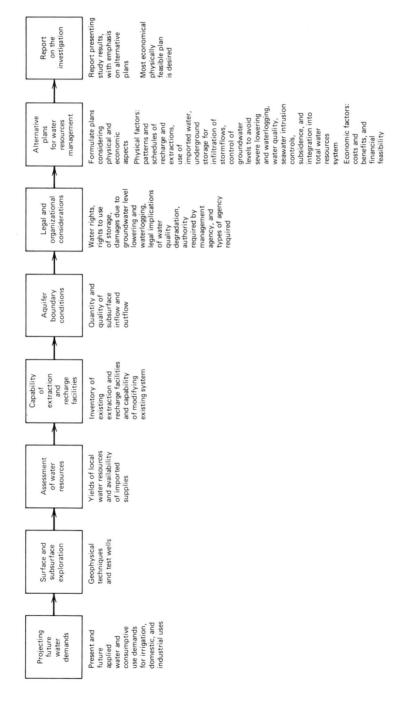

Fig. 9.3 Sequence of activities during a feasibility investigation for groundwater management (after Amer. Soc. Civil Engrs.[2]).

Data Collection and Fieldwork

Topographic Data. Contour maps, aerial photographs, and benchmarks related to a leveling network are basic requirements. They are directly applicable for locating and identifying wells, measuring groundwater levels, conducting crop and land use surveys, and plotting areal data.

Geologic Data. Surface and subsurface geologic mapping provides the framework for the occurrence and movement of groundwater and hence is essential for feasibility studies. Subsurface information is gained from a drilling program, including classification and analysis of well logs, and geophysical surveys (see Chapter 12). As part of the drilling program, pumping tests of wells are conducted to evaluate storage coefficients and transmissivities of aquifers, while samples of groundwater are collected and analyzed for quality. From interpretation of subsurface geologic data, principal aquifers and their extent are mapped together with regions of confined and unconfined groundwater. Location of faults, dikes, and other structures that may significantly affect groundwater is also a part of the geologic program.

Hydrologic Data. The principle purpose of hydrologic data collection is to evaluate the equation of hydrologic equilibrium. The following outline summarizes types of basic data required and methods of their analysis.

Surface Inflow and Outflow; Imported and Exported Water. These quantities are measurable by standard hydrographic and hydraulic procedures. Where complete data on surface flows to and from the basin are not available, supplemental stream gaging stations should be installed.

Precipitation. Records of precipitation in the area should be assembled. Gages should be well distributed over the basin to provide a good estimate of the annual precipitation from the isohyetal or Thiessen methods.* If gages are not so located, supplemental stations should be established.

Consumptive Use. All water, surface and subsurface, released into the atmosphere by processes of evaporation and transpiration is consumptive use, or evapotranspiration. To compute this discharge from a given basin, it is first necessary to make a land use, or cultural, survey to yield the amount of each type of water-consuming area. Aerial photographs are helpful for this task. Unit values of consumptive use must then be determined. For crops and native vegetation, methods based on available heat (such as the

---

*See, for example: R. K. Linsley, et al., *Hydrology for Engineers,* 2nd ed., McGraw-Hill, New York, 482 pp., 1975.

Thornthwaite or Blaney-Criddle method) are generally satisfactory. For water surfaces local evaporation records should be employed. Urban and industrial areas require careful estimates from samples of representative areas using metered deliveries and sewage outflows. Multiplying the unit value of consumptive use by the corresponding acreage gives the water consumption for each area; the sum of these products yields the total consumptive use over the basin.

Changes in Surface Storage. These can be computed directly from changes in water levels of surface reservoirs and lakes.

Changes in Soil Moisture. The moisture content of the soil can be measured by devices embedded in the soil or by a neutron probe (see Chapter 2). In practice, however, the variability of soil moisture both in time and place makes it difficult to obtain an accurate basin-wide measurement. The problem can be minimized by selecting periods of storage change in which the amount of water in unsaturated storage at the beginning and end of the period is nearly equal. In irrigated areas period limits should correspond to the beginning or ending of the irrigation season.

Changes in Groundwater Storage. From geologic data on aquifers and measurement of groundwater levels, changes in groundwater storage can be determined. Antecedent information on groundwater levels, pumping records, pumping tests, and artificial recharge should be collected. Specific yields of unconfined aquifers are determined by laboratory tests of samples and/or by classifications of well logs; storage coefficients are best determined from pumping tests of wells.

Select a grid of measuring wells distributed over the basin. Supplement with test holes where required. Water levels in these wells should be measured under conditions as nearly static as possible, preferably after the season of heavy draft and again after the season of recharge. A few control wells should be equipped with automatic water-level recorders or have their water levels measured monthly to facilitate detailed study of groundwater fluctuations. A basin map showing lines of equal change in groundwater level is then prepared. The product of change in water level times storage coefficient times area gives the change of groundwater storage for each aquifer within the basin.

Subsurface Inflow and Outflow. These items of the equation are the most difficult to evaluate because they cannot be directly measured. Often one of them, or the difference, is fixed by being the only unknown in the equation. From geologic investigation it may be found that either subsurface inflow or outflow is lacking, or both. Many times after study, subsurface inflow may be estimated to equal that of subsurface outflow so that the items cancel.

Difficulties arise in situations where underground flows from one basin to another occur. The direction of flow can be established from water table or piezometric gradients. Knowing groundwater slopes and transmissivities, subsurface flows can be computed from Darcy's law. Where surface streams and subsurface drainage systems control groundwater levels, better estimates of subsurface flow are usually possible because more data are available.

Alternative Basin Yields

The maximum quantity of water that is actually available from a groundwater basin on a perennial basis is limited by the possible deleterious side effects that can be caused by pumping and by the operation of the basin. As a result, several concepts of basin yield are generally recognized.[2] These are briefly defined in the following subsections together with comments as to their consequences.

Mining Yield. If groundwater is withdrawn at a rate exceeding the recharge, a *mining yield* exists.[17] As a consequence, this yield must be limited in time until the aquifer storage is depleted. Many groundwater basins today are being mined; if mining continues, the local economy served by this pumping may change, evolving into other forms that use less water or involve importations of water into the basin. The Salt River Valley of southern Arizona and the High Plains of western Texas are classic examples of such situations.

Various valid arguments, economic and other, have been advanced to justify mining of groundwater. One is that water in storage is of no value unless it is used.[46] In arid areas, such as the Sahara Desert, where groundwater represents the only available water resource, almost any development of groundwater constitutes a mining yield. But the needs are there and the benefits are great so that such exploitation will continue. With proper management plus water conservation, such groundwater resources can be made to last from several decades to a few centuries.

Perennial Yield. The *perennial yield* of a groundwater basin defines the rate at which water can be withdrawn perennially under specified operating conditions without producing an undesired result.* An undesired result is an adverse situation such as (1) progressive reduction of the water resource, (2) development of uneconomic pumping conditions, (3) degradation of groundwater quality,

---

*In the past the term *safe yield,* implying a fixed quantity of extractable water basically limited to the average annual basin recharge, has been widely used. The term has now fallen into disfavor because a never-changing quantity of available water depending solely on natural water sources and a specified configuration of wells is essentially meaningless from a hydrologic standpoint.

(4) interference with prior water rights, or (5) land subsidence caused by lowered groundwater levels.[29,30,54] Evaluation of perennial yield is discussed in a subsequent section. Any draft in excess of perennial yield is referred to as *overdraft*. Existence of overdraft implies that continuation of present water management practices will result in significant negative impacts on environmental, social, or economic conditions.

A schematic diagram of a groundwater basin developed to less than perennial yield is shown in Fig. 9.4a. Here a portion of the natural recharge is lost by subsurface outflow from the basin. But Fig. 9.4b suggests a minimum perennial yield situation in which extractions balance recharge so that no groundwater is lost.

Deferred Perennial Yield. The concept of a deferred perennial yield consists of two different pumping rates. The initial rate is larger and exceeds the perennial yield, thereby reducing the groundwater level. This planned overdraft furnishes water from storage at low cost and without creating any undesirable effects. In fact, reducing storage eliminates wasteful subsurface outflow of groundwater and losses to the atmosphere by evapotranspiration from high water table areas. After the groundwater level has been lowered to a predetermined depth, a second rate, comparable to that of perennial yield, is established so that a balance of water entering and leaving the basin is maintained thereafter. With a larger available storage volume, more water can be recharged and a larger perennial yield can be obtained. Figure 9.4c indicates this situation schematically.

Maximum Perennial Yield. The *maximum perennial yield,* as the name suggests, means the maximum quantity of groundwater perennially available if all possible methods and sources are developed for recharging the basin. In effect, this quantity depends on the amount of water economically, legally, and politically available to the organization or agency managing the basin. Clearly, the more water that can be recharged both naturally and artificially to a basin, the greater the yield.

To achieve the maximum perennial yield the aquifer should be managed as a unit. Thus, efficient and economic production of water requires that all pumping, importations, and distributions of water be done for the benefit of the largest manageable system. Where surface water is available in addition to groundwater, these two sources are operated conjunctively. Such a conjunctive use scheme provides a larger and more economic yield of water than can be obtained

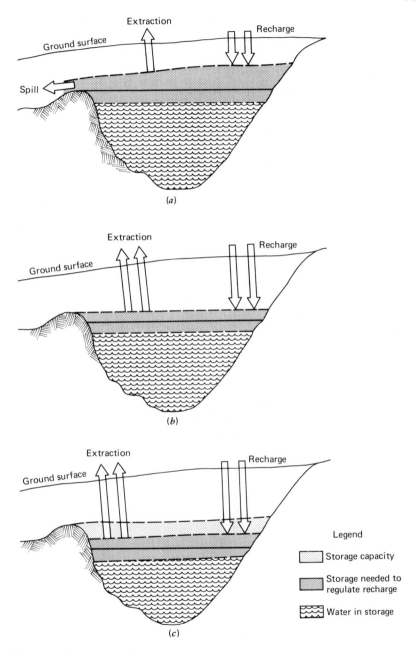

Fig. 9.4 Schematic diagram showing storage relations in a groundwater basin for three stages of development. (a) Less than perennial yield. (b) Minimum perennial yield. (c) Increased perennial yield (after Peters[39]).

from the two sources operated independently. The limit to such an operation is governed by the ability to import and distribute water and also by the storage available for surface water and groundwater.

Evaluation of Perennial Yield

Consideration of the above definitions of perennial yield reveals that there can be more than one "undesired result" from pumping a groundwater basin, that perennial yield may be limited to an amount less than the net amount of water supplied to the basin, and that perennial yield can vary with different patterns of recharge, development, and use of water in a basin.

If groundwater is regarded as a renewable natural resource, then only a certain quantity of water may be withdrawn annually from a groundwater basin. The maximum quantity of water that can be extracted from an underground reservoir, and still maintain that supply unimpaired, depends on the perennial yield. Overdraft areas constitute the largest potential groundwater problem in the United States.[49] Until overdrafts are reduced to perennial yields in these basins, permanent damage or depletion of groundwater supplies must be anticipated.

Factors Governing Perennial Yield. Determination of the perennial yield of a groundwater basin requires analysis of the undesired results that may accrue if the extraction rate is exceeded. The recharge* criterion (progressive reduction of the water resource) is the most important because exceeding this factor is normally responsible for introducing other undesired results. Water supplied to a basin may be limited either by the storage volume of the underground basin or by the rate of water movement through the basin from the recharge area to the withdrawal area. The quantity concept is usually applicable to unconfined aquifers where supply and disposal areas are near, whereas the rate concept applies more to confined aquifers where supply and disposal areas are widely separated.

Economic considerations can govern perennial yield in basins where the cost of pumping groundwater becomes excessive. Excessive costs may be associated with lowered groundwater levels, necessitating deepening wells, lowering pump bowls, and installing larger pumps. Where pumpage is largely for irrigation, power costs, crop prices, or government farm subsidies may establish an eco-

---

*Recharge here refers to water reaching the saturated zone of an aquifer, where it is available for extraction.

nomic limit for pumping groundwater; alternatively, other uses that can support higher pumping costs may evolve.

Water quality can govern perennial yield if draft on a basin produces groundwater of inferior quality. Possibilities include: (1) pumping in a coastal aquifer could induce seawater intrusion into the basin (see Chapter 14); (2) lowered groundwater levels could lead to pumping of underlying connate brines; (3) polluted water from nearby areas might be drawn into a pumped aquifer. A quality limitation on perennial yield depends on the minimum acceptable standard of water quality, which in turn depends on the use made of the pumped water. Therefore, by lowering the quality requirement, the perennial yield can be increased.

Legal considerations affect perennial yield if pumpage interferes with prior water rights.[32] Finally, if pumpage is responsible for land subsidence, a limitation on perennial yield can result.

Calculation of Perennial Yield. In general, the basin recharge criterion will govern perennial yield because, as mentioned earlier, one or more of the other undesired results will often be induced by pumpage exceeding this rate. Quantitative determination of perennial yield where recharge is the limiting factor can be made under specified conditions if adequate knowledge of the hydrology of the basin is available. Methods are based on the equation of hydrologic equilibrium or approximations thereto.[31] Basically, this implies that perennial yield is defined in terms of a rate at which groundwater can be withdrawn from a basin over a representative time period without producing a significant change in groundwater storage.

Variability of Perennial Yield. It is important to recognize that perennial yield of a groundwater basin tends to vary with time. Any quantitative determination is based on specified conditions, either existing or assumed, and any changes in these conditions will modify the perennial yield. This fact applies to the degree and pattern of groundwater development within a basin as well as to the other factors that govern safe yield.

Investigations of the availability of groundwater within a basin are typically not initiated until basin development has produced an overdraft. Yet this is almost necessary in order to obtain a reasonable estimate for perennial yield. In a virgin basin, where a balance exists between natural inflow and outflow and there is no pumping, the absence of hydrogeologic data may not justify the cost of a management investigation. Similarly, estimating future perennial yield of a basin under greater development than at present requires careful evaluation of all items in the equation of hydrologic equilibrium.

Perennial yield may vary with the level of groundwater within a basin. Thus, if levels are lowered, subsurface inflow will be increased and subsurface outflow will be decreased, recharge from losing streams will be increased and discharge from gaining streams will be decreased, and uneconomic evapotranspiration losses will be reduced. Conversely, a rise in water levels will have the opposite effects. Therefore, where recharge is sufficient, the greater the utilization of underground water, the larger the perennial yield. The maximum perennial yield will be controlled by economic or legal constraints.

An unconfined basin fed by an adequate recharge source can increase its perennial yield, not only by increasing pumpage but also by rearrangement of the pumping pattern. If the concentration of wells is shifted to near the recharge source, greater inflow can be induced. The rearrangement has the additional advantage that a greater supply may be obtained without necessarily increasing pumping lifts. For example, in the cross section shown in Fig. 9.5a, it is assumed that the stream is the principal recharge source. By moving the well field nearer to the stream as in Fig. 9.5b, the water

Fig. 9.5 Example of increased groundwater yield for same pumping depths obtained by shifting wells nearer to a recharge source.

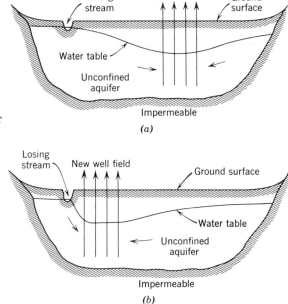

table slope is increased and a greater yield for equal pumping depths results.

For a confined aquifer with its recharge area located some distance from the pumping area, the rate of flow through the aquifer will govern the perennial yield. In large confined aquifers, pumpage of water from storage can be carried on for many years without establishing an equilibrium with basin recharge. Although the slope of the piezometric surface will increase, the permeability of the aquifer is seldom sufficient to maintain a compensating flow into the basin.[49]

Besides operational changes perennial yield can also vary due to gradual and subtle modifications occurring within a basin. Changes in vegetation and even in crops, particularly where root depth is affected, may influence surface infiltration and subsequent percolation to the water table. Urbanization of an area, accompanied by greater surface runoff and installation of sewer systems, can be expected to reduce recharge. Changes in the purpose of pumping groundwater, such as from irrigation to municipal or industrial use, may—from an economic viewpoint—permit greater pumping lifts; consequently, perennial yield can be increased. Other economic factors include, among others, changes in value of irrigated crops, increased efficiency of new wells and pumps, treatment to meet revised water quality standards, and power costs.

Salt Balance

Maintenance of a usable groundwater basin requires that the salinity of the groundwater not increase with time to a point where it destroys the value of the resource. A dynamic balance of total salts entering and leaving a basin is desired, so that on a long-term basin

$$\sum_{n} (CQ)_i = 0 \tag{9.2}$$

where $(CQ)_i$ is the salt concentration times the discharge of one of n flow components to or from the basin. In practice this condition seldom exists because most uses of water add dissolved solids to water, which is subsequently recharged to groundwater.

Salt may be added to groundwater by solution of aquifer materials, from rainfall and surface and subsurface inflows, and in special circumstances from connate brines and seawater. Evapotranspiration removes water, leaving higher salt concentrations behind. Domestic and industrial uses of water add salts, as do fertilizers, soil conditioners, pesticides, and other chemicals in agricultural areas. Salts leave a groundwater basin by natural outflow, drainage, and pumped extractions.[11]

The salt problem becomes most important for irrigated land in arid and semiarid regions.[23,37] If a high water table persists with inadequate drainage, evapotranspiration of irrigation water and groundwater gradually increases the salt content of the soil, leading to destruction of the land for agricultural purposes. The solution depends on local conditions, but, in general, the requirements are that the water table be lowered, that soil salinity be reduced by leaching, and that a drainage system to transport saline water out of the basin be constructed.

It should be recognized that excellent groundwater can be found in a basin with an adverse salt balance, and vice versa. Groundwater is rarely uniformly mixed. Typically, good- and poor-quality ground-waters are segregated both horizontally and vertically within a basin. Thus, an unfavorable salt balance poses a serious long-term threat but seldom concerns the current usability of groundwater.

An illustration of salt balance for a semiarid region in the Central Valley of California is shown in Table 9.2. Under 1970 conditions it can be seen that input of salt exceeds output by 1629 tons. During the decade 1970–1980 an increased volume of imported water entered the basin, newly irrigated lands were leached, and a drainage system was started. The effect of these changes by 1980 is a projected

TABLE 9.2 Calculated and Projected Salt Balances for the Tulare Lake Basin, California (after Schmidt[47]) (values in 1000 tons)

Item	1970	1980
Input		
Precipitation	23	25
Streamflow	357	357
Imported water	326	846
Soil amendments	476	543
Fertilizers	176	190
Animal waste	51	55
Leaching new lands	0	771
Municipal wastewater	38	44
Urban runoff	7	8
Industrial waste	28	30
Oil field waste	182	5
Subtotal	1664	2874
Output		
Streamflow and groundwater	35	102
Subsurface drainage export	0	545
Subtotal	35	647
Net accretion	1629	2227

net input of 2227 tons of salt, representing an increase of 73 percent in salt accumulation. Assuming this salt is mixed within the principal aquifers, the salinity of the groundwater will increase 1 to 4 mg/l per year in the eastern zone of higher precipitation and 10 to 30 mg/l per year in the arid western zone.

Basin Management by Conjunctive Use

In basins approaching full development of water resources, optimal benefical use can be obtained by *conjunctive use,* which involves the coordinated and planned operation of both surface water and groundwater resources to meet water requirements in a manner whereby water is conserved.* The basic difference between the usual surface water development with its associated groundwater development and a conjunctive operation of surface water and groundwater resources is that the separate firm yields of the former can be replaced by the larger and more economic joint yields of the latter.

The concept of conjunctive use of surface water and groundwater is predicated on surface reservoirs impounding streamflow, which is then transferred at an optimum rate to groundwater storage. Surface storage in reservoirs behind dams supplies most annual water requirements, while the groundwater storage can be retained primarily for cyclic storage to cover years of subnormal precipitation. Thus, groundwater levels would fluctuate, being lowered during a cycle of dry years and being raised during an ensuing wet period. Figure 9.6 depicts how groundwater levels might vary under such a system of conjunctive use.

During periods of above-normal precipitation, surface water is utilized to the maximum extent possible and also artificially recharged into the ground to augment groundwater storage and raise groundwater levels (see Chapter 13). Conversely, during drought periods limited surface water resources are supplemented by pumping groundwater, thereby lowering groundwater levels. The feasibility of the conjunctive-use approach depends on operating a groundwater basin over a range of water levels; that is, there must be space to store recharged water, and, in addition, there must be water in storage for pumping when needed.

Management by conjunctive use requires physical facilities for water distribution, for artificial recharge, and for pumping. The procedure does require careful planning to optimize use of avail-

---

*Coordinated use of surface water and groundwater does not preclude importing water, as required, to meet growing needs. In fact, to store and distribute additional water economically may require more intensive use of groundwater storage space.

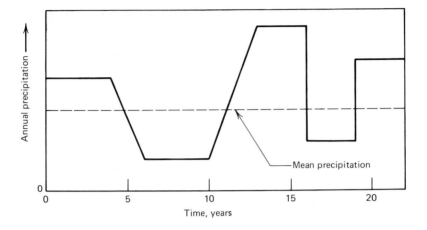

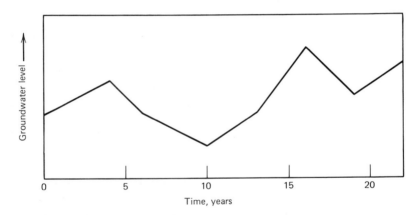

Fig. 9.6 Illustrative example of variation in groundwater levels in relation to annual precipitation under conjunctive use management.

able surface-water and groundwater resources. Such operations can be complex and highly technical; they require competent personnel, detailed knowledge of the hydrogeology of the basin, records of pumping and recharge rates, and continually updated information on groundwater levels and quality. A schematic diagram of a systematic approach for a conjunctive use analysis is illustrated in Fig. 9.7.

A conjunctive use management study requires data on surface water resources, groundwater resources, and geologic conditions;

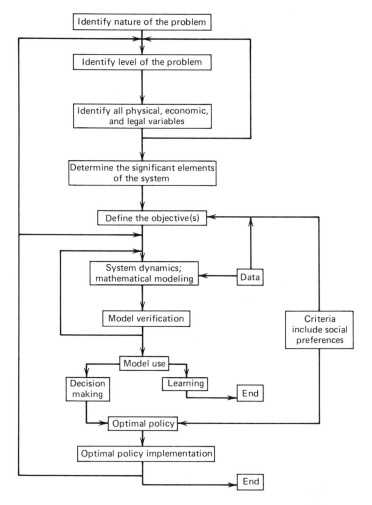

Fig. 9.7 Schematic diagram of a systematic approach for studying conjunctive use problems (after Maknoon and Burges[27]; reprinted from *Journal American Water Works Association,* Vol. 70, by permission of the Association; copyright © 1978 by American Water Works Association, 6666 West Quincy Avenue, Denver, Col. 80235).

data on water distribution systems, water use, and wastewater disposal are also necessary.[24,41] Figure 9.8 shows a simplified flowchart of the various phases and steps involved for a basin management study in California. This suggests the diversity of data and effort required in order to determine an optimal basin management plan.

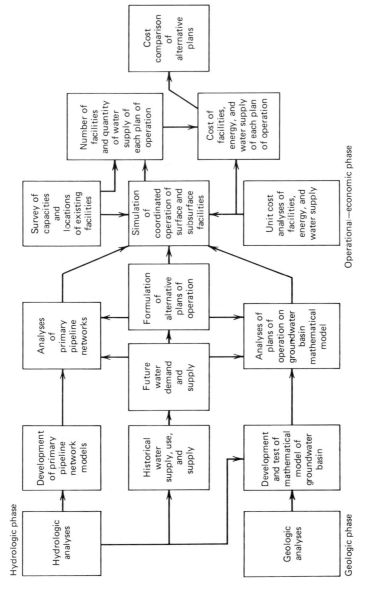

Fig. 9.8 Flow diagram of a management study for the San Gabriel Valley, California, groundwater basin (after Amer. Soc. Civil Engrs.[2]).

It should be noted from Figs. 9.7 and 9.8 that mathematical models are usually incorporated in such studies (see Chapter 10). A basin model simulates the responses of a basin to variations in variables such as natural and artificial recharge and pumping so that the best operating procedures for basin management can be practiced. In effect, this will optimize the water supply obtained from the basin.[16, 22, 44]

Because every water development project is unique, it is impossible to present economic considerations generally for conjunctive operations and have them apply specifically to any given situation. Nevertheless, the advantages and disadvantages, mostly economic, are summarized in Table 9.3. The tabulation compares a conjunctive-use operation relative to development of surface-water resources only, assuming irrigation to be the principal water use in a semiarid region.

Total usable water supply can be increased by coordinated operation of surface and underground water resources. With an optimum coordinated operation the unit cost of water supply storage and distribution can be minimized. The basic principles of groundwater

TABLE 9.3   Conjunctive Use of Surface Water and Groundwater Resources (after Clendenen[13])

Advantages	Disadvantages
1. Greater water conservation	1. Less hydroelectric power
2. Smaller surface storage	2. Greater power consumption
3. Smaller surface distribution system	3. Decreased pumping efficiency
4. Smaller drainage system	4. Greater water salination
5. Reduced canal lining	5. More complex project operation
6. Greater flood control	6. More difficult cost allocation
7. Ready integration with existing development	7. Artificial recharge is required
8. Stage development facilitated	8. Danger of land subsidence
9. Smaller evapotranspiration losses	
10. Greater control over outflow	
11. Improvement of power load and pumping plant use factors	
12. Less danger from dam failure	
13. Reduction in weed seed distribution	
14. Better timing of water distribution	

basic operation that will produce an optimum water resources management scheme include, as reported by Fowler:[19]

1. The surface and underground storage capacities must be integrated to obtain the most economical utilization of the local storage resources and the optimum amount of water conservation.

2. The surface distribution system must be integrated with the groundwater basin transmission characteristics to provide the minimum cost distribution system.

3. An operating agency must be available with adequate power to manage surface-water resources, groundwater recharge sites, surface-water distribution facilities, and groundwater extractions.

The procedure for developing a sound conjunctive-use operation within a basin requires estimation of the various elements of water supply and distribution. The optimum use of surface-water and groundwater resources is determined for assumed conditions, usually those during the most critical drought period of record. Examples of coordinated basin management include studies for basins in California,[5,33,38,53] Colorado,[6,34,48] Idaho,[42] Maryland,[21] New York,[20] England,[18] and India.[45]

Examples of Groundwater Management

Los Angeles Coastal Plain, California. This 1240 km² basin supplies approximately one-half of the water supply for the Los Angeles metropolitan area. In the recent past the basin has been critically overdrawn, resulting in declining groundwater levels and seawater intrusion. Detailed management studies[10,12] were undertaken to formulate the most economic plan for operating the groundwater basin in coordination with surface-water storage and transmission facilities to: (1) meet the growing and fluctuating water demands of the area, (2) conserve the maximum amount of locally available water, and (3) minimize the undesirable effects of overdraft. A schematic representation of the coordinated use of surface-water and groundwater resources in shown in Fig. 9.9.

Because of the vast increases in imported water to the Los Angeles area,* the study concentrated, first, on evaluating the dynamic response of the basin to recharging and pumping so that maximum use of the underground reservoir could be made, and, second, on determining the most economic plan for operating the basin—taking into

*It should be noted that the Los Angeles coastal plain is served by a network of water sources, including local surface water and groundwater, reclaimed wastewater, and imported water from the Owens, Colorado, and Feather rivers.

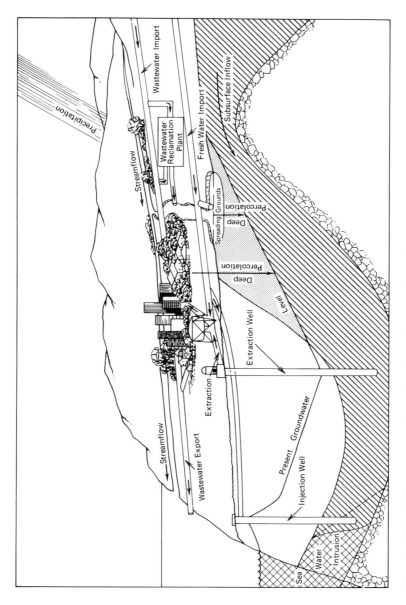

Fig. 9.9 Pictorial representation of conjunctive use of surface water and groundwater resources, Los Angeles Coastal Plain, California (after Calif. Dept. Water Resources[10]).

account patterns and rates of water extraction, rates of artificial recharge, and methods for controlling seawater intrusion.

High Plains, Texas and New Mexico. The High Plains straddling the Texas–New Mexico border define the boundaries of the Ogallala Formation, an aquifer containing approximately 250×10^9 m^3 of water in 1958. A phenomenal increase in pumping of groundwater for irrigation began in the 1940s resulting in an increase of 362 percent of irrigated acreage in the 1948–1958 decade together with an increase in wells from 8,356 to 45,522 during the same period. Already by 1958 some 50×10^9 m^3 of water had been extracted and the pumping rate amounted to 9×10^9 m^3/yr, which was more than 100 times the recharge rate. Effects on wells are apparent in Fig. 9.10 in terms of changes in yields and pumping lifts. Although rapid mining of groundwater is underway, restricting pumpage to the rate of recharge would also essentially stop all extraction and permit a large volume of water to remain unused.

Recognizing the economic and political constraints on obtaining imported water, the only feasible solution involves managed mining of the groundwater.[7,24,43] Because the aquifer has a low transmis-

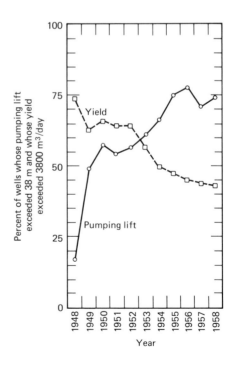

Fig. 9.10 Graph showing decreased well yield and increased pumping lift for wells in the High Plains of Texas and New Mexico as mining of groundwater occurred during the 1948–1958 decade (after Nace[36]).

sivity, depressions in the water table reflect centers of concentrated pumping. By conservative pumping from adequately spaced wells, optimum development based on long-term economic considerations can be achieved.* An example of such management has been established in the New Mexico portion of the High Plains.[15] Here groundwater extraction is limited, on a township basis, to what will provide a firm minimum supply for a period of ^0 years. With time it is anticipated that improved conservation measures, changed water uses, and possible imported supplemental water will enable needs to be met further into the future. Stable development over a long term permits amortization of capital expenditures and enhances opportunities for measures to secure a more permanent water supply.

Indus River Valley, Pakistan. With the advent of canal irrigation before 1900, the Indus Plain of Pakistan developed gradually into the largest single irrigated region on the earth.† Leakage from the canal network, however, brought the water table close to ground surface over large areas (see Fig. 6.16) so that about 2.6×10^6 ha have reduced fertility caused by salinity and waterlogging. With inadequate drainage, minimal application of irrigation water, and a high evaporation rate, salts continue to accumulate and expand the problem areas. Considering Pakistan's dependence on agriculture and its continued population growth, the problem is really twofold: to eliminate salinity and waterlogging and to increase agricultural production.[40]

The solution that has been undertaken[28,35] involves drilling a large network of deep (60–100 m) high-capacity wells spaced about 1.6 km apart. Water pumped from the wells is released into existing local canals for irrigation use. In effect, the wells serve three complementary purposes by (1) lowering the water table, (2) providing supplemental irrigation water for agricultural production, and (3) furnishing water for leaching of saline soils.** Figure 9.11 illustrates the

*Also, as a result of strong local cooperation, a program for conserving all available water is effectively underway. This involves recharging ponded surface runoff and irrigation tailwater into wells.

†The irrigated area amounts to some 9×10^6 ha, comparable to the total irrigated acreage in the entire United States. The Indus River, which supplies the water, has an average discharge twice that of the Nile and more than ten times that of the Colorado River.

**It should be noted that the wells disperse the salinity of the upper soil layer throughout the body of the groundwater but do not remove the salt. The salinity of the groundwater will slowly increase; when necessary in the future, a fraction of the groundwater can be pumped to waste to achieve a salt balance.

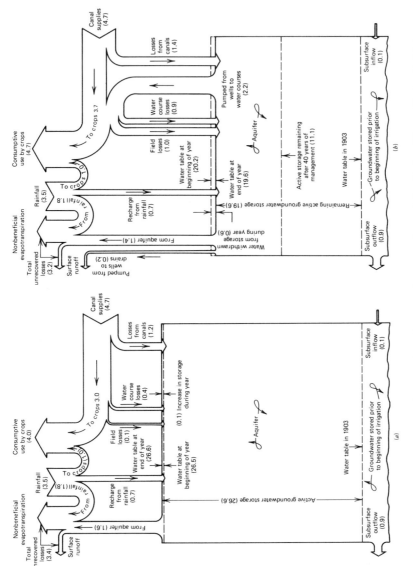

Fig. 9.11 Schematic water balances for Chaj Doab, a portion of the Indus River Plain, Pakistan. (*a*) Premanagement conditions. (*b*) Ten years after start of management program. All values are in 10⁹ m³/yr (after Tipton and Kalmbach, Inc., *Feasibility Rept. on Salinity Control and Reclamation, Project No. 2, West Pakistan, Denver, Col.*, 1960).

hydrologic balances before and after the wells were in operation. By this system of basin management, together with improvements in agricultural techniques, agricultural productivity is being increased several fold.

References

1. Ambroggi, R. P., Underground reservoirs to control the water cycle, *Sci. Amer.,* v. 236, no. 5, pp. 21–27, 1977.
2. Amer. Soc. Civil Engrs., *Ground water management,* Manual Engrng. Practice 40, 216 pp., 1972.
3. Banks, H. O., Utilization of underground storage reservoirs, *Trans. Amer. Soc. Civil Engrs.,* v. 118, pp. 220–234, 1953.
4. Bear, J., and O. Levin, The optimal yield of an aquifer, *Intl. Assoc. Sci. Hydrology Publ.* 72, pp. 401–412, 1967.
5. Beaver, J. A., and M. L. Frankel, Significance of ground-water management strategy—a systems approach, *Ground Water,* v. 7, no. 3, pp. 22–26, 1969.
6. Bittinger, M. W., The problem of integrating ground-water and surface water use, *Ground Water,* v. 2, no. 3, pp. 33–38, 1964.
7. Brown, R. F., et al., *Artificial ground-water recharge as a water-management technique on the Southern High Plains of Texas and New Mexico,* Texas Dept. Water Resources Rept. 220, 32 pp., 1978.
8. Buras, N., Conjunctive operation of dams and aquifers, *Jour. Hydraulics Div.,* Amer. Soc. Civil Engrs., v. 89, no. HY6, pp. 111–131, 1963.
9. Burt, O., Temporal allocation of groundwater, *Water Resources Research,* v. 3, pp. 45–56, 1967.
10. California Dept. Water Resources, *Planned utilization of ground water basins: coastal plain of Los Angeles County,* Bull. 104, Sacramento, 25 pp., plus apps., 1968.
11. Casey, H. E., *Salinity problems in arid lands irrigation,* Office of Arid Lands Studies, Univ. of Arizona, Tucson, 300 pp., 1972.
12. Chun, R. Y. D., et al., Ground-water management for the nation's future—optimum conjunctive operation of ground-water basins. *Jour. Hydraulics Div.,* Amer. Soc. Civil Engrs., v. 90, no. HY4, pp. 79–95, 1964.
13. Clendenen, F. B., Economic utilization of ground water and surface water storage reservoirs, Paper presented before meeting of Amer. Soc. Civil Engrs., San Diego, Feb. 1955.
14. Conkling, H., Utilization of ground-water storage in stream system development, *Trans. Amer. Soc. Civil Engrs.,* v. 111, pp. 275–354, 1946.
15. Conover, C. S., Ground-water resources—development and management, *U.S. Geological Survey Circ.* 442, 7 pp., 1961.
16. Domenico, P. A., Economic aspects of conjunctive use of water, Smith Valley, Nevada, USA, *Intl. Assoc. Sci. Hydrology Publ.* 72, pp. 474–482, 1967.

17. Domenico, P. A., et al., Optimal ground-water mining, *Water Resources Research*, v. 4, pp. 247–255, 1968.

18. Downing, R. A., et al., Regional development of groundwater resources in combination with surface water, *Jour. Hydrology*, v. 22, pp. 155–177, 1974.

19. Fowler, L. C., Ground-water management for the nation's future—ground-water basin operation, *Jour. Hydraulics Div.*, Amer. Soc. Civil Engrs., v. 90, no. HY4, pp. 51–57, 1964.

20. Franke, O. L., and N. E. McClymonds, Summary of the hydrologic situation on Long Island, New York, as a guide to water-management alternatives, *U.S. Geological Survey Prof. Paper 627-F*, 59 pp., 1972.

21. Hansen, H. J., Zoning plan for managing a Maryland coastal aquifer, *Jour. Amer. Water Works Assoc.*, v. 62, pp. 286–292, 1970.

22. Hartman, L. M., Economics and ground-water development, *Ground Water*, v. 3, no. 2, pp. 4–8, 1965.

23. Helweg, O. J., A nonstructural approach to control salt accumulation in ground water, *Ground Water*, v. 15, pp. 51–57, 1977.

24. Johnson, A. I., et al., Symposium on planning and design of groundwater data programs, *Water Resources Research*, v. 8, pp. 177–241, 1972.

25. Jones, O. R., and A. D. Schneider, Groundwater management on the Texas High Plains, *Water Resources Bull.*, v. 8, pp. 516–522, 1972.

26. Mack, L. E., *Ground water management in development of a national policy on water*, Rept. NWC-EES-71-004, Natl. Water Comm., Washington, D.C., 179 pp., 1971.

27. Maknoon, R., and S. J. Burges, Conjunctive use of ground and surface water, *Jour. Amer. Water Works Assoc.*, v. 70, pp. 419–424, 1978.

28. Malmberg, G. T., Reclamation by tubewell drainage in Rechna Doab and adjacent areas, Punjab region, Pakistan, *U.S. Geological Survey Water-Supply Paper 1608-O*, 72 pp., 1975.

29. Mann, J. F., Jr., Safe yield changes in groundwater basins, *Intl. Geological Congress, XXI Session*, Pt. XX, pp. 17–23, 1960.

30. Mann, J. F., Jr., Factors affecting the safe yield of ground-water basins, *Trans. Amer. Soc. Civil Engrs.*, v. 128, Pt. III, pp. 180–190, 1963.

31. Mann, J. F., Jr., Ground-water management in the Raymond Basin, California, *Engrng, Geol. Case Histories 7*, Geol. Soc. Amer., pp. 61–74, 1969.

32. McCleskey, G. W., Problems and benefits in ground-water management, *Ground Water*, v. 10, no. 2, pp. 2–5, 1972.

33. Moore, C. V., and J. H. Snyder, Some legal and economic implications of sea water intrusion—a case study of ground water management, *Natural Resources Jour.*, vol. 9, pp. 401–419, 1969.

34. Morel-Seytoux, H. J., A simple case of conjunctive surface-ground-water management, *Ground Water*, v. 13, pp. 506–515, 1975.

35. Mundorff, M. J., et al., Hydrologic evaluation of salinity control and reclamation projects in the Indus Plain, Pakistan—a summary, *U.S. Geological Survey Water-Supply Paper 1608-Q*, 59 pp., 1976.

36. Nace, R. L., Water management, agriculture, and ground-water supplies, *U.S. Geological Survey Circ.* 415, 12 pp., 1960.
37. Nightingale, H. I., and W. C. Bianchi, Ground-water chemical quality management by artificial recharge, *Ground Water,* v. 15, pp. 15-22, 1977.
38. Orlob, G. T., and B. B. Dendy, Systems approach to water quality management, *Jour. Hydraulics Div.,* Amer. Soc. Civil Engrs., v. 99, no. HY4, pp. 573-587, 1973.
39. Peters, H. J., Groundwater management, *Water Resources Bull,* v. 8, pp. 188-197, 1972.
40. Peterson, D. F., Ground water in economic development, *Ground Water,* v. 6, no. 3, pp. 33-41, 1968.
41. Pfannkuch, H. O., and B. A. Labno, Design and optimization of groundwater monitoring networks for pollution studies, *Ground Water,* v. 14, pp. 455-462, 1976.
42. Ralston, D. R., Administration of ground water as both a renewable and nonrenewable resource, *Water Resources Bull.,* v. 9, pp. 908-917, 1973.
43. Rayner, F. A., Ground-water basin management on the High Plains of Texas, *Ground Water,* v. 10, no. 5, pp. 12-17, 1972.
44. Renshaw, E. F., The management of ground water reservoirs, *Jour. Farm Economics,* v. 45, pp. 285-295, 1963.
45. Revelle, R., and V. Lakshminarayana, The Ganges water machine, *Science,* v. 188, pp. 611-616, 1975.
46. Sasman, R. T., and R. J. Schicht, To mine or not to mine groundwater, *Jour. Amer. Water Works Assoc.,* v. 70, pp. 156-161, 1978.
47. Schmidt, K. D., Salt balance in groundwater of the Tulare Lake Basin, California, *Hydrology and Water Resources in Arizona and the Southwest,* v. 5, Proc. Amer. Water Resources Assoc., Arizona Section, Tempe, pp. 177-184, 1975.
48. Taylor, O. J., and R. R. Luckey, Water management studies of a stream-aquifer system, Arkansas River Valley, Colorado, *Ground Water,* v. 12, pp. 22-38, 1974.
49. Thomas, H. E., *The conservation of ground water,* McGraw-Hill, New York, 327 pp., 1951.
50. Thomas, H. E., Cyclic storage, where are you now?, *Ground Water,* v. 16, pp. 12-17, 1978.
51. U.S. Bureau of Reclamation, *Ground water manual,* U.S. Dept. Interior, 480 pp., 1977.
52. Water Resources Dev. Centre, *Large-scale ground-water development,* United Nations, New York, 84 pp., 1960.
53. Weschler, L. F., *Water resources management: the Orange County experience,* California Govt. Ser. no. 14, Inst. Govt. Affairs, Univ. California, Davis, 67 pp., 1968.
54. Young, R. A., Safe yield of aquifers: an economic reformulation, *Jour. Irrig. Drain. Div.,* Amer. Soc. Civil Engrs., v. 96, no. IR4, pp. 377-385, 1970.

Groundwater
Modeling Techniques

CHAPTER 10 •••••••••••••••••••••••••••••••••••

Because groundwater is essentially a hidden resource, studies of groundwater under both natural and artificial boundary conditions have employed modeling techniques. A variety of types of models have been developed.[2, 72, 74] These can be categorized as porous media models, miscellaneous analog models, electrical analog models based on the similarity between Ohm's law and Darcy's law, and digital computer models for numerical solution of aquifer flow equations. In recent years the development and application of digital computer models has done much to improve the efficient management of extensive groundwater resources.

In a comprehensive article on groundwater modeling, Prickett[52] lists applications of various types of groundwater models, including aquifer features, purpose of the models, and schematic sketches of the models. Detailed lists of references on the subject can be found in Prickett.[53,54]

Porous Media Models

Sand Tank Models. A sand tank model is a scale model of an aquifer with the boundaries scaled down and the permeability—absolute value and spatial distribution—modified. Sand models have

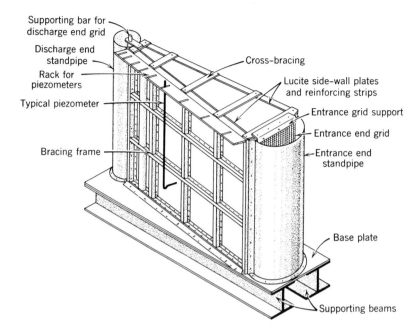

Fig. 10.1 Sand tank model of a well sector to study steady flow to a well penetrating an unconfined aquifer (after Hall[33]).

been constructed in watertight boxes of various shapes—rectangular forms, columns, and sectors are most common. An example of a well sector design appears in Fig. 10.1. From the standpoint of aquifer type, unconfined aquifers can be modeled with the water table serving as the upper boundary; confined aquifers are reproduced by providing an impermeable cover so that pressure can be applied.

Visual identification of a water table is difficult; consequently, water table and piezometric levels can be obtained best from piezometers tapped into the model. Piezometer tubes should be small in order to minimize modifications of the flow pattern. The flow field can be revealed by dye streams from points of dye added to the sand.

Coarse sand placed in small quantities under water and compacted consistently to remove air will yield a uniform permeability within the model. Anisotropic permeabilities can be achieved by layers of different sands.

Capillary rise in a sand model is disproportionately large compared to that occurring under field conditions. For studies of flow patterns, especially those involving confined aquifers, the effects are unimportant; however, in unconfined aquifer investigations corrections for the large capillary rise are necessary. The rise can be minimized by using a coarse-grained porous medium and, if necessary, a more viscous liquid.

Geometric similarity is defined by the model-prototype length ratio

$$L_r = \frac{L_m}{L_p} \qquad (10.1)$$

where the subscripts r, m, and p refer to ratio, model, and prototype, respectively. Because Darcy's law applies to both model and prototype, the velocity ratio for the isotropic case can be expressed as

$$v_r = \frac{v_m}{v_p} = \frac{K_m i_m}{K_p i_p} \qquad (10.2)$$

where K is hydraulic conductivity and i is hydraulic gradient. With equal slopes, the prototype velocity is given by

$$v_r = K_r \qquad (10.3)$$

and the flow rate by

$$Q_r = K_r L_r^2 \qquad (10.4)$$

For unconfined aquifers the time scale can be obtained from

$$Q_r = \frac{V_m/t_m}{V_p/t_p} = \frac{V_r}{t_r} \qquad (10.5)$$

where V is the volume of fluid and t is time. With a specific yield S_y, it follows that

$$S_{yr} L_r^3 = V_r \qquad (10.6)$$

so that from Eqs. 10.4 and 10.5,

$$K_r L_r^2 = \frac{S_{yr} L_r^3}{t_r} \qquad (10.7)$$

and hence

$$t_r = \frac{S_{yr} L_r}{K_r} \qquad (10.8)$$

Sand tank models have been widely employed for investigating a variety of groundwater flow problems.* Examples include studies of well flow, seepage, artificial recharge, dispersion, and seawater intrusion.[14, 16, 39, 40]

---

*The first groundwater model is attributed to P. Forchheimer, who constructed a sand tank model for study of well flow in Graz, Austria, in 1898.

Transparent Models. It is possible to create a transparent porous media model by matching the index of refraction of the medium with that of the fluid. Crushed Pyrex glass and a mineral oil have been successfully combined to achieve the effect inside a glass-walled container. With the transparency it is possible to follow the movement of the fluid by means of dye streams; the phenomenon of dispersion can be particularly well displayed. The model is most useful as a demonstration and educational tool.

Analog Models

Flow through porous media obeys laws that govern other physical systems including laminar flow of fluids, heat, and electricity. These similarities make available a variety of techniques for studying the movement of groundwater. Nonelectrical analog models are described in this section.

Viscous Fluid Models. The movement of a viscous liquid flowing between two closely spaced parallel plates is analogous to that of groundwater flow in two dimensions. The first viscous fluid model based on this principle was developed by Hele-Shaw[35] in England in 1897 to demonstrate flow patterns around variously shaped boundaries.*

With laminar flow between two parallel plates, it can be shown that the flow lines form a two-dimensional potential flow field. The derivation follows from the generalized Navier-Stokes equations of motion. For vertical plates (see Fig. 10.2) the mean velocity of flow with the Dupuit assumptions is given by

$$v_m = -\frac{b^2 \rho_m g}{12 \mu_m} \frac{dh}{dx} \qquad (10.9)$$

where b is the spacing between the plates; ρ_m and μ_m are the model fluid density and viscosity, respectively; g is the acceleration of gravity; and dh/dx is the hydraulic gradient. From the analogy to Darcy's law, it follows that the hydraulic conductivity K_m of the model is

$$K_m = \frac{b^2 \rho_m g}{12 \mu_m} \qquad (10.10)$$

which indicates that the plate spacing and fluid can be selected to correspond to a desired permeability. The velocity ratio

$$v_r = \frac{v_m}{v_p} = \frac{\rho_m b^2 \mu_p}{12 \rho_p k \mu_m} = \frac{b^2 \rho_r}{12 k \mu_r} \qquad (10.11)$$

*The viscous fluid model is also referred to as a Hele-Shaw or parallel plate model.

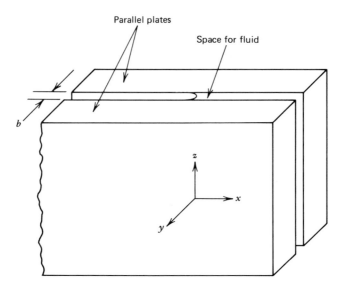

Fig. 10.2 Sketch of viscous fluid flow between two closely spaced vertical and parallel plates. To ensure laminar flow conditions, spacing is usually about one millimeter.

where k is intrinsic permeability; this equation together with a given scale factor (length ratio) L_r of the model enables the time ratio to be found from

$$t_r = \frac{L_r}{v_r} \tag{10.12}$$

Details on model scales are available in the literature.[6,53]

Models of this type are constructed from two sheets of glass or plastic spaced a fixed distance apart. Reservoirs to control fluid flow between the plates are attached at the sides or ends of the model. Oil or glycerin functions satisfactorily for the fluid; for small spacings even water flows in the laminar range. Dye added to the fluid defines the free surface for unconfined flows, while point sources of dye along inflow boundaries reveal flow lines.

The vertical viscous fluid model representing a vertical cross-section through an aquifer has been employed to study bank storage adjacent to streams, seepage and drainage phenomena, seawater intrusion, artificial recharge, and irregular aquifer boundary problems.[18,19,44,73,86] Figure 10.3 shows the design of a vertical model

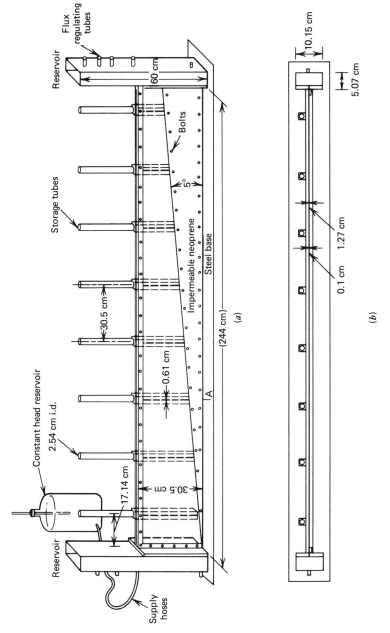

Fig. 10.3 Design of a vertical viscous fluid model for studying flow in a wedge-shaped aquifer. (*a*) General view. (*b*) Plan view (after Columbus[19]).

for flow through a wedge-shaped aquifer.[84] A horizontal model can represent a two-dimensional aquifer for study of regional effects of pumping, recharge, evapotranspiration, drainage, and even flow in multiple-layered aquifers. The space between the plates is analogous to aquifer transmissivity, while vertical tubes represent aquifer storage. Recharge to or discharge from the aquifer is simulated by supplying or withdrawing fluid at appropriate locations.[60,80]

An important advantage of the viscous fluid model is its ability to define the free surface and flow directions for steady and non-steady flow conditions in two dimensions. Care is required in the model construction because the flow rate varies with the cube of the model width; to ensure laminar flow, the plate spacing should be only about one millimeter. Temperature control is usually necessary to maintain fluid properties constant, and corrections for capillary effects should be considered.

Membrane Models. Another model analog for groundwater flow can be constructed with a stretched thin rubber membrane. Small slopes of the membrane surface can be expressed in polar coordinates as

$$\frac{d^2z}{dr^2} + \frac{1}{r}\frac{dz}{dr} = -\frac{W_m}{T_m} \tag{10.13}$$

where dz is the deflection at a radial distance dr from a central deflecting point, W_m is the weight of membrane per unit area, and T_m is a uniform membrane tension. For steady radial flow to a well in a homogeneous and isotropic confined aquifer,

$$\frac{d^2h}{dr^2} + \frac{1}{r}\frac{dh}{dr} = 0 \tag{10.14}$$

When the weight of the membrane is made very small, or when it is placed in a vertical position, the analogy between Eq. 10.13 and 10.14 becomes apparent.

To study the shape of a free surface around a well, a rubber membrane is clamped under uniform tension, while a central probe, representing a pumping well, deflects the membrane (see Fig. 10.4). Deflections, analogous to drawdowns, can be measured by micrometer or by optical techniques.[22] The technique can be adapted for unconfined well flow,[88] for multiple-well systems as shown in Fig. 10.4, and for complex aquifer boundary conditions.[34]

Fig. 10.4 Membrane model representing the piezometric surface in the vicinity of a multiple-well system with equal drawdowns in each well (courtesy V. E. Hansen).

Moiré Pattern Models. Moiré patterns* can be employed to display equipotentials or streamlines of two-dimensional groundwater flow situations. A moiré pattern is an optical phenomenon that occurs when two sets of grid lines are superposed, the pattern being the loci of points of intersection of the two grids. Figure 10.5 shows a moiré pattern representing the streamline pattern about an interface between two fluids of different density in a confined aquifer.

The moiré pattern technique can be adapted to simulate a variety of source-sink situations and is most useful for demonstration purposes.[22,26]

Thermal Models. A thermal analogy can be developed for studdies of flow in homogeneous and isotropic confined aquifers. The flow of heat in a uniform body of material satisfies the Laplace equa-

---

Moiré refers to the irregular wavy finish created on fabric by pressing between engraved rollers; the term is derived from the Frenchman, M. de Moiré, who introduced this texture on silk.

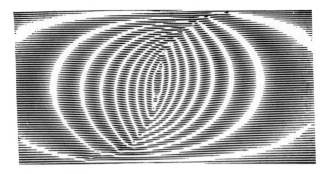

Fig. 10.5 Example of a moiré pattern analogous to stream-lines about an interface between fluids of different density. The illustration represents an instantaneous picture after an initially unstable vertical interface has begun to rotate to a stable horizontal position (courtesy G. de Josselin de Jong).

tion and hence moves as a potential flow system in the same manner as groundwater. The analogies are as follows:

Aquifer	Thermal Model
Hydraulic conductivity	Thermal conductivity
Storage coefficient	Model thickness × density × specific heat
Flow rate	Flow of heat
Head	Temperature

One application of the technique involved nonsteady flow of groundwater to a partially penetrating well.[37] A thick slab of steel, representing the aquifer, was contained between styrofoam insulation, representing confining layers. A heater was imbedded in the steel slab, while temperatures at various distances were measured with thermocouples. In general, the method has not been widely applied because of practical problems of model design and instrumentation.

Blotting Paper Models. A convenient demonstration model for simulating vertical cross sections of groundwater flow under saturated or unsaturated conditions can be constructed from ordinary blotting paper. The paper is cut to any desired shape, hung vertically, and water is introduced by other saturated sheets at appropriate boundary edges. Figure 10.6 illustrates an experimental setup for simulating furrow irrigation. A network of ink dots reveals the pattern of flow. Impermeable zones can be added by painting the paper with quick-drying lacquer, which seals the pores.

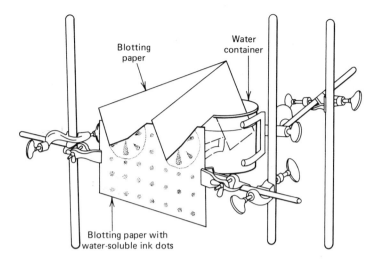

Fig. 10.6 Experimental setup of a blotting paper model for simulating furrow irrigation. Water movement changes ink dots to streaks, thereby revealing flow directions (after Sevenhuysen[62]).

Electric Analog Models

The flow of an electric current can be expressed by Ohm's law

$$I = -\sigma \frac{dE}{dx} \tag{10.15}$$

where I is the electric current per unit area through a material of specific conductivity σ, and dE/dx is the voltage gradient. Equation 10.15 satisfies the Laplace equation and when compared with the Darcy equation

$$v = -K \frac{dh}{dx} \tag{10.16}$$

the similarity between the two equations becomes apparent: velocity v is analogous to electric current I, hydraulic conductivity K to specific conductivity σ, and head h to voltage E. This correspondence serves as the basis for electric analog models of groundwater flow in aquifers.

There are two basic categories of electric analog models. Continuous systems are those in which aquifer properties are modeled by an electric conductive medium that is continuous in space. For these a conductive liquid or a conductive solid serves as the analogous aquifer. The other category includes discrete systems in which aquifer properties are modeled by an assemblage of discrete electric elements forming a network. Resistance-capacitance and resistance networks belong in this group.

Conductive Liquid Models. A conductive liquid model is formed by an insulated tank filled with an electrolyte such as a dilute solution of copper sulfate. Boundaries of the tank are scaled to represent aquifer boundaries. Copper electrodes are immersed in the tank to create equipotential surfaces. Lines of constant potential drop can be traced by a probe connected to an oscilloscope, a voltmeter circuit, and a pantograph. Figure 10.7 illustrates the electrical circuit for a conductive liquid model.

Both equipotential lines and flow lines can be mapped by reversing the conducting and nonconducting boundary surfaces (see Fig.

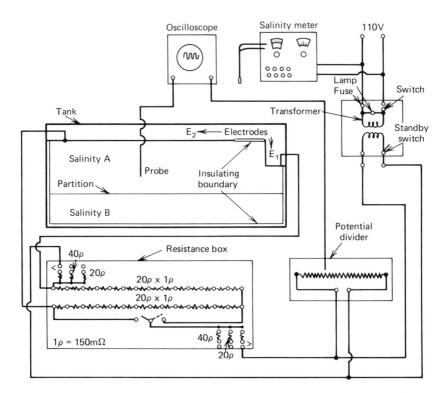

Fig. 10.7 Electrical circuit for a conductive liquid model designed to study seepage from a river into a two-layered anisotropic aquifer (after Todd and Bear[75]).

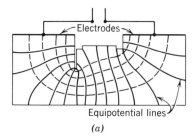

Equipotential lines

(a)

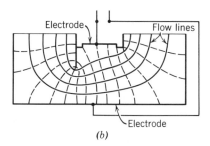

(b)

Fig. 10.8 Electrode arrangments in a conductive liquid model for determining (a) equipotential lines and (b) flow lines of seepage beneath a masonry dam with a cutoff on a pervious foundation.

10.8). Such models are usually restricted to two-dimensional steady-state situations; however, by appropriate model modifications, multiple aquifers, anisotropic permeabilities, and even three-dimensional cases can be studied.

Because there is no analogous force of gravity to produce a water table, free surface boundaries must be determined on a trial-and-error basis. The criterion for locating a water table is based on the fact that the decrease in head is proportional to the decrease in electric potential along the surface.

Conductive liquid models have been applied to investigate a variety of seepage conditions, drawdowns near a well field, and regional groundwater flows.[21,75]

Conductive Solid Models. Conductive solid models operate on the same principle as conductive liquid models except that the medium is a solid instead of a liquid. Materials employed include a narrow wedge of pressed carbon (for radial flow to a well), a saline gelatin mixture (for three-dimensional flow to a partially penetrating well, copper sheets, and carbonized (Teledeltos) paper (for two-dimensional flow). The models can be shaped to represent desired aquifer boundaries under study and are best adapted for studies of steady flow in confined aquifers.[65,66]

Resistance-Capacitance Networks. Resistance-capacitance networks are typically employed to evaluate confined aquifers under
nonsteady, two-dimensional flow conditions.[67,83] The technique
enables an aquifer to be modeled together with the effects of pumping wells. In essence the aquifer is represented by a large array of
individual electric elements that form a scaled-down version of the
aquifer.[9,13] Appropriate electric voltage and current sources are
connected to the individual junctions, or *nodes,* of the model network to create sources, sinks, and external boundaries for the aquifer. Voltage-measuring devices determine the voltage distribution
over the network; this is equivalent after a scale conversion to the
head distribution over the aquifer.

An example of a resistance-capacitance network is shown in Fig.
10.9. Here an electronic analyzer is coupled to the analog model.
Electric resistors are made inversely proportional to aquifer transmissivity, while electric capacitors are made directly proportional
to the aquifer storage. The correspondence of electric and hydraulic
units is described by the scale factors K_1 to K_5 in Fig. 10.9.

The pulse and waveform generators of the electronic analyzer
cause electric current to flow in the analog model at the appropriate
times and in proportion to aquifer flow rates. The oscilloscope measures the time variations of potential levels in the model. Traces of
the oscilloscope, actually time-voltage graphs, are analogous to time-
drawdown graphs.

The analog is based on a finite-difference grid superposed over
a map of the aquifer, as shown in Fig. 10.10a. The elemental area
a^2 should be small compared to the total area of the aquifer so that
the discrete model can represent the continuous aquifer. The finite-
difference form of the nonsteady, two-dimensional equation describing the flow of groundwater is

$$T(h_2 + h_3 + h_4 + h_5 - 4h_1) = a^2 S \frac{\partial h_1}{\partial t} \qquad (10.17)$$

where subscripts on the heads h refer to the node locations of the
aquifer shown in Fig. 10.10a. A corresponding resistor-capacitor
network with a square spacing ρ is shown in Fig. 10.10b. A typical
network node consists of four resistors of equal value R and one
capacitor of value C connected to electric ground. For the node numbered 1, the relation from Kirchoff's current law of

$$\frac{1}{R}(V_2 + V_3 + V_4 + V_5 - 4V_1) = C \frac{\partial V_1}{\partial t} \qquad (10.18)$$

applies where the V terms, with subscripts, refer to voltages at the
corresponding numbered nodes shown in Fig. 10.10b.

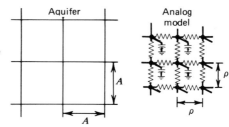

Discretize aquifer with square grid of lengths A, in meters. Model the discretized aquifer with a scaled down array of electrical resistors and capacitors as outlined

Assign numerical values to the following scale factors based on capabilities of available electronic equipment, the particular aquifer simulation desired, and the physical size of the model preferred

$$K_1 = \frac{q\ \text{m}^3}{Q\ \text{coul}} \qquad K_2 = \frac{h\ \text{m}}{V\ \text{volt}} \qquad K_3 = \frac{Q\ \text{m}^3/\text{day}}{I\ \text{amp}} \qquad K_4 = \frac{t_d\ \text{days}}{t_s\ \text{sec}} \qquad K_5 = \frac{A\ \text{m}}{\rho\ \text{cm}}$$

Compute values of resistors and capacitors needed for simulating each portion of the aquifer with the following formulas:

$$R = \frac{K_2}{K_3 T}\ \text{ohms} \qquad\qquad C = A^2 S \frac{K_2}{K_1}\ \text{farads}$$

where T is the local aquifer transmissivity in m^2/day and S is the local aquifer storage coefficient

Use the scale factor equations given below to calculate the electric currents needed to simulate individual pumping rates

$$I_1 = \frac{Q_1}{K_3}, \qquad I_2 = \frac{Q_2}{K_3}, \ldots\ldots\ldots\ldots I_n = \frac{Q_n}{K_3}\ \text{amps}$$

Construct the resistor–capacitor network on a suitable frame to form the analog model. Interconnect the analog model, waveform generator, pulse generator, and oscilloscope according to the wiring diagram below to form the analog simulator

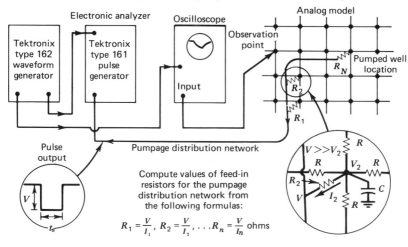

Compute values of feed-in resistors for the pumpage distribution network from the following formulas:

$$R_1 = \frac{V}{I_1}, \quad R_2 = \frac{V}{I_2}, \ldots R_n = \frac{V}{I_n}\ \text{ohms}$$

Install pumpage distribution network, adjust t_s to coincide with the desired length of pumping t_d through the use of scale factor K_4, adjust waveform generator for repetitive control of pulse generator and oscilloscope

Simulator output is in the form of time-voltage traces on the oscilloscope for individual observation points within the aquifer. The time-voltage traces are converted to time-head graphs with the scale factors K_4 and K_2

Fig. 10.9 Example design of a resistance-capacitance network in which an electronic analyzer is coupled to an analog model (after Prickett and Lonnquist[55]).

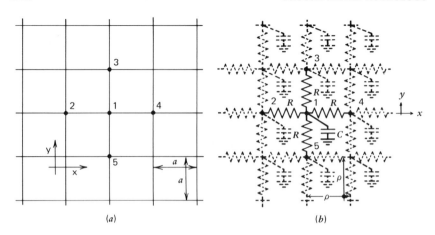

Fig. 10.10   Nodal configuration for an electrical analog of an aquifer.
(*a*) Finite-difference grid superposed over the aquifer. (*b*) Model
schematic of the electrical circuit (after Prickett[53]).

Comparing Eqs. 10.17 and 10.18 demonstrates the analogy between
the two-dimensional flow of groundwater in a homogeneous, isotropic, confined aquifer and the flow of electric current in a resistance-capacitance network. Further analogies include quantity of
water and electric charge, head loss and voltage drop; from Fig. 10.9
the electric pulse generator is analogous to a water pump, while the
oscilloscope functions as a water-level recorder.

Resistance-capacitance network analogs are versatile in that they
can readily study a variety of aquifer conditions as well as extensive
aquifers requiring a large number of nodes.[1,7,25,38,69,78] The technique
can even be extended to the three-dimensional case. Nonhomogeneous aquifer properties are readily incorporated. The only significant limitations on application of the analog involve nonlinear
conditions of varying transmissivity in unconfined aquifers and
two-fluid flow problems.

Resistance Networks.   The design of the resistance type of analog model is basically similar to that already described for the
resistance-capacitance model. The resistance model is composed of
an electronic analyzer coupled to an analog model; however, the
analog consists of an array of resistors only.[5,8,36,70] Capacitors are
eliminated; therefore, no storage elements are included, reducing
the analog to steady-state applications only. Electric and hydraulic
units correspond as before.

The finite difference form of the Laplace equation, which governs steady two-dimensional flow in a homogeneous and isotropic confined aquifer, is given by (from Eq. 10.17)

$$h_2 + h_3 + h_4 + h_5 - 4h_1 = 0 \qquad (10.19)$$

Digital Computer Models

With the widespread availability of digital computers has come the development of mathematical models of aquifers.[58] Applications are expanding, programming techniques are steadily improving, and computer capabilities are growing so that it is safe to say that almost any type of groundwater situation can be studied by means of a digital computer model.[41,45,57] Finite-difference methods, similar to those for electric analog simulation, are well developed; more recently, finite-element methods have emerged as promising alternative techniques. Finally, hybrid computer models combine a resistance network with a digital computer.

Finite-Difference Methods. The finite-difference method is a computational procedure based on dividing an aquifer into a grid and analyzing the flows associated within a single zone of the aquifer. The flow equation is based on the equation of continuity

$$\text{inflow} - \text{outflow} = \text{change of storage} \qquad (10.21)$$

which for a small portion of an aquifer can be restated as

$$\text{sum of subsurface} + \text{net flow to or} = \text{change in storage} \qquad (10.22)$$
$$\text{flows} \qquad \text{from surface}$$

This relation plus Darcy's law for the equation of motion yields the equation[79]

$$\frac{\partial}{\partial x}\left(T\frac{\partial h}{\partial x}\right) + \frac{\partial}{\partial y}\left(T\frac{\partial h}{\partial y}\right) - Q = S\frac{\partial h}{\partial t} \qquad (10.23)$$

where T and S are the aquifer transmissivity and storage coefficient, respectively, Q is the net external inflow, h is head, and t is time.

In finite-difference form Eq. 10.23 can be expressed as

$$\sum_i \frac{W_{iB}T_{iB}}{L_{iB}}(h_i^{j+1} - h_B^{j+1}) - A_B Q_B^{j+1} = \frac{A_B S_B}{\Delta t}(h_B^{j+1} - h_B^j) \qquad (10.24)$$

where W, T, and L are the zonal boundary width, transmissivity, and flow path length, respectively (see Fig. 10.11); A is the area of a single zone; the superscript j denotes points along the time coordinate with Δt being one time step; the subscripts i and B refer to a contiguous zone and the zone in question, respectively (see Fig. 10.11).

Fig. 10.11 Geometry of
an aquifer zone for a
finite-difference solution
to flow within an
aquifer. Zones may have
any polygonal form;
however, squares are
often adopted for
convenience of
computation (after Tyson
and Weber[79]).

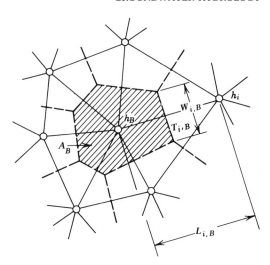

The quantity Q represents the algebraic sum of extraction flows
(pumpage) and replenishment flows (including precipitation, excess
irrigation, imported water, stream percolation, and artificial recharge).

With the zonal configurations defining values of W, L, and A, and
estimates from hydrogeologic data for S, T, and Q, time variations
of h over the aquifer can then be computed from solution of the system of simultaneous equations. For verification of the model, a period
of past records of groundwater levels is selected. Adjustments of
the physical constants S, T, and perhaps Q are made, as needed,
until a satisfactory agreement is reached between the computed
water-level responses and the historical data. Once the model has
been calibrated, it can be applied to study the dynamic behavior of
the basin for a variety of alternative future operational conditions.

One of the earliest digital computer models employing the finite-
difference method was developed by the California Department of
Water Resources[79] to study the dynamic behavior of the Los Angeles
coastal plain groundwater basin, comprising an area of 1240 km².
Since then several computer methods have been developed for solving the simultaneous equations.[49,56] Detailed computer programs
are available in the literature.[10,27,28,29,77]

Applications of numerical modeling employing finite-difference
methods cover a wide range of groundwater topics: groundwater
and well flow,[15,20,59,71,76] unsaturated flow,[31] flow with surface water
bodies,[85] dispersion,[64] saltwater intrusion,[50,63] land subsidence,[30]
mass transport (quality models),[11,17,42,61] and management.[12,43,79,87]

Finite-Element Methods. The finite-element technique involves solving a differential equation for groundwater flow by means of variational calculus.[32,48,51,52] The equation for two-dimensional nonsteady groundwater flow in a nonhomogeneous aquifer can be expressed as

$$\frac{\partial}{\partial x}\left(K_x b \frac{\partial h}{\partial x}\right) + \frac{\partial}{\partial y}\left(K_y b \frac{\partial h}{\partial y}\right) + Q_s = S \frac{\partial h}{\partial t} \qquad (10.25)$$

where K_x and K_y are hydraulic conductivities in the coordinate directions, L is head, b is aquifer thickness, and Q_s is a source or sink function. The solution to this equation is equivalent to finding a solution for h that minimizes the variational function

$$F = \iint \left[\frac{K_x}{2}\left(\frac{\partial h}{\partial x}\right)^2 + \frac{K_y}{2}\left(\frac{\partial h}{\partial y}\right)^2 + \left(S\frac{\partial h}{\partial t} - Q_s\right)h\right] dx\, dy \qquad (10.26)$$

To obtain a numerical solution to Eq. 10.26, the aquifer is subdivided into "finite elements." Figure 10.12 shows an example of such an element within an aquifer. The size and shape of the finite elements are arbitrary, typically being triangular or quadrilateral. In fact, the elements can be disordered and nonuniform and should be smallest where flow is concentrated, such as near a well. The parameters K_x, K_y, S, and Q_s are kept constant for a given element, but they may vary from element to element. To minimize Eq. 10.26,

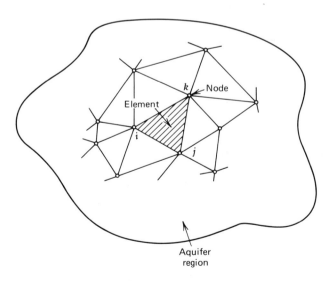

Fig. 10.12 Example of a triangular finite element within an aquifer (after Prickett[53]).

the differential $\partial F/\partial h$ is evaluated for each node and equated to zero. The resulting system of simultaneous equations can then be readily solved by a digital computer.[52]

The choice of whether a finite-element or a finite-difference method is better for aquifer modeling depends on variables such as (1) complexity of the flow system, (2) computer time required for solution, (3) problems of stability and truncation error, and (4) applicability of computer programs. The finite-element method is relatively new; the literature contains applications[23,24,82,89] as well as comparisons of the technique with the finite-difference method.

Hybrid Computer Models. A combination of a digital model and a resistance network analog, known as a hybrid computer model, has been developed to reduce the lengthy computer time sometimes required for iterative finite-difference solutions.[46,81] The digital computer provides the input data, such as sources, sinks, and aquifer properties and boundaries; these are expressed in electric form by a digital-analog converter and connected with the resistance network by means of a distributor. After the analog relaxes the system, the node voltages are fed back to the digital computer through a multiplexer and an analog-digital converter. This approach is most advantageous for solving iteration-intensive problems such as nonsteady flows in unconfined aquifers.

Modeling for Groundwater Management. The development and use of improved mathematical tools are necessary to foster more efficient groundwater management. Digital computer models serve as tools with considerable capability for aiding in decision making related to the various uses, both actual and potential, of groundwater systems.

Numerical modeling of groundwater is a relatively new field; it was not extensively pursued until the mid-1960s, when digital computers with adequate capacity became generally available.[3] Since then significant progress has been made in the development and application of such techniques to groundwater management. A recent survey,[4] however, pointed out gaps that exist between the need for and the actual use of groundwater models in management. Specifically, it was pointed out that:

1. Difficulties in the accessibility of existing models to potential users form a serious impediment; documentation including descriptions of models, listings of codes, and user's manuals would help alleviate this problem.

2. There is need for improved communications between water managers and technical personnel responsible for modeling.

3. Because of inadequacies of input data, the reliability of model output is often seriously questioned; hence, more cost-effective means of data collection are required.

4. Improvements in modeling are needed to make computer codes more understandable and easier to use.

5. Further model development is needed to handle problems in the following areas:

- Flow in media of secondary porosity.
- Flow of immiscible fluids.
- Fully integrated surface and subsurface flows.
- Pollutant transport with chemical and biological reactions.
- Socioeconomic aspects.
- Ecological aspects.
- Consideration of stochasticity.
- Parameter identification.

At present the majority of groundwater modeling is concerned with flows; however, mass-transport models for handling groundwater quality, pollution, and dispersion are increasingly in evidence. The future, in Prickett's words,[54] "looks exciting as more models will be developed, more investigators at the grass-roots level will be effectively using models, and low-priced computer equipment will become commonplace for nearly everyone's use."

References

1. Anderson, T. W., Electrical-analog analysis of the hydrologic system, Tucson Basin, Southeastern Arizona, *U.S. Geological Survey Water-Supply Paper* 1939-C, 34 pp., 1972.
2. Intl. Assoc. Sci. Hydrology, *The use of analog and digital computers in hydrology,* Publs. 80 and 81, 755 pp., 1968.
3. Appel, C. A., and J. D. Bredehoeft, Status of ground-water modeling in the U.S. Geological Survey, *U.S. Geological Survey Circ. 737,* 9 pp., 1976.
4. Bachmat, Y., et al., *Utilization of numerical groundwater models for water resource management,* Rept. EPA-600/8-78-012, U.S. Environmental Protection Agency, Ada, Okla., 177 pp., 1978.
5. Baturic-Rubcic, J., The study of non-linear flow through porous media by means of electrical models, *Jour. Hydr. Research,* v. 7, pp. 31–65, 1969.
6. Bear, J., Scales of viscous analogy models for ground water studies, *Jour. Hydraulics Div.,* Amer. Soc. Civil Engrs., v. 86, no. HY2, pp. 11–23, 1960.
7. Bedinger, M. S., et al., Methods and applications of electrical simulation in ground-water studies in the Lower Arkansas and Verdigris River Valleys, Arkansas and Oklahoma, *U.S. Geological Survey Water-Supply Paper* 1971, 71 pp., 1970.
8. Bouwer, H., Analyzing ground-water mounds by resistance network,

Jour. Irrig. Drain. Div., Amer. Soc. Civil Engrs., v. 88, no. IR3, pp. 15–36, 1962.

9. Bouwer, H., Analyzing subsurface flow systems with electric analogs, *Water Resources Research*, v. 3, pp. 897–907, 1967.

10. Bredehoeft, J. D., and G. F. Pinder, Digital analysis of areal flow in multi-aquifer groundwater systems: a quasi three-dimensional model, *Water Resources Research*, v. 6, pp. 883–888, 1970.

11. Bredehoeft, J. D., and G. F. Pinder, Mass transport in flowing ground-water, *Water Resources Research*, v. 9, pp. 194–210, 1973.

12. Bredehoeft, J. D., and R. A. Young, The temporal allocation of ground-water—a simulation approach, *Water Resources Research*, v. 6, pp. 3–21, 1970.

13. Bredehoeft, J. D., et al., Inertial and storage effects in well-aquifer systems, an analog investigation, *Water Resources Research*, v. 2, pp. 697–707, 1966.

14. Bruch, J. C., Jr., Two-dimensional dispersion experiments in a porous medium, *Water Resources Research*, v. 6, pp. 791–800, 1970.

15. Brutsaert, W. F., et al., Computer analysis of free surface well flow, *Jour. Irrig. Drain. Div.*, Amer. Soc. Civil Engrs., v. 97, no. IR3, pp. 405–420, 1971.

16. Cahill, J. M., Hydraulic sand-model studies of miscible-fluid flow, *U.S. Geological Survey Jour. Research*, v. 1, pp. 243–250, 1973.

17. California Dept. Water Resources, *Mathematical modeling of water quality for water resources management*, 2 vols., The Resources Agency, Sacramento, 304 pp., 1974.

18. Collins, M. A., et al., Hele-Shaw model of Long Island aquifer system, *Jour. Hydraulics Div.*, Amer. Soc. Civil Engrs., v. 98, no. HY9, pp. 1701–1714, 1972.

19. Columbus, N., The design and construction of Hele-Shaw models, *Ground Water*, vol. 4, no. 2, pp. 16–22, 1966.

20. Cooley, R. L., A finite difference method for unsteady flow in variably saturated porous media: application to a single pumping well, *Water Resources Research*, v. 7, pp. 1607–1625, 1971.

21. Debrine, B. E., Electrolytic model study for collector wells under river beds, *Water Resources Research*, v. 6, pp. 971–978, 1970.

22. de Josselin de Jong, G., Moiré patterns of the membrane analogy for ground-water movement applied to multiple fluid flow, *Jour. Geophysical Research*, v. 66, pp. 3625–3629, 1961.

23. Durbin, T. J., Calibration of a mathematical model of the Antelope Valley ground-water basin, California, *U.S. Geological Survey Water-Supply Paper* 2046, 51 pp., 1978.

24. Fang, C. S., et al., Groundwater flow in a sandy tidal beach, 2. Two-dimensional finite element analysis, *Water Resources Research*, v. 8, pp. 121–128, 1972.

25. Fidler, R. E., Potential development and recharge of ground water in Mill Creek Valley, Butler and Hamilton Counties, Ohio, based on analog

model analysis, *U.S. Geological Survey Water-Supply Paper* 1893, 37 pp., 1970.

26. Freeze, R. A., Moiré pattern techniques in groundwater hydrology, *Water Resources Research,* v. 6, pp. 634–641, 1970.

27. Freeze, R. A., Three-dimensional, transient, saturated-unsaturated flow in a groundwater basin, *Water Resources Research,* v. 7, pp. 347–366, 1971.

28. Freeze, R. A., and R. L. Harlan, Blueprint for a physically-based, digitally-simulated hydrologic response model, *Jour. Hydrology,* v. 9, pp. 237–258, 1969.

29. Freeze, R. A., and P. A. Witherspoon, Theoretical analysis of regional groundwater flow: 1. Analytical and numerical solutions to the mathematical model, *Water Resources Research,* v. 2, pp. 641–656, 1966.

30. Gambolati, G., and R. A. Freeze, Mathematical simulation of the subsidence of Venice, *Water Resources Research,* v. 9, pp. 721–733, 1973, and v. 10, pp. 563–577, 1974.

31. Green, D. W., et al., Numerical modeling of unsaturated groundwater flow and comparison of the model to a field experiment, *Water Resources Research,* v. 6, pp. 862–874, 1970.

32. Guymon, G. L., et al., A general numerical solution of the two-dimensional diffusion-convection equation by the finite element method, *Water Resources Research,* v. 6, pp. 1611–1617, 1970.

33. Hall, H. P., An investigation of steady flow toward a gravity well, *La Houille Blanch,* v. 10, pp. 8–35, 1955.

34. Hansen, V. E., Complicated well problems solved by the membrane analogy, *Trans. Amer. Geophysical Union,* v. 33, pp. 912–916, 1952.

35. Hele-Shaw, H. S., Experiments on the nature of the surface resistance in pipes and on ships, *Trans. Inst. Naval Architects,* v. 39, pp. 145–156, 1897.

36. Herbert, R., Time variant ground water flow by resistance network analogues, *Jour. Hydrology,* v. 6, pp. 237–264, 1968.

37. Javandel, I., and P. A. Witherspoon, Use of thermal model to investigate the theory of transient flow to a partially penetrating well, *Water Resources Research,* v. 3, pp. 591–597, 1967.

38. Jorgenson, D. G., *Analog-model studies of groundwater hydrology in the Houston District, Texas,* Texas Water Dev. Board Rept. 190, Austin, 84 pp., 1975.

39. Kraijenhoff van de Leur, Some effects of the unsaturated zone on nonsteady free-surface groundwater flow as studied in a scaled granular model, *Jour. Geophysical Research,* v. 67, pp. 4347–4362, 1962.

40. Kimbler, O. K., Fluid model studies of the storage of freshwater in saline aquifers, *Water Resources Research,* v. 6, pp. 1522–1527, 1970.

41. Kleinecke, D., Use of linear programming for estimating geohydrologic parameters of ground water basins, *Water Resources Research,* v. 7, pp. 367–374, 1971.

42. Konikow, L. F., Modeling chloride movement in the alluvial aquifer at

the Rocky Mountain Arsenal, Colorado, *U.S. Geological Survey Water-Supply Paper* 2044, 43 pp., 1977.

43. Maddaus, W. O., and M. A. Aaronson, A regional groundwater resource management model, *Water Resources Research,* v. 8, pp. 231–237, 1972.

44. Marino, M. A., Hele-Shaw model study of the growth and decay of ground-water ridges, *Jour. Geophysical Research,* vol. 72, pp. 1195–1205, 1967.

45. Meyer, C. F., Using experimental models to guide data gathering, *Jour. Hydraulics Div.,* Amer. Soc. Civil Engrs., v. 97, no. HY10, pp. 1681–1697, 1971.

46. Morris, W. J., et al., Combined surface water-groundwater analysis of hydrological systems with the aid of the hybrid computer, *Water Resources Bull.,* v. 8, pp. 63–76, 1972.

47. Mundorff, M. J., et al., Electric analog studies of flow to wells in the Punjab aquifer of West Pakistan, *U.S. Geological Survey Water-Supply Paper* 1608-N, 28 pp., 1972.

48. Neuman, S. P., and P. A. Witherspoon, Analysis of nonsteady flow with a free surface using the finite element method, *Water Resources Research,* v. 7, pp. 611–623, 1971.

49. Pinder, G. F., and J. D. Bredehoeft, Application of digital computer for aquifer evaluation, *Water Resources Research,* v. 4, pp. 1069–1093, 1968.

50. Pinder, G. F., and H. H. Cooper, Jr., A numerical technique for calculating the transient position of the salt-water front, *Water Resources Research,* v. 6, pp. 876–882, 1970.

51. Pinder, G. F., and E. O. Frind, Application of Galerkin's procedure to aquifer analysis, *Water Resources Research,* v. 8, pp. 108–120. 1972.

52. Pinder, G. F., and W. G. Gray, *Finite element simulation in surface and subsurface hydrology,* Academic Press, New York, 295 pp., 1977.

53. Prickett, T. A., Modeling techniques for groundwater evaluation, *in* Chow, V. T. (ed.), *Advances in hydroscience,* Academic Press, v. 10, pp. 1–143, 1975.

54. Prickett, T. A., Ground-water computer models—state of the art, *Ground Water,* v. 17, pp. 167–173, 1979.

55. Prickett, T. A., and C. G. Lonnquist, Comparison between analog and digital simulation techniques for aquifer evaluation, *Intl. Assoc. Sci. Hydrology Publ.* 81, pp. 625–634, 1968.

56. Prickett, T. A., and C. G. Lonnquist, *Selected digital computer techniques for groundwater resource evaluation,* Bull. 55, Illinois State Water Survey, Urbana, 62 pp., 1971.

57. Prickett, T. A., and C. G. Lonnquist, *Aquifer simulation model for use on disk supported small computer systems,* Circ. 114, Illinois State Water Survey, 21 pp., 1973.

58. Remson, I., et al., *Numerical methods in subsurface hydrology,* John Wiley & Sons, New York, 389 pp., 1971.

59. Rushton, K. R., and S. C. Redshaw, *Seepage and groundwater flow, numerical analysis by analog and digital methods,* John Wiley & Sons, New York, 339 pp., 1979.

60. Santing, G., A horizontal scale model based on the viscous flow analogy for studying groundwater flow in an aquifer having storage, *Intl. Assoc. Sci. Hydrology Publ.* 43, pp. 105–114, 1958.
61. Schwartz, F. W., and P. A. Domenico, Simulation of hydrochemical patterns in regional groundwater flow, *Water Resources Research*, v. 9, pp. 707–720, 1973.
62. Sevenhuysen, R. J., Blotting paper models simulating groundwater flow, *Jour. Hydrology*, v. 10, pp. 276–281, 1970.
63. Shamir, U., and G. Dagan, Motion of the seawater interface in coastal aquifers: a numerical solution, *Water Resources Research*, v. 7, pp. 644–657, 1971.
64. Shamir, U. Y., and D. R. F. Harleman, Numerical solutions for dispersion in porous mediums, *Water Resources Research*, v. 3, pp. 557–581, 1967.
65. Sherwood, C. B., and H. Klein, Use of analog plotter in water-control problems, *Ground Water*, v. 1, no. 1, pp. 8–15, 1963.
66. Shestakov, V. M., On the technique for solving hydrological problems using solid and network models, *Intl. Assoc. Sci. Hydrology Publ.* 77, pp. 353–360, 1968.
67. Skibitzke, H. E., The use of analogue computers for studies in ground-water hydrology, *Jour. Inst. Water Engineers*, v. 17, pp. 216–230, 1963.
68. Skibitzke, H. E., and J. A. DaCosta, The ground-water flow system in the Snake River Plain, Idaho—an idealized analysis, *U.S. Geological Survey Water-Supply Paper* 1536-D, pp. 47–67, 1962.
69. Spieker, A. M., Effect of increased pumping of ground water in the Fairfield-New Baltimore area, Ohio—a prediction by analog-model study, *U.S. Geological Survey Prof. Paper* 605-C, 34 pp., 1968.
70. Stallman, R. W., Electric analog of three-dimensional flow to wells and its application to unconfined aquifers, *U.S. Geological Survey Water-Supply Paper* 1536-H, pp. 205–242, 1963.
71. Taylor, G. S., and J. N. Luthin, Computer methods for transient analysis of water-table aquifers, *Water Resources Research*, v. 5, pp. 144–152, 1969.
72. Thomas, R. G., *Groundwater models*, Irrig. and Drainage Paper 21, Food and Agriculture Organization, United Nations, Rome, Italy, 192 pp., 1973.
73. Todd, D. K., Unsteady flow in porous media by means of a Hele-Shaw viscous fluid model, *Trans. Amer. Geophysical Union*, v. 35, pp. 905–916, 1954.
74. Todd, D. K., Laboratory research with ground-water models, *Intl. Assoc. Sci. Hydrology Publ.* 41, pp. 199–206, 1956.
75. Todd, D. K., and J. Bear, Seepage through layered anisotropic porous media, *Jour. Hydraulics Div.*, Amer. Soc. Civil Engrs., v. 87, no. HY3, pp. 31–57, 1961.
76. Todsen, M., On the solution of transient free-surface flow problems in porous media by finite difference methods, *Jour. Hydrology*, v. 12, pp. 177–210, 1971.
77. Trescott, P. C., et al., Finite-difference model for aquifer simulation in

two dimensions with results of numerical experiments, *U.S. Geological Survey Techniques of Water-Resources Investigations,* Bk. 7, Chap. C1, 116 pp., 1976.

78. Tyley, S. J., Analog model study of the ground-water basin of the Upper Coachella Valley, California, *U.S. Geological Survey Water-Supply Paper 2027,* 77 pp., 1974.

79. Tyson, H. N., Jr., and E. M. Weber, Groundwater management for the nation's future—computer simulation of groundwater basins, *Jour. Hydraulics Div.,* Amer. Soc. Civil Engrs., v. 90, no. HY4, pp. 59–77, 1964.

80. Varrin, R. D., and H. Y. Fang, Design and construction of a horizontal viscous flow model, *Ground Water,* v. 5, no. 3, pp. 35–41, 1967.

81. Vemuri, V., and W. J. Karplus, Identification of nonlinear parameters of ground water basins by hybrid computation, *Water Resources Research,* v. 5, pp. 172–185, 1969.

82. Volker, R. E., Nonlinear flow in porous media by finite elements, *Jour. Hydraulics Div.,* Amer. Soc. Civil Engrs., v. 95, no. HY6, pp. 2093–2114, 1969.

83. Walton, W. C., and T. A. Prickett, Hydrogeologic electric analog computers, *Jour. Hydraulics Div.,* Amer. Soc. Civil Engrs., v. 89, no. HY6, pp. 67–91, 1963.

84. Williams, D. E., Viscous model study of ground-water flow in a wedge-shaped aquifer, *Water Resources Research,* v. 2, pp. 479–486, 1966.

85. Winter, T. C., Numerical simulation analysis of the interaction of lakes and ground water, *U.S. Geological Survey Prof. Paper 1001,* 45 pp., 1976.

86. Yen, B. C., and C. H. Hsie, Viscous flow model for groundwater movement, *Water Resources Research,* v. 8, pp. 1299–1306, 1972.

87. Young, R. A., and J. D. Bredehoeft, Digital computer simulation for solving management problems of conjunctive groundwater and surface water systems, *Water Resources Research,* v. 8, pp. 533–556, 1972.

88. Zee, C. H., et al., Flow into a well by electric and membrane analogy, *Trans. Amer. Soc. Civil Engrs.,* v. 122, pp. 1088–1112, 1957.

89. Zienkiewicz, O., et al., Solutions of anisotropic seepage by finite elements, *Jour. Engrng. Mechs. Div.,* Amer. Soc. Civil Engrs., v. 92, no. EM1, pp. 111–120, 1966.

Surface Investigations
of Groundwater

Although groundwater cannot be seen on the earth's surface, a variety of techniques can provide information concerning its occurrence and—under certain conditions—even its quality from surface or above-surface locations. Surface investigations of groundwater are seldom more than partially successful in that results usually leave the hydrogeologic picture incomplete; however, such methods are normally less costly than subsurface investigations. Geologic methods, involving interpretation of geologic data and field reconnaissance, represent an important first step in any groundwater investigation. Remote sensing from aircraft or satellite has become an increasing valuable tool for understanding subsurface water conditions. Finally, geophysical techniques, especially electric resistivity and seismic refraction methods, provide only indirect indications of groundwater so that underground hydrologic data must be inferred from surface data. Correct interpretation requires supplemental data from subsurface investigations (described in Chapter 12) to substantiate surface findings.

Geologic Methods

Geologic studies enable large areas to be rapidly and economically appraised on a preliminary basis as to their potential for groundwater development. A geologic investigation begins with the collection, analysis, and hydrogeologic interpretation of existing topographic maps, aerial photographs, geologic maps and logs, and other pertinent records. This should be supplemented, when possible, by geologic field reconnaissance and by evaluation of available hydrologic data on: streamflow and springs; well yields; groundwater recharge, discharge, and levels; and water quality. Such an approach should be regarded as a first step in any investigation of subsurface water because no expensive equipment is required; furthermore, information on geologic composition and structure defines the need for field exploration by other methods.

Knowledge of the depositional and erosional events in an area may indicate the extent and regularity of water-bearing formations. The type of rock formation will suggest the magnitude of water yield to be expected; one formation may be adequate for a domestic supply but entirely unsatisfactory for an industrial or municipal supply. Stratigraphy and geologic history of an area may reveal aquifers beneath unsuitable upper strata, the continuity and interconnection of aquifers, or important aquifer boundaries. The nature and thickness of overlying beds, as well as the dip of water-bearing formations, will enable estimates of drilling depths to be made. Similarly, confined aquifers may be noted and the possibility of flowing wells or low pumping lifts foretold. Landforms can often reveal nearsurface unconsolidated formations serving as aquifers, such as glacial outwash, eskers, terraces, and sand dunes. Faults, which may form impermeable barriers to subsurface flow, frequently can be mapped from surface traces.

The fundamental relationships between geology and groundwater are presented elsewhere: geologic formations as aquifers in Chapter 2, the quality of groundwater as affected by geologic sources in Chapter 7, and geologic logs in Chapter 12.

Remote Sensing

Photographs of the earth taken from aircraft or satellite at various electromagnetic wavelength ranges can provide useful information regarding groundwater conditions.[5] The technology of remote sensing has developed rapidly in recent years, while its applications to water resources are still being discovered.[22,23,36] Furthermore, the ready availability of photographs from commercial firms and government agencies has stimulated their use.*

---

*Satellite photographs covering the entire globe are available to the public on a worldwide basis; inquiries should be addressed to EROS Data Center, U.S. Geological Survey, Souix Falls, SD 57198, USA.

Stereoscopic examination of black-and-white aerial photographs has gained steadily in importance. Observable patterns, colors, and relief make it possible to distinguish differences in geology, soils, soil moisture, vegetation, and land use.[3] Thus, photogeology can differentiate between rock and soil types and indicate their permeability and areal distribution—and hence areas of groundwater recharge and discharge.[45] Maps classifying an area into good, fair, and poor groundwater yields can be prepared.[21] Table 11.1 summarizes the role aerial photographs can play as interpretive aids in groundwater studies.

Aerial photographs also reveal fracture patterns in rocks, which can be related to porosity, permeability, and ultimately well yield.[29,47] Springs and marshy areas indicate relatively shallow depths to groundwater. Hydrobotanical studies of vegetation in photographs can be productive.[10,52] Phreatophytes, which transpire water from shallow water tables, define depths to groundwater; Fig. 11.1 illustrates vegetation on an alluvial fan. Halophytes, plants with a high tolerance for soluble salts, and white efflorescences of salt at ground surface indicate the presence of shallow brackish or saline groundwater. Xerophytes, desert plants subsisting on minimal water, suggest a considerable depth to the water table.

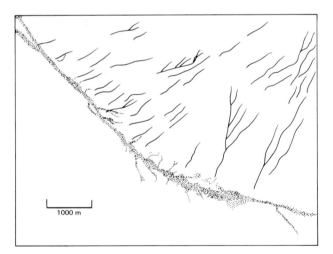

1000 m

Fig. 11.1 Tracing of an aerial photograph showing a strip of phreatophytes along the toe of an alluvial fan in a desert area. There should be a good supply of shallow groundwater along the upslope portion of the strip (after Mann[31]).

**TABLE 11.1   Surficial Features Identified on Aerial Photographs
that Aid in Evaluating Groundwater Conditions
(after Heath and Trainer,** *Introduction to Ground-Water
Hydrology,* **John Wiley, 1968, and Mollard**[36]**)**

Topography
  Appraisal of regional relief setting
  Appraisal of local relief setting
Phreatophytes and aquatic plants
Geologic land forms likely to contain relatively permeable strata
  Modern alluvial terraces and floodplains
  Stratified valley-fill deposits in abandoned meltwater and spillway
  channels
  Glacial outwash and glacial deltas
  Kames and kame-moraine complexes
  Eskerine-kame complexes
  Alluvial fans
  Beach ridges
  Partly drift-filled valleys marked by a chain of elongate closed
  depressions
  Largely masked bedrock valleys cutting across modern valleys, indicated
  by local nonslumping of weak shale strata in valley sides
  Local drift-filled valleys in extensive bedrock-exposed terrains
  Sand dunes assumed to overlie sandy glaciofluvial sediments
Lakes and streams
  Drainage density of stream network
  Localized gain or loss of streamflow
  Nearby small perennial and intermittent lakes (e.g., lakes in outwash,
  elongate saline lakes in inactive drainage systems)
  Perennial rivers and larger creeks in valleys having inactive floodplains
  Small intermittent drainages (including misfit creeks in abandoned glacial
  spillways and meltwater channels)
  No defined drainage channel in former glacial spillways and meltwater
  channels
Moist depressions and seepages
  Moist depressions, marshy environments, and seepages (significance
  depends on interpretation of associated phenomena)
  String of alkali flats or lakes (playas, salinas) along inactive drainage
  systems
  Salt precipitates (e.g., salt crusts), localized anomalous-looking "burn-
  out" patches in the soil, and vegetation associated with salt migration
  and accumulation
Springs (types tentatively inferred from aerial photographs)
  Depression springs (where land surface locally cuts the water table or the
  upper surface of the zone of saturation)
  Contact springs (permeable water-bearing strata overlying relatively

impermeable strata—usually along the sides of valleys that cut across the interface between different strata)
Artesian springs occurring on undulating upland till plains (permeable water-bearing bed between relatively impermeable confining beds, and with enough head to discharge water at the ground surface)
Artesian springs occurring on or near the base of hillsides, valley slopes, and local scarps
Springs where the type could not be reasonably inferred from aerial photographs
Artificial water features
  Wells
  Developed springs
  Reservoirs
  Canals

Other nonvisible portions of the electromagnetic spectrum hold promise for a whole array of imaging techniques that can contribute to hydrogeologic surveys. Infrared imagery, which records differences in apparent surface temperatures, enables information on soil moisture, groundwater circulation, and faults functioning as aquicludes to be obtained.[44] Near-infrared imaging has outlined seepage patterns from canals.[1] One of the most interesting results of infrared aerial imaging has been mapping coastal submarine springs—both hot and cold—in regions of basalt or limestone.[15] Figure 11.2 shows swirls of groundwater, colder than seawater, emerging around the island of Hawaii.* Radar imagery can provide information on the presence of moisture on or at shallow depths below ground surface. Finally, low frequency electromagnetic aerial surveys have outlined buried channels and zones of seawater intrusion.[38]

Geophysical Exploration

Geophysical exploration is the scientific measurement of physical properties of the earth's crust for investigation of mineral deposits or geologic structure.[12,18] With the discovery of oil by geophysical methods in 1926, economic pressures for locating petroleum and mineral deposits stimulated the development and improvement of

---

*The detection of submarine springs should prove most beneficial in the eastern Mediterranean area where such springs are common in karstic limestone. One such spring 2 km offshore of Chekka, Lebanon, discharges an estimated 17 m³/sec at a depth of 43 m and produces potable water at the sea surface.

Fig. 11.2 Infrared aerial photograph showing submarine springs
emerging along the coast of Hilo Bay, Hawaii. Dark ocean areas contain
cooler fresh water (courtesy U.S. Geological Survey).

many geophysical methods and equipment. Application to groundwater investigations was slow because the commercial value of oil
overshadows that of water. In recent years, however, refinement of
geophysical techniques—as well as an increasing recognition of the
advantages of the methods for groundwater study—has changed the
situation. Today, many organizations concerned with groundwater
employ geophysical methods. The methods are frequently inexact
or difficult to interpret, and they are most useful when supplemented
by subsurface investigations.

   Geophysical methods detect differences, or anomalies, of physical
properties within the earth's crust. Density, magnetism, elasticity,
and electrical resistivity are properties most commonly measured;
these are described in the following sections. Experience and research have enabled pronounced differences in these properties to
be interpreted in terms of geologic structure, rock type and porosity,
water content, and water quality.

Electric Resistivity Method

   The electric resistivity of a rock formation limits the amount of
current passing through the formation when an electric potential is
applied. It may be defined as the resistence in ohms between opposite faces of a unit cube of the material. If a material of resistance R

has a cross-sectional area A and a length L, then its resistivity can be expressed as

$$\rho = \frac{RA}{L} \qquad (11.1)$$

Units of resistivity are ohm-m^2/m, or simply ohm-m.

Resistivities of rock formations vary over a wide range, depending on the material, density, porosity, pore size and shape, water content and quality, and temperature.[24] There are no fixed limits for resistivities of various rocks; igneous and metamorphic rocks yield values in the range 10^2 to 10^8 ohm-m; sedimentary and unconsolidated rocks, 10^0 to 10^4 ohm-m. Figure 11.3 provides a representative guide to electric resistivity ranges of various sediments and rocks. In relatively porous formations, the resistivity is controlled more by water content and quality within the formation than by the rock resistivity. For aquifers composed of unconsolidated materials, the resistivity decreases with the degree of saturation and the salinity of the groundwater. Clay minerals conduct electric current through their

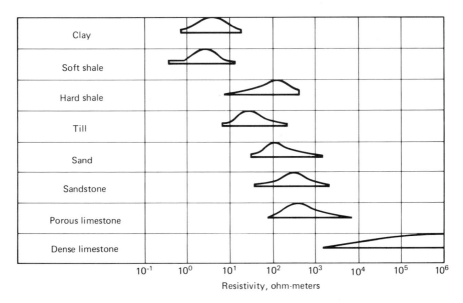

Fig. 11.3 Representative ranges of electrical resistivity for various sediments and rocks. Values assume presence of fresh groundwater; saline water will shift values at least an order of magnitude to the left (after Amer. Soc. Civil Engrs.[1]).

matrix; therefore, clayey formations tend to display lower resistivities than do permeable alluvial aquifers.*

Actual resistivities are determined from apparent resistivities, which are computed from measurements of current and potential differences between pairs of electrodes placed in the ground surface. The procedure involves measuring a potential difference between two electrodes (P in Fig. 11.4) resulting from an applied current through two other electrodes (C in Fig. 11.4) outside but in line with the potential electrodes. If the resistivity is everywhere uniform in the subsurface zone beneath the electrodes, an orthogonal network of circular arcs will be formed by the current and equipotential lines, as shown in Fig. 11.4. The measured potential difference is a

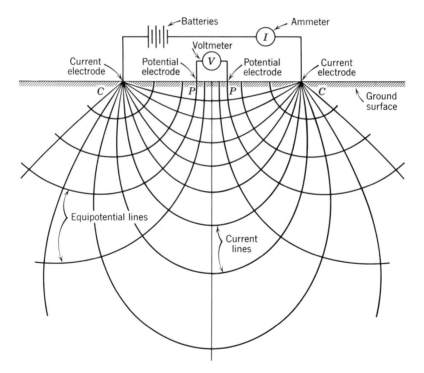

Fig. 11.4 Electrical circuit for resistivity determination and electrical field for a homogeneous subsurface stratum.

*Clay and till when wet typically have low resistivities of 5–30 ohm-m whereas wet sand and gravel have resistivities five to ten times higher; therefore, relatively high resistivity zones are of interest as shallow aquifers.

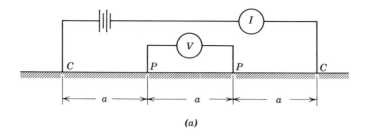

(a)

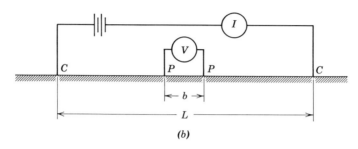

(b)

Fig. 11.5 Common electrode arrangements for resistivity determination. (*a*) Wenner. (*b*) Schlumberger.

weighted value over a subsurface region controlled by the shape of the network. Thus, the measured current and potential differences yield an apparent resistivity over an unspecified depth. If the spacing between electrodes is increased, a deeper penetration of the electric field occurs and a different apparent resistivity is obtained. In general, actual subsurface resistivities vary with depth; therefore, apparent resistivities will change as electrode spacings are increased, but not in a like manner. Because changes of resistivity at great depths have only a slight effect on the apparent resistivity compared to those at shallow depths, the method is seldom effective for determining actual resistivities below a few hundred meters.

Electrodes consist of metal stakes driven into the ground.* In practice, various standard electrode spacing arrangements have been adopted; most common are the Wenner and Schlumberger arrangements.

The Wenner[56] arrangement, shown in Fig. 11.5*a*, has the potential electrodes located at the third points between the current electrodes.

*For the potential electrodes, porous cups filled with a saturated solution of copper sulfate are sometimes employed to inhibit electric fields from forming around them.

The apparent resistivity is given by the ratio of voltage to current times a spacing factor. For the Wenner arrangement, the apparent resistivity

$$\rho_a = 2\pi a \frac{V}{I} \qquad (11.2)$$

where a is the distance between adjacent electrodes, V is the voltage difference between the potential electrodes, and I is the applied current.

The Schlumberger arrangement, shown in Fig. 11.5b, has the potential electrodes close together. The apparent resistivity is given by

$$\rho_a = \pi \frac{(L/2)^2 - (b/2)^2}{b} \frac{V}{I} \qquad (11.3)$$

where L and b are the current and potential electrode spacings, respectively (Fig. 11.5b). Theoretically, $L >> b$, but for practical application good results can be obtained if $L \geq 5b$.[11]

When apparent resistivity is plotted against electrode spacing (a for Wenner, and $L/2$ for Schlumberger) for various spacings at one location, a smooth curve can be drawn through the points. The interpretation of such a resistivity-spacing curve in terms of subsurface conditions is a complex and frequently difficult problem. The solution can be obtained in two parts: (1) interpretation in terms of various layers of actual (as distinguished from apparent) resistivities and their depths; (2) interpretation of the actual resistivities in terms of subsurface geologic and groundwater conditions. Part (1) can be accomplished with theoretically computed resistivity-spacing curves of two-, three-, and four-layer cases for various ratios of resistivities.[4] Curves and explanations of curve-matching techniques have been published for the Wenner configuration[37,51] and the Schlumberger configuration.[11,41] Part (2) depends on supplemental data. Comparing actual resistivity variations with depth to data from a nearby logged test hole enables a correlation to be established with subsurface geologic and groundwater conditions. This information can then be applied for interpretation of resistivity measurements in surrounding areas.

Figure 11.6 illustrates the interpretation of a two-layer situation from measurements with the Schlumberger electrode spacing. The field curve, plotted on logarithmic transparent paper to the same scale as published master curves, is superposed on the two-layer master set. By keeping the coordinate axes parallel, the sheet is moved until a best fit of the field and theoretical curves is obtained. The abscissa of the cross, which is the origin of the theoretical curve,

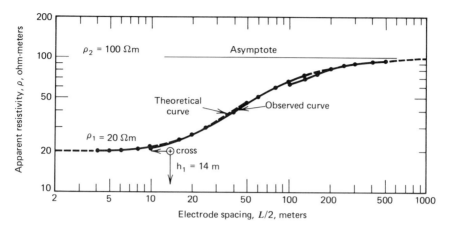

Fig. 11.6 Interpretation of a two-layer electrical resistivity measurement from Schlumberger electrode spacings (after Zohdy[59]).

equals the thickness of the first layer, while the ordinate of the cross defines the actual resistivity ρ_1 of the first layer. The asymptote of the end of the curve with the largest spacing defines the actual resistivity ρ_2 of the second layer. Physically, such a curve might represent a clay layer overlying a sandy aquifer at a depth of 14 m.

Resistivity surveys can cover vertical variations (soundings) at selected locations by varying electrode spacings.* More generally, they are conducted to obtain horizontal profiles of apparent resistivity or apparent-resistivity maps of an area by adopting a constant electrode spacing. Figure 11.7 shows a horizontal resistivity profile across a shallow gravel deposit together with its geologic interpretation. Areal resistivity changes can be interpreted in terms of aquifer limits and changes in groundwater quality, whereas sounding surveys may indicate aquifers, water tables, salinities, impermeable formations, and bedrock depths.

Any factors that disturb the electric field in the vicinity of electrodes may invalidate the resistivity measurements. These may include lateral geologic inhomogeneities; in addition, buried pipelines, cables, and wire fences are common hazards.

Of all surface geophysical methods, electric resistivity has been applied most widely for groundwater investigations. Its portable

---

*It is often assumed that a given electrode spacing represents the depth of resistivity measurement. Although this rule-of-thumb is untrue, the greater the current electrode separation, the greater the amount of current that penetrates a given depth.[59]

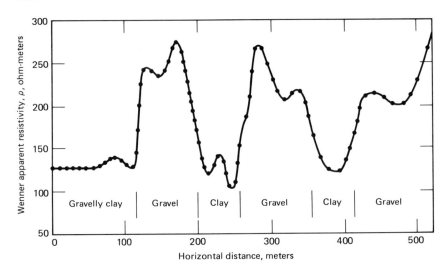

Fig. 11.7 Horizontal profile by surface resistivity measurements over a shallow gravel deposit in California and its interpretation (after Zohdy[59]).

equipment and ease of operation facilitate rapid measurement. The method frequently aids in planning efficient and economic test drilling programs.[39] It is especially well adapted for locating subsurface saltwater boundaries because the decrease in resistance when salt water is encountered becomes apparent on a resistivity-spacing curve. Where subsurface conditions are relatively homogeneous, the technique can be employed to detect the water table as the top of a relatively conductive layer. In California, locations of highly permeable zones for groundwater recharge were aided by resistivity measurements.[42] The method has also been employed for delineating geothermal areas[20] and estimating aquifer permeability.[27]

An important new application of resistivity surveys involves defining areas and magnitudes of polluted groundwater. Results correlate best with groundwater samples where a highly conductive pollutant, such as soluble salt, is moving in a relatively shallow zone with uniform geologic conditions.[7,28] Studies of pollution from landfills,[8,26,49] wastewater disposal,[55] industrial wastes,[19] and acid mine drainage[35] have demonstrated the feasibility of the technique.

Seismic Refraction Method

The seismic refraction method involves the creation of a small shock at the earth's surface either by the impact of a heavy instrument or by a small explosive charge and measuring the time required for the resulting sound, or shock, wave to travel known distances. Seismic waves follow the same laws of propagation as light rays and may be reflected or refracted at any interface where a velocity change occurs. Seismic reflection methods provide information on geologic structure thousands of meters below the surface, whereas seismic refraction methods—of interest in groundwater studies—go only about 100 meters deep.[40] The travel time of a seismic wave depends on the media through which it is passing; velocities are greatest in solid igneous rocks and least in unconsolidated materials.

Characteristic seismic velocities for a variety of geologic materials are shown in Fig. 11.8; these can be employed to identify the nature of alluvium or bedrock. In coarse alluvial materials, seismic velocity increases markedly from unsaturated to saturated zones; consequently, the depth to water table can be mapped, often to an accuracy of 10 percent, where geologic conditions are relatively uniform. Changes in seismic velocities are governed by changes in elastic properties of the formations. The greater the contrast of these properties, the more clearly the formations and their boundaries can be identified. In sedimentary rocks, the texture and geologic history are more important than the mineral composition. Porosity tends to decrease wave velocity, but water content increases it.

For consolidated formations with a uniform distribution of small pores, such as a sandstone, velocity and porosity can be related by[50]

$$\frac{1}{v} = \frac{\alpha}{v_L} + \frac{1 - \alpha}{v_S} \tag{11.4}$$

where v is the measured velocity, v_L is the velocity in the liquid saturating the rock,* v_S is the velocity of the solid rock matrix, and α is porosity of the rock.

A spherical wave expands outward from a shock point, as shown in Fig. 11.9a. It travels at a speed governed by the material through which it is passing. Assume, for example, a homogeneous unconsolidated material with a water table; when the wave reaches the water table it will travel along the interface. As it travels, a series of waves is propagated back into the unsaturated layer. Positions of the wave front drawn at intervals of a few milliseconds in Fig. 11.9a illustrate this refraction. At any location on the surface, the first wave will arrive either directly from the shot point or from a refracted path. By measuring the time interval of the first arrival at varying distances from the shot point, a time-distance graph can be

---

*Seismic velocity in water under typical groundwater conditions approximates 1460 m/s.

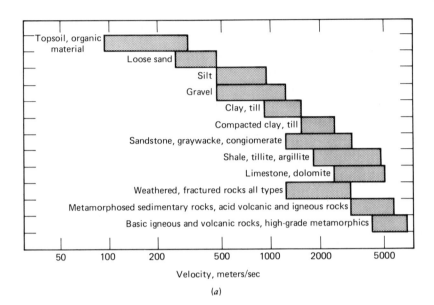

(a)

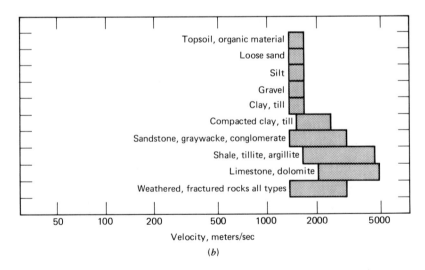

(b)

Fig. 11.8 Seismic velocity of geologic materials. (*a*) Unsaturated materials. (*b*) Saturated materials. (after Amer. Soc. Civil Engrs.[1])

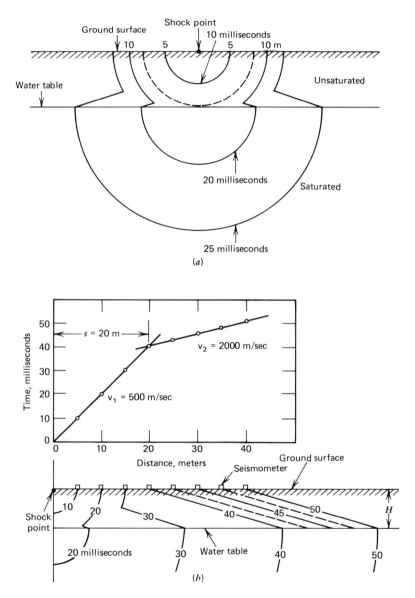

Fig. 11.9 Seismic refraction method applied to determine depth to water table. (*a*) Wave front advance. (*b*) Time-distance graph.

plotted. Reciprocals of the slopes in the time-distance graph of Fig. 11.9b give 500 m/s for the velocity v_1 above the water table and 2000 m/s for the velocity v_2 below. For the horizontal two-layer case here described, the depth H to the water table can be computed from the velocities v_1 and v_2 and the distance s to the intersection on the graph, as shown in Fig. 11.9b. The equation is

$$H = \frac{s}{2} \sqrt{\frac{v_2 - v_1}{v_2 + v_1}} \qquad (11.5)$$

which, when the values of the example are substituted, gives 8 m.

Multilayered problems can be solved in a similar manner, often aided by nomographs. Different surface elevations, sloping formations, faults, and changes in the interfacial configuration require special analysis.[34,40,59] Computational procedures are described in textbooks of geophysics.[12,17,18]

The field procedure for seismic refraction investigations has been simplified with the help of compact and efficient instruments. A small charge of dynamite is placed in a hand-augered hole about one meter deep, and the hole is backfilled. Seismometers, also known as geophones, detectors, or pickups, are spaced in a line from the shock point 3 to 15 m apart. They receive the shock wave and convert the vibration into electric impulses. An electric circuit connects the seismometers to an amplifier and a recording oscillograph, which automatically records the instant of firing and the various first arrivals of the shock wave. Depth determinations to 60 to 100 m are typical with this equipment, although satisfactory work to depths of 300 m has been accomplished. For investigations of depths less than about 20 m, a sledgehammer blow on the ground surface can produce a recordable shock wave.

Interpretation of seismic refraction data assumes homogeneous layers bounded by interfacial planes. Where no distinct boundary exists, but rather a gradual transition zone, a curve replaces the break in slope on the time-distance graph. Fortunately, water tables approximate planes, so that many of the problems imposed by irregular configurations of geologic structure are avoided.[13] Efficient application of the method requires skill in proper interpretation in terms of rock materials, depths, and irregularities. Other knowledge of subsurface conditions aids in proper analysis of field records. The actual presence of groundwater is difficult to determine without supplemental information because velocities overlap in saturated and unsaturated zones. Seismic velocities must increase with depth in order to obtain satisfactory results; as a result, a dense layer overlying an unconsolidated aquifer can mask the presence of the aquifer.

The seismic refraction method in applicable areas can eliminate rapidly and economically areas unfavorable for test drilling. It is not readily adapted to small areas. Minimum distances of a few hundred meters are needed for seismic profiles in different directions. Local noise or vibrations from sources such as highways, airports, and construction sites interfere with seismic work.

The seismic method, because it requires special equipment and trained technicians for operation and interpretation, has been applied to only a relatively limited extent for groundwater investigations. It is commonly employed to map cross sections of alluvial valleys so that variations in thickness of unconfined aquifers can be determined. The successful work of Linehan and Keith[30] in locating groundwater supplies in New England provides notable examples of the applicability of the seismic method. Lateral seismic velocity variations in unsaturated sediments can represent lithology differences that correlate with water well yields.

Gravity and Magnetic Methods

The gravity method measures differences in density on the earth's surface that may indicate geologic structure. Because the method is expensive and because differences in water content in subsurface strata seldom involve measurable differences in specific gravity at the surface, the method has little application to groundwater prospecting. Under special geologic conditions, such as a large buried valley, the gross configuration of an aquifer can be detected from gravity variations.[48,57]

The magnetic method enables magnetic fields of the earth to be mapped. Because magnetic contrasts are seldom associated with groundwater occurrence, the method has little relevance. Indirect information pertinent to groundwater studies, such as dikes that form aquifer boundaries or limits of a basaltic flow, has been obtained with the method.[59]

Water Witching

Using a forked stick to locate water is known as water witching, or dowsing.[14] Although lacking scientific justification for the method, water witches diligently practice the art wherever people can be persuaded of its potential value. Commonly, the method consists of holding a forked stick in both hands and walking over the local area until the butt end is attracted downward—ostensibly by subsurface water.

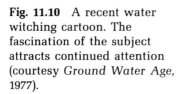

Fig. 11.10 A recent water witching cartoon. The fascination of the subject attracts continued attention (courtesy *Ground Water Age*, 1977).

"I doubt he'll find water, but you gotta admire his confidence"

It is amazing that the idea of supernatural powers has such a continued fascination for people.* Literature on the subject is extensive and spans four centuries. The U.S. Geological Survey advises inquirers not to employ water witches, yet the practice continues and receives frequent publicity (see Fig. 11.10), whereas routine scientific procedures seldom do.[9] During the height of the 1977 drought in California, a San Francisco newspaper reported completion of a suburban flowing well located by means of a bent coat hanger.

An intriguing and well-documented analysis of water witching has been written by an anthropologist and a psychologist.[53] To learn of the practice in the United States they conducted a survey of county agricultural extension agents. This revealed an average of 181 water witches per one million population. Witching proved to be more common in rural areas and where groundwater was difficult to find. The authors concluded water witching to be "magical divination," meaning an irrational system of decision-making in which the signs have no demonstrable connection to the anticipated outcome. The authors present a convincing psychological and social rationale for witching based on the fact that it provides certain and specific answers that tend to relieve anxiety about groundwater resources.

---

*A perpetrator of the legend was the novelist Kenneth Roberts. His novels advocating water witching (including *Henry Gross and His Dowsing Rod*, 1951, *The Seventh Sense*, 1953, and *Water Unlimited*, 1957) are interesting to read. For a humorous and sometimes scathing rebuttal to Roberts, see the paper by Riddick.[46]

References

1. American Soc. Civil Engrs., *Ground water management,* Manual Engrng. Practice 40, New York, 216 pp., 1972.
2. Anon., Geophysics and ground water, *Water Well Jour.,* v. 25, no. 7, pp. 43–60, no. 8, pp. 35–50, 1971.
3. Avery, T. E., *Interpretation of aerial photographs,* Burgess Publishing, Minneapolis, 392 pp., 1977.
4. Bhattacharya, P. K., and H. P. Patra, *Direct current geoelectric sounding—principles and interpretation,* Elsevier, Amsterdam, 135 pp., 1968.
5. Bowden, L. W., and E. L. Pruit (eds.), *Manual of remote sensing,* vol. II, Interpretation and applications, Amer. Soc. Photogrammetry, Falls Church, Virginia, pp. 869–2144, 1975.
6. Breusse, J. J., Modern geophysical methods for subsurface water exploration, *Geophysics,* v. 28, pp. 633–657, 1963.
7. Bugg, S. F., and J. W. Lloyd, A study of fresh water lens configuration on the Cayman Islands using resistivity methods, *Quarterly Jour. Engrng. Geol.,* v. 9, pp. 291–302, 1976.
8. Cartwright, K., and M. R. McComas, Geophysical surveys in the vicinity of sanitary landfills in Northeastern Illinois, *Ground Water,* v. 6, no. 5, pp. 23–30, 1968.
9. Chadwick, D. G., and L. Jensen, *The detection of magnetic fields caused by groundwater and the correlation of such fields with water dowsing,* Utah Water Research Lab., Logan, 57 pp., Jan. 1971.
10. Chikishev, A. G. (ed.), *Plant indicators of soils, rocks, and subsurface waters,* Consultants Bureau, New York, 210 pp., 1965.
11. Compagnie Générale de Géophysique, *Master curves for electrical sounding,* 2nd rev. ed., European Assoc. Exploration Geophysicists, The Hague, Netherlands, 49 pp., 1963.
12. Dobrin, M. B., *Introduction to geophysical prospecting,* 3rd ed., McGraw-Hill, New York, 630 pp., 1976.
13. Duguid, J. O., Refraction determination of water table depth and alluvium thickness, *Geophysics,* v. 33, pp. 481–488, 1968.
14. Ellis, A. J., The divining rod—a history of water witching, *U.S. Geological Survey Water-Supply Paper* 416, 59 pp., 1917.
15. Fischer, W. A., et al., Fresh-water springs of Hawaii from infrared images, *U.S. Geological Survey Hydrologic Atlas* 218, 1966.
16. Flathe, H., Five-layer master curves for the hydrogeological interpretation of geoelectrical resistivity measurements above a two-storey aquifer, *Geophys. Prospecting,* v. 11, pp. 471–508, 1963.
17. Grant, F. S., and G. F. West, *Interpretation theory in applied geophysics,* McGraw-Hill, New York, 583 pp., 1965.
18. Griffiths, D. H., and R. F. King, *Applied geophysics for engineers and geologists,* Pergamon, Oxford, 223 pp., 1965.
19. Hackbarth, D. A., Field study of subsurface spent sulfite liquor movement using earth resistivity measurements, *Ground Water,* v. 9, no. 3, pp. 11–16, 1971.

20. Hatherton, T., et al., Geophysical methods in geothermal prospecting in New Zealand, *Bull. Volcanol.*, v. 29, pp. 485–498, 1966.

21. Howe, R. H., et al., Application of air photo interpretation in the location of ground water, *Jour. Amer. Water Works Assoc.*, vol. 48, pp. 1380–1390, 1956.

22. Idso, S. B., et al., Detection of soil moisture by remote surveillance, *Amer. Sci.*, vol. 63, pp. 549–557, 1975.

23. Inland Waters Branch, *Instrumentation and observation techniques*, Proc. Hydrology Symposium no. 7, Dept. of Energy, Mines and Resources, Ottawa, 343 pp., 1969.

24. Keller, G. V., and F. C. Frischknecht, *Electrical methods in geophysical prospecting*, Pergamon, Oxford, 517 pp., 1966.

25. Kelly, S. F., Geophysical exploration for water by electrical resistivity, *Jour. New England Water Works Assoc.*, v. 76, pp. 118–189, 1962.

26. Kelly, W. E., Geoelectric sounding for delineating ground-water contamination, *Ground Water*, vol. 14, pp. 6–10, 1976.

27. Kelly, W. E., Geoelectric sounding for estimating aquifer hydraulic conductivity, *Ground Water*, v. 15, pp. 420–425, 1977.

28. Klefstad, G., et al., Limitations of the electrical resistivity method in landfill investigations, *Ground Water*, v. 13, pp. 418–427, 1975.

29. Lattman, L. H., and Parizek, R. R., Relationships between fracture traces and the occurrence of groundwater in carbonate rocks, *Jour. Hydrology*, v. 2, pp. 73–91, 1964.

30. Linehan, D., and S. Keith, Seismic reconnaissance for ground-water development, *Jour. New England Water Works Assoc.*, v. 63, pp. 76–95, 1949.

31. Mann, J. F., Jr., Estimating quantity and quality of ground water in dry regions using airphotos, *Intl. Assoc. Sci. Hydrology Publ.* 44, 125–134, 1958.

32. McDonald, H. R., and D. Wantland, Geophysical procedures in ground water study, *Trans. Amer. Soc. Civil Engrs.*, v. 126, pt. III, pp. 122–135, 1961.

33. McGinnis, L. D., and J. P. Kempton, *Integrated seismic, resistivity and geologic studies of glacial deposits*, Circ. 323, Illinois Geol. Survey, Urbana, 23 pp., 1961.

34. Meidav, T., A multilayer seismic refraction nomogram, *Geophysics*, v. 33, pp. 524–526, 1968.

35. Merkel, R. H., The use of resistivity techniques to delineate acid mine drainage in ground water, *Ground Water*, v. 10, no. 5, pp. 38–42, 1972.

36. Mollard, J. D., The role of photo-interpretation in finding groundwater sources in Western Canada, *Proc. 2nd Seminar on Air Photo Interpretation in the Development of Canada*, The Queen's Printer, Ottawa, pp. 57–75, 1968.

37. Mooney, H. M., and W. W. Wetzel, *The potentials about a point electrode and apparent resistivity curves for a two-, three-, and four-layered earth*, Univ. Minnesota Press, Minneapolis, 146 pp. + set of curves, 1956.

38. Morley, L. W. (ed.), *Mining and groundwater geophysics/1967*, Econ. Geol. Rept. 26, Geological Survey of Canada, Ottawa, 722 pp., 1970.

39. Morris, D. B., The application of resistivity methods to ground-water exploration of alluvial basins in semi-arid areas, *Jour. Instn. Water Engrs.*, v. 18, pp. 59–65, 1964.

40. Musgrave, A. W. (ed.), *Seismic refraction prospecting*, Soc. Explor. Geophysists, Tulsa, 604 pp., 1967.

41. Orellana, E., and H. M. Mooney, *Master tables and curves for vertical electrical sounding over layered structures*, Interciencia, Madrid, 150 pp., 66 tables, 1966.

42. Page, L. M., Use of the electrical resistivity method for investigating geologic and hydrologic conditions in Santa Clara County, California, *Ground Water*, v. 6, no. 5, pp. 31–40, 1968.

43. Paver, G. L., The geophysical interpretation of underground water supplies; a geological analysis of observed resistivity data, *Jour. Instn. Water Engrs.*, v. 4, pp. 237–266, 1950.

44. Pluhowski, E. J., Hydrologic interpretations based on infrared imagery of Long Island, New York, *U.S. Geological Survey Water-Supply Paper 2009-B*, 20 pp., 1972.

45. Ray, R. G., Aerial photographs in geologic interpretation and mapping, *U.S. Geological Survey Prof. Paper 373*, 227 pp., 1960.

46. Riddick, T. M., Dowsing—an unorthodox method of locating underground water supplies or an interesting facet of the human mind, *Proc. Amer. Philosophical Soc.*, v. 96, pp. 526–534, 1952.

47. Setzer, J., Hydrologic significance of tectonic fractures detectable on airphotos, *Ground Water*, v. 4, no. 4, pp. 23–27, 1966.

48. Spangler, D. P., and F. J. Libby, Application of the gravity survey methods to watershed hydrology, *Ground Water*, v. 6, no. 6, pp. 21–26, 1968.

49. Stollar, R., and P. Roux, Earth resistivity surveys—a method for defining ground-water contamination, *Ground Water*, v. 13, no. 2, pp. 145–150, 1975.

50. Wyllie, M. R. J., *The fundamentals of well log interpretation*, 3rd ed., Academic Press, New York, 238 pp., 1963.

51. Van Nostrand, R. G., and K. L. Cook, Interpretation of resistivity data, *U.S. Geological Survey Prof. Paper 499*, 310 pp., 1966.

52. Viktorov, S. V., et al., *Short guide to geo-botanical surveying*, Pergamon, Oxford, 158 pp., 1964.

53. Vogt, E. Z., and R. Hyman, *Water witching U.S.A.*, Univ. of Chicago Press, Chicago, 248 pp., 1959.

54. Volker, A., and J. Dijkstra, Détermination des salinités des eaux dans le sous-sol du Zuiderzee par prospection géophysique, *Geophysical Prospecting*, v. 3, pp. 111–125, 1955.

55. Warner, D., Preliminary field studies using earth resistivity measurements for delineating zones of contaminated water, *Ground Water*, v. 7, no. 1, pp. 9–16, 1969.

56. Wenner, F., A method of measuring earth-resistivity, *Bull. Bureau Standards*, v. 12, Washington, D.C., pp. 469–478, 1916.

57. West, R. E., and J. S. Sumner, Ground-water volumes from anomalous mass determinations for alluvial basins, *Ground Water*, v. 10, no. 3, pp. 24–32, 1972.

58. Woollard, G. P., and G. F. Hanson, *Geophysical methods applied to geologic problems in Wisconsin*, Wis. Geological Survey Bull. 78, Madison, 255 pp., 1954.

59. Zohdy, A. A., et al., Application of surface geophysics to ground-water investigations, *U.S. Geological Survey Techniques of Water-Resources Investigations*, Chap. D1, Bk. 2, 116 pp., 1974.

Subsurface Investigations of Groundwater

Detailed and comprehensive study of groundwater and conditions under which it occurs can only be made by subsurface investigations. Whether the information needed concerns an aquifer (its location, thickness, composition, permeability, and yield) or groundwater (its location, movement, and quality), quantitative data can be obtained from subsurface examinations. It should be emphasized that all work classed as subsurface investigations is conducted entirely by personnel on the surface who operate equipment extending underground. Test drilling furnishes information on substrata in a vertical line from the surface. Geophysical logging techniques provide information on physical properties of geologic formations, water quality, and well construction.

Test Drilling

Test drilling of small-diameter holes to ascertain geologic and groundwater conditions is useful in verifying other means of investigation and to obtain assurance of underground conditions prior to well drilling. Many times, if a test hole proves fruitful, it is redrilled or reamed to a larger diameter to form a pumping well. Test holes also serve as observation wells for measuring water levels or for conducting pumping tests.

Almost any well-drilling method can be employed for test drilling; however, in unconsolidated formations, cable tool and hydraulic rotary methods are most common (see Chapter 5). The former is slower but provides more accurate samples from the bailer; the latter is faster, but it is sometimes difficult to determine the exact character of the formations. This fact is particularly true where fine-grained materials are encountered because these mix with the drilling fluid. For great depths and fairly uniform sands, the hydraulic rotary method is quicker and cheaper. Accurate samples can be obtained by pulling the drill stem and using a sampler at the bottom of the hole, or if intact cores are desired in exploratory drilling, a hollow-stem cutter head can be affixed to cut cylindrical cores. For test holes in soft ground and shallow depths, drilling with an auger is quick and economic. Jetting has proved to be an economic method of drilling shallow, small-diameter holes for investigational purposes. The rapidity of the jetting operation combined with its lightweight portable equipment gives it important advantages, but the lack of good samples is a disadvantage. The choice of method for test drilling depends on what information is necessary, the type of material encountered, drilling depth, and location.

Geologic Log. A geologic log is constructed from sampling and examination of well cuttings collected at frequent intervals during the drilling of a well or test hole. Such logs furnish a description of the geologic character and thickness of each stratum encountered as a function of depth, thereby enabling aquifers to be delineated. Figure 12.1 shows a geologic log for a well in unconsolidated alluvium.

Considering all types of logs, the geologic log is probably the most important, but preparation of a good geologic log can be difficult. One problem is that well cuttings are small and mixed with mud. Particularly in rotary drilling, the drilling mud masks the presence of material in the silt and clay ranges. Another problem stems from the fact that most geologic logs are prepared by well drillers, who are busy with many activities during drilling operations and sometimes lack formal geologic training; therefore, logs often tend to be prepared in a perfunctory manner. But experienced drillers, who recognize the value of geologic logs, make diligent efforts to prepare them carefully and completely.

It is good practice to store samples of well cuttings systematically. These not only permit detailed geologic logs to be prepared but also enable grain size analyses and correlations with other nearby wells to be made after drilling is finished. Figure 12.2 shows well cuttings

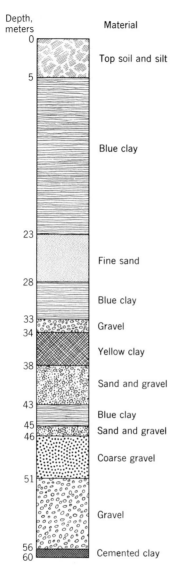

Depth, meters

Material

Fig. 12.1 Geologic log of a well prepared from well cuttings.

collected by a bucket sampler during drilling; they have been arranged in order on the ground, bagged, and labeled for storage.

Drilling-Time Log. A drilling-time log is a useful supplement to test drilling. It consists of an accurate record of the time, in minutes and seconds, required to drill each unit depth of the hole.[29] The technique is most practical with hydraulic rotary drilling although it

Fig. 12.2 Well cuttings, taken at regular intervals from a rotary-drilled well in Joufrah, Libya, being tagged and labeled for subsequent storage in compartmented boxes (photo by David K. Todd).

is applicable to other methods as well. Because the texture of a stratum being penetrated largely governs the drilling rate, a drilling-time log may be readily interpreted in terms of formation types and depths. A portion of one obtained by the hydraulic rotary method is shown in Fig. 12.3 together with the log of the test hole based on cuttings.

Water Level Measurement

One of the most common measurements in groundwater investigations is the determination of the depth to groundwater. In both existing and new wells such data are needed to define groundwater flow directions, changes in water levels over time, and effects of pumping tests.

A simple and accurate method for obtaining water depth is lowering a steel tape into a well. By adding chalk to the end of the tape, the length of submersion becomes apparent, thus giving the distance from the top of the well to the water surface.

For repeated measurements and for depths exceeding 50 m, an electric water-level sounder is preferable. A sounder consists of a battery, a voltmeter, a calibrated two-wire cable on a reel, and an electrode. When the electrode contacts water, the circuit is com-

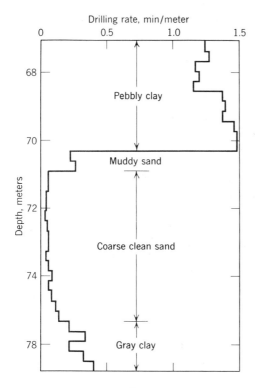

Fig. 12.3 Drilling-time log and strata penetrated (after Kirby[29]; courtesy Johnson Div., UOP Inc.).

pleted and the voltmeter shows a deflection. The depth is read directly from graduations along the cable.

Another widely employed technique is the air-line method.[15] A small-diameter tube is placed in the annular space between the pump column and the casing. Commonly, the tube is fastened to the pump column so that the two are installed simultaneously. The tube extends below the water surface and is connected to a tire pump or small air compressor with a pressure gage. Air is pumped into the tube until a maximum pressure is observed; this pressure, converted to depth of water, indicates the distance from the lower end of the tube to the water surface. Although the air-line method is less accurate than the above methods, it is especially applicable in pumped wells where water splash and turbulence may invalidate other techniques.

A unique and convenient method for measuring water levels in deep wells is the rock technique developed by Stewart.[49] He determined empirically the time required for a common 1.55-cm glass marble or a standard BB (air rifle shot) to fall to the water surface plus the time for the sound of the splash to return to ground surface. Measuring the elapsed time by stopwatch, the depth to water can be read directly from Table 12.1. For water depths exceeding 57 m, the sphere reaches a constant terminal velocity; therefore, for depths greater than those listed in Table 12.1, the equation

$$d = 27.3\, t - 47.6 \tag{12.1}$$

can be employed where d is the depth to water in meters and t is the time interval in seconds. The method is accurate to within 1.5 m.*

Automatic water level recorders are installed in observation wells where short-term fluctuations are of interest such as near intermittently operating wells. A typical recorder consists of a float and counterweight, a gear linkage that rotates a chart drum, and a recording pen driven across the chart by a clock mechanism.

Where multiple aquifers exist with differing water levels, individual observation wells screened in only one aquifer are often drilled. Alternatively, individual small piezometer tubes extending down to the levels of the various aquifers are placed inside a large single perforated casing. The casing is backfilled with sand and sealed by grout between adjoining aquifers.

TABLE 12.1   Depth to Water Surface in a Well as a Function of the Time Interval of a Falling Sphere (after Stewart[49])

Time, s	Distance, m	Time, s	Distance, m	Time, s	Distance, m
0.0	0.0	1.4	9.6	2.8	33.3
0.1	0.0	1.5	11.0	2.9	35.4
0.2	0.2	1.6	12.4	3.0	37.6
0.3	0.4	1.7	13.9	3.1	39.8
0.4	0.8	1.8	15.3	3.2	42.1
0.5	1.2	1.9	16.8	3.3	44.4
0.6	1.8	2.0	18.4	3.4	46.9
0.7	2.4	2.1	20.0	3.5	49.2
0.8	3.1	2.2	21.7	3.6	51.6
0.9	4.0	2.3	23.5	3.7	54.1
1.0	4.9	2.4	25.4	3.8	56.7
1.1	5.9	2.5	27.3	3.9	59.1
1.2	7.1	2.6	29.3	4.0	61.6
1.3	8.3	2.7	31.3	4.1	64.3

*Table 12.1 applies only to the two specified spheres; ordinary pebbles give erratic results because of their irregular shapes. Stewart[49] also pointed out that BBs make a sharp "ping" sound, while marbles cause a short "blurred" sound. Furthermore, he noted, BBs are deflected by spiderwebs, but marbles are not.

Geophysical Logging

Geophysical logging involves lowering sensing devices in a bore-hole and recording a physical parameter that may be interpreted in terms of formation characteristics; groundwater quantity, quality, and movement; or physical structure of the borehole.[23,27,39] A wide variety of logging techniques is available;[24,25] Table 12.2 lists the types of information that can be obtained from various logging techniques described in this chapter.

TABLE 12.2 Summary of Logging Applications to Groundwater Hydrology (after Keys and MacCrary[27])

Required Information	Possible Logging Techniques
Lithology and stratigraphic correlation of aquifers and associated rocks	Resistivity, sonic, or caliper logs made in open holes; radiation logs made in open or cased holes
Total porosity or bulk density	Calibrated sonic logs in open holes; calibrated neutron or gamma-gamma logs in open or cased holes
Effective porosity or true resistivity	Calibrated long-normal resistivity logs
Clay or shale content	Natural gamma logs
Permeability	Under some conditions long-normal resistivity logs
Secondary permeability—fractures, solution openings	Caliper, sonic, or television logs
Specific yield of unconfined aquifers	Calibrated neutron logs
Grain size	Possible relation to formation factor derived from resistivity logs
Location of water level or saturated zones	Resistivity, temperature, or fluid conductivity logs; neutron or gamma-gamma logs in open or cased holes
Moisture content	Calibrated neutron logs
Infiltration	Time-interval neutron logs
Dispersion, dilution, and movement of waste	Fluid conductivity or temperature logs; natural gamma logs for some radioactive wastes
Source and movement of water in a well	Fluid velocity or temperature logs
Chemical and physical characteristics of water, including salinity, temperature, density and viscosity	Calibrated fluid conductivity or temperature logs; resistivity logs

TABLE 12.2   Continued

Required Information	Possible Logging Techniques
Construction of existing wells, diameter and position of casing, perforations, screens	Gamma-gamma, caliper, casing, or television logs
Guide to screen setting	All logs providing data on the lithology, water-bearing characteristics, and correlation and thickness of aquifers
Cementing	Caliper, temperature, or gamma-gamma logs; acoustic logs for cement bond
Casing corrosion	Under some conditions caliper, casing, or television logs
Casing leaks and/or plugged screen	Fluid velocity logs

Geophysical logs furnish continuous records of subsurface conditions that can be correlated from one well to another.[20] They serve as valuable supplements to geologic logs. Data from geophysical logs can be digitized, stored on magnetic tape, or transmitted by radio or telephone for interpretation. Graphic displays of log data permit rapid visual interpretations and comparisons in the field so decisions regarding completion and testing of wells can be made immediately.

The application of geophysical logging to groundwater hydrology lags far behind its comparable use in petroleum exploration. It is doubtful if more than a few percent of the new water wells drilled each year are logged by geophysical equipment. The primary reason for this is cost. Most water wells are shallow, small-diameter holes for domestic water supply; logging costs would be relatively large and usually unnecessary. But for deeper and more expensive wells, such as for municipal, irrigation, or injection purposes, logging can be economically justified in terms of improved well construction and performance. Another deterrent to geophysical logging is the lack of experience among drillers, engineers, and geologists in the interpretation of logs. As logging techniques become more sophisticated, the data they produce become more complex. The interpretation of many logs is more of an art than a science; log responses are governed by numerous environmental factors, making quantitative analysis difficult. In general, best results are obtained with experience and with supplemental hydrogeologic information.

In the following sections geophysical logging techniques are described that are most important in groundwater hydrology. Emphasis

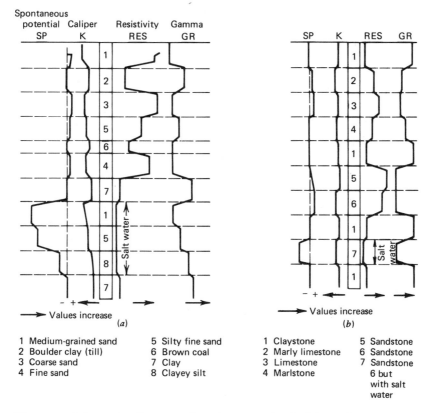

Spontaneous
potential Caliper   Resistivity   Gamma
SP   K   RES   GR

1 Medium-grained sand   5 Silty fine sand
2 Boulder clay (till)   6 Brown coal
3 Coarse sand   7 Clay
4 Fine sand   8 Clayey silt

1 Claystone   5 Sandstone
2 Marly limestone   6 Sandstone
3 Limestone   7 Sandstone
4 Marlstone   6 but
                    with salt
                    water

Fig. 12.4 Schematic diagram of various geophysical logs showing their relative responses in (*a*) unconsolidated rocks and (*b*) consolidated rocks (after Brown, et al.[7]; reproduced by permission of UNESCO).

is on concepts and applications. Figure 12.4 is a schematic diagram showing several of the logs and their typical relative responses in various unconsolidated and consolidated geologic formations.

Resistivity Logging

Within an uncased well, current and potential electrodes can be lowered to measure electric resistivities of the surrounding media and to obtain a trace of their variation with depth. The result is a resistivity (or electric) log. Such a log is affected by fluid within a well, by well diameter, by the character of surrounding strata, and by groundwater.

Of several possible methods for measuring underground resistivities, the multielectrode method is most commonly employed, because it minimizes effects of the drilling fluid and well diameter and also makes possible a direct comparison of several recorded resistivity curves.[31] Four electrodes, two for emitting current and two for potential measurement, constitute the system. Recorded curves are termed *normal* or *lateral,* depending on the electrode arrangement, as shown in Fig. 12.5. In the normal arrangement the effective spacing is considered to be the distance *AM* (Fig. 12.5*a*) and the recorded curve is designated *AM.* Sometimes a long normal curve (*AM'*) is recorded based on the same electrode arrangement as the normal but with a larger *AM* distance (Fig. 12.5*b*). The spacing for lateral (*AO*) curves is taken as the distance *AO,* measured between *A* and a point midway between the electrodes *M* and *N* (Fig. 12.5*c*). Boundaries of formations having different resistivities are located most readily with a short electrode spacing, whereas information on fluids in thick permeable formations can be obtained best with long spacings.

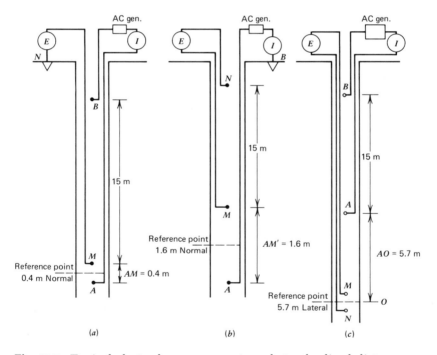

Fig. 12.5 Typical electrode arrangements and standardized distances for resistivity logs. (*a*) Short normal. (*b*) Long normal. (*c*) Lateral (after Keys and MacCrary[27]).

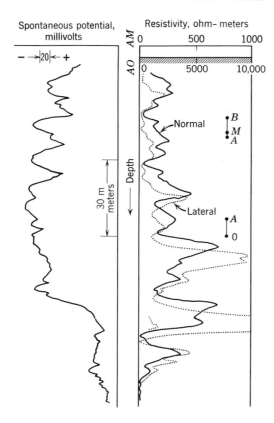

Fig. 12.6 Spontaneous potential and resistivity logs of a well (courtesy Schlumberger Well Surveying Corp.).

An electric log of a well usually consists of vertical traverses that record the short and long normals, the lateral, and the spontaneous potential curves (see following section). An illustration of an electric log is given in Fig. 12.6. Accurate interpretation of resistivity logs is difficult, requires careful analysis, and is best done by specialists.[18,32,55]

Resistivity curves indicate the lithology of rock strata penetrated by the well and enable fresh and salt waters to be distinguished in the surrounding material.[28,30] In old wells exact locations of casings can be determined. Resistivity logs may be used to determine specific resistivities of strata, or they may indicate qualitatively changes of importance. As mentioned in the previous chapter, resistivity of an unconsolidated aquifer is controlled primarily by porosity, packing, water resistivity, degree of saturation, and temperature. Although specific resistivity values cannot be stated for different aquifers, on a relative basis shale, clay, and saltwater sand give low values, freshwater sand moderate to high values, and cemented sandstone and nonporous limestone high values (see Fig. 11.3). Casings and metallic objects will indicate very low resistivities. Correlation of rock samples, taken from wells during drilling, with resistivity curves furnishes a sound basis for interpretation of curves measured in nearby wells without available samples.

Resistivity of groundwater depends on ionic concentration and ionic mobility of the salt solution. This mobility is related to the molecular weight and electrical charge, so that differences exist for various compounds. For example, the ion mobility of a sodium chloride solution is several times that of a comparable calcium carbonate solution. Relationships between resistivity and total dissolved solids for several salt solutions and natural groundwaters are shown in Fig. 12.7.

As the temperature of a groundwater increases, it has a greater ionic mobility, associated with a decrease in viscosity. Hence, an inverse relation exists between resistivity and temperature. The

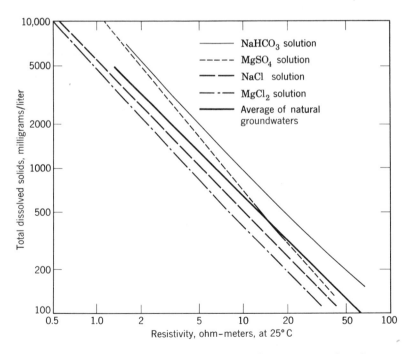

Fig. 12.7 Resistivity-concentration curves for various salt solutions and natural groundwaters (after *Agric. Handbook* 60, U.S. Dept. of Agriculture).

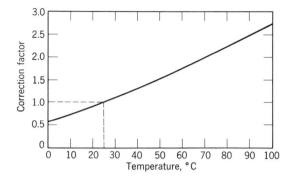

Fig. 12.8 Correction factor to convert resistivities at other temperatures to resistivities at 25°C (after Jones and Buford[22]).

relation, expressed as a correction factor, is shown in Fig. 12.8. Resistivity at the measurement temperature when multiplied by the correction factor for that temperature yields the resistivity at the standard temperature of 25° C.

One of the most common uses of an electric log is to determine the proper place to set well screens. A log provides a basis for selecting proper lengths of screens and for setting them opposite the best formations. Because of this application, many well drillers have their own "loggers" for this purpose.

Investigations of Louisiana aquifers by Jones and Buford[22] and later by Turcan[51] have extended the applicability of resistivity logs to the estimation of groundwater quality. First, a field-formation factor F for an aquifer is determined from previous data by

$$F = \frac{\rho_o}{\rho_w} \tag{12.2}$$

where ρ_o is the resistivity of the saturated aquifer and ρ_w is the resistivity of the groundwater in the aquifer.* Second, specific conductance is related to chloride content or total dissolved solids for the aquifer, as in Fig. 12.7. Finally, with these relationships known, ρ_o is read from the long-normal resistivity curve in an aquifer; this enables ρ_w and then the salinity of the groundwater to be calculated. The method yields best results in uniform clastic aquifers such as sand and sandstone, consisting mainly of intergranular pores saturated with water.

Another application of long-normal curves has been suggested by

---

*It should be noted that the resistivity of groundwater is the reciprocal of its specific conductance (see Chapter 7). The relation has the form

$$\rho_w = 10^4/E_c \tag{12.3}$$

where ρ_w is in ohm-m and E_c is specific conductance in $\mu S/cm$.

Croft[11] to estimate permeability. A value of F is determined as above; then from a previously established relationship, permeability is calculated directly from F. Again, the method should be limited to clastic rock formations.

Resistivity logs can also aid in identifying wells that intersect both fresh and saline zones. Circulation within such a well under nonpumping conditions depends on the relative hydrostatic heads, water densities, aquifer locations and thicknesses, and the physical structure and condition of the well. Various hydrologic conditions for pumping and nonpumping wells are shown diagrammatically in Fig. 12.9 together with corresponding resistivity curves. Resistivity logs are also employed for locating aquifers, determining bed sequences, correlating aquifers, and estimating changes in groundwater quality.

Spontaneous Potential Logging

The spontaneous potential method measures natural electrical potentials found within the earth.* Measurements, usually in millivolts, are obtained from a recording potentiometer connected to two like electrodes. One electrode is lowered in an uncased well and the other is connected to the ground surface, as illustrated by electrodes M and N in Fig. 12.5a. The potentials are primarily produced by electrochemical cells formed by the electrical conductivity differences of drilling mud and groundwater where boundaries of permeable zones intersect a borehole.[33] In some instances electrokinetic effects of fluids moving through permeable formations are also responsible for spontaneous potentials. Therefore, potential logs indicate permeable zones but not in absolute terms; they can also aid in determining casing lengths[14] and in estimating total dissolved solids in groundwater.[53] Where no sharp contrasts occur in permeable zones, as often happens in shallow alluvial formations, potential logs lack relief and contribute little. In urban and industrial areas, spurious earth currents may occur, such as from electric railroads, which interfere with potential logging.

Potential values range from zero to several hundred millivolts. By convention potential logs are read in terms of positive and negative deflections from an arbitrary baseline, usually associated with an impermeable formation of considerable thickness. The sign of the potential depends on the ratio of the salinity (or resistivity) of the drilling mud to the formation water.[13]

---

*Potentials are also referred to as self-potentials or simply SP.

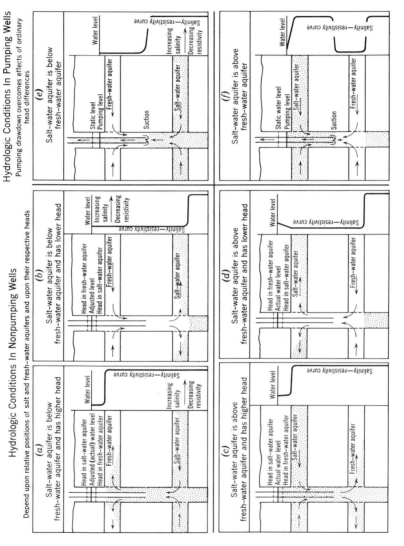

Fig. 12.9 Hydrologic conditions and resistivity curves for wells penetrating two aquifers of different salinities (after Poland and Morrison[45]).

Spontaneous potentials resulting from electrochemical potentials can be expressed by

$$SP = -(64.3 + 0.239T) \log \frac{\rho_f}{\rho_w} \tag{12.4}$$

where ρ_f is the drilling fluid resistivity in ohm-m, ρ_w is the groundwater resistivity in ohm-m, and T is the borehole temperature in $^\circ$ C. Therefore, for measured SP, ρ_f, and T values, the resistivity and hence salinity of groundwater can be determined.[27,53] It should be noted, however, that the formula applies only where the groundwater is very saline, NaCl is the predominant salt, and the drilling mud contains no unusual additives.

In practice, potential and resistivity logs are usually recorded together as shown in Fig. 12.6. The two logs often indicate the same subsurface conditions and thereby supplement each other; however, occasionally the two types of logs will furnish information not available directly from either alone.

Radiation Logging

Radiation logging, also known as nuclear or radioactive logging, involves the measurement of fundamental particles emitted from unstable radioactive isotopes. Logs having application to groundwater are natural gamma, gamma-gamma, and neutron;[54] these are promising but not widely used hydrogeologic tools. An important advantage of these logs over most others is that they may be recorded in either cased or open holes that are filled with any fluid.

Natural-Gamma Logging. Because all rocks emit natural-gamma radiation, a record of this constitutes a natural-gamma log. The radiation originates from unstable isotopes of potassium, uranium, and thorium. In general, the natural-gamma activity of clayey formations is significantly higher than that of quartz sands and carbonate rocks. The most important application to groundwater hydrology is identification of lithology, particularly clayey or shale-bearing sediments, which possess the highest gamma intensity.[36,43] Because most of the gamma rays detected originate within 15–30 cm of the borehole wall, logs run before and after well development can reveal zones where clay and fine-grained material were removed. Borehole dimensions and fluid, casing, and gravel pack all exert minor influences on gamma probe measurements.

Figure 12.10 shows the natural-gamma log of a test hole in unconsolidated sediments together with its geologic interpretation.

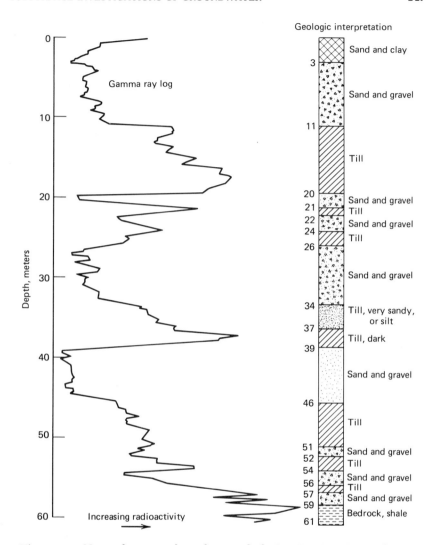

Fig. 12.10 Natural-gamma log of a test hole in Moraine City, Ohio, together with its geologic interpretation (after *USGS Water-Supply Paper* 1808).

Gamma-Gamma Logging. Gamma radiation originating from a source probe and recorded after it is backscattered and attenuated within the borehole and surrounding formation constitutes a gamma-gamma log. The source probe generally contains cobalt-60 or cesium-137, which is shielded from a sodium iodide detector built into the probe. Primary applications of gamma-gamma logs are for identification of lithology and measurement of bulk density and porosity of rocks. The porosity α can be determined by

$$\alpha = \frac{\rho_G - \rho_B}{\rho_G - \rho_F} \tag{12.5}$$

where ρ_G is grain density (obtained from cuttings or cores), ρ_B is bulk density (measured from a calibrated log), and ρ_F is the fluid density. Also, within the same geologic formation it should be possible to derive specific yield from the difference in bulk density measured above and below a water table. Finally, gamma-gamma logs can assist in locating casing, collars, grout, and zones of hole enlargement. As with natural-gamma logs, borehole and fluid conditions affect readings.

Neutron Logging.   Neutron logging is accomplished by a neutron source and detector arranged in a single probe, which produces a record related to the hydrogen content of the borehole environment. In most formations the hydrogen content is directly proportional to the interstitial water; therefore, neutron logs can measure moisture content above the water table and porosity below the water table.[41] Neutrons have a relative mass of 1 and no electric charge; therefore, the loss of energy when passing through matter is by elastic collisions. Neutrons are slowed most effectively by collisions with hydrogen because the nucleus of a hydrogen atom has approximately the same mass as a neutron.* Several designs of neutron probes are currently available utilizing sources of beryllium combined with radium-226, plutonium-239, or americium-241. Probes for measurement of soil moisture (moisture meters) are compact and designed to fit snugly in a small-diameter access tube for accurate quantitative results. For porosity determination in large-diameter holes, larger probes are employed. By measuring moisture contents above and below the water table, the specific yield of unconfined aquifers can be determined.[34] The lateral penetration of neutron logs is in the range 0.2–0.6 m. Neutron log results are influenced by hole size; therefore, in large uncased holes information on hole diameter is required for proper interpretation.

   Figure 12.11 shows a neutron log of a shallow well in unconsolidated alluvium together with the geologic log. This log is calibrated in moisture content as a percentage of bulk volume. Note that the capillary fringe is apparent above the water table as well as high porosity clay layers in the saturated zone.

---

*Dynamically, the slowing of a neutron by a hydrogen atom is analogous to a golf ball losing energy upon collision with another golf ball but rebounding elastically with little energy loss from a large mass such as a concrete wall.

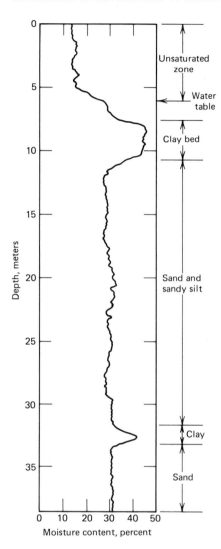

Fig. 12.11 Neutron log of a shallow well in unconsolidated alluvium near Garden City, Kansas, together with its geologic interpretation (courtesy U.S. Geological Survey).

Temperature Logging

A vertical traverse measurement of groundwater temperature in a well can be readily obtained with a recording resistance thermometer. Such data can be of value in analyzing subsurface conditions.[3,47,50] Ordinarily, temperatures will increase with depth in accordance with the geothermal gradient, amounting to roughly 3° C for each 100 m in depth. Departures from this normal gradient may provide information on circulation or geologic conditions in the well.[5,9,48] Abnormally cold temperatures may indicate the presence of gas or, in deep wells, may suggest recharge from ground surface. Likewise, abnormally warm water may occur from water of deep-seated origin. Temperatures may indicate waters from different aquifers intersected by a well. In a few instances temperature logs have aided the location of the approximate top of new concrete behind a casing,[1] because the heat generated during setting produces a marked temperature increase of the water within the casing.

Caliper Logging

A caliper log provides a record of the average diameter of a borehole. Caliper tools are designed either with arms hinged at the upper end and pressed against the hole wall by springs or with bow springs fastened at both ends. These logs aid in the identification of lithology and stratigraphic correlation, in the location of fractures and other rock openings,[38] and in correcting other logs for hole-diameter effects. During well construction caliper logs indicate the size of casing that can be fitted into the hole and enable the annular volume for gravel packing to be calculated. Other applications include measuring casing diameters in old wells and locating swelling and caving zones. A hole caliper and the resulting log are shown in Fig. 12.12.

Fluid-Conductivity Logging

A continuous record of the conductivity of fluid in a borehole is a fluid-conductivity log. The probe measures the AC-voltage drop across two closely spaced electrodes and is governed by the resistivity of the fluid between the electrodes. Fluid resistivity is generally measured in ohm-m; its reciprocal, conductivity, is measured in μS/cm (see Eq. 12.3). Use of the term *fluid-conductivity log* avoids confusion with a resistivity log, which measures rock and fluid conditions outside a borehole. Temperature logs should be made in conjunction with fluid-conductivity logs so that values can be corrected to a standard temperature.

Fluid-conductivity logs enable saline water zones to be located, furnish information on fluid flow within a well, and provide a means to extrapolate water-sample data from a well. Figure 12.13 illustrates how a fluid-conductivity log can define the location and transition zone of saline water underlying fresh water within a well.

Fluid-Velocity Logging

Measurement of fluid movement within a borehole constitutes a fluid-velocity log. Such data reveal strata contributing water to a well, flow from one stratum to another within a well, hydraulic differences between aquifers intersected by a well, and casing leaks. Several flowmeter designs have been developed for boreholes; it is

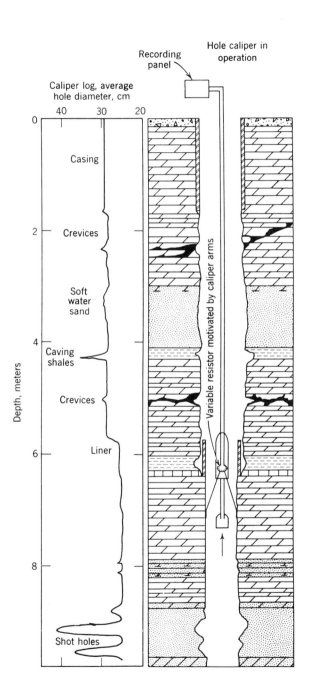

Fig. 12.12 Hole caliper and corresponding caliper log (after Bays and Folk[1]).

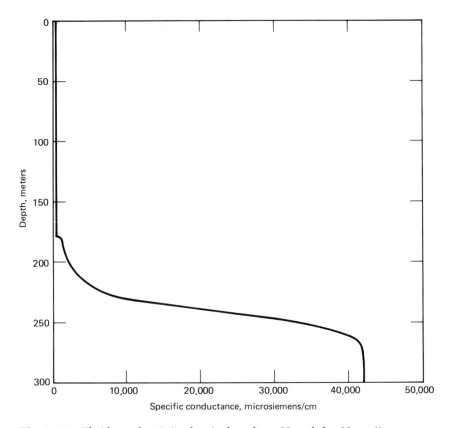

Fig. 12.13 Fluid conductivity log in basalt at Honolulu, Hawaii, showing fresh water overlying saline water and separated by a transition zone (courtesy Honolulu Board of Water Supply).

important that they be compact and sensitive to small water movements and directions.[40] Figure 12.14 illustrates a fluid-velocity log for a pumping well.

Miscellaneous Logging Techniques

Television Logging. A convenient tool with increasing use is a television camera lowered in a well. Specially designed wide-angle cameras, typically less than 7 cm in diameter, are equipped with lights and provide continuous visual inspection of a borehole; with videotape a record of the interior can be preserved. Among the variety of applications are locating changes in geologic strata, pin-pointing large pore spaces, inspecting the condition of well casing and screen, checking for debris in wells, locating zones of sand

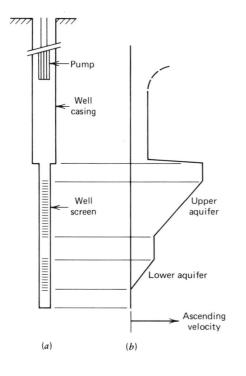

Fig. 12.14 Example of a fluid-velocity log for a well tapping two confined aquifers. (*a*) Well. (*b*) Fluid-velocity log.

entrance, and searching for lost drilling tools.[8,16] Photographs taken within a well at close intervals, termed a photolog,[30] can be employed for the same purposes. As an example, photographs of cased and uncased portions of a well in Honolulu appear in Fig. 12.15.

Acoustic Logging. Acoustic, or sonic, logging measures the velocity of sound through the rock surrounding an uncased, fluid-filled hole. Sound velocity in rock is governed by the velocity of the rock matrix and the fluid filling the pore space (see Chapter 11); therefore, the greater the porosity, the closer the measured sound velocity approaches that of the fluid. Chief applications of the acoustic log include determining the depth and thickness of porous zones,[44] estimating porosity, identifying fracture zones, and determining the bonding of cement between the casing and the formation.

Casing Logging. A casing-collar locator is a useful device for recording locations of casing collars, perforations, and screens.[27] The instrument consists of a magnet wrapped with a coil of wire; voltage fluctuations caused by changes in the mass of metal cutting the lines of flux from the magnet are recorded to form the log.

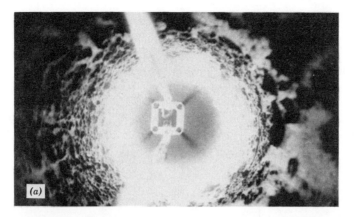

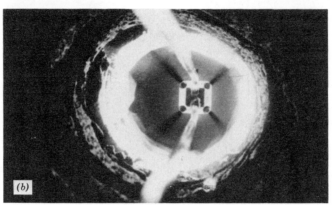

Fig. 12.15 Photographs inside a water supply well in Honolulu, Hawaii. (*a*) Tuberculated condition within a wrought iron casing 83 years old. (*b*) Fractured and cavernous lava formation below the casing (courtesy Honolulu Board of Water Supply).

Other Subsurface Methods

Besides the array of logging techniques described in this chapter, it should be recognized that other subsurface methods can yield important information about hydrogeologic conditions. These have been described in earlier chapters and by way of summary are listed here:

1. Tracer tests for groundwater flow (Chapter 3).
2. Groundwater level measurements for flow directions and aquifer conditions (Chapter 3).
3. Pumping tests of wells for aquifer characteristics (Chapter 4).

4. Groundwater level fluctuation measurements for aquifer characteristics (Chapter 6).

5. Groundwater samples for water quality determination (Chapter 7).

References

1. Bays, C. A., and S. H. Folk, *Developments in the application of geophysics to ground-water problems,* Illinois Geological Survey Circ. 108, Urbana, 25 pp., 1944.

2. Bennett, G. D., and E. P. Pattern, Jr., Borehole geophysical methods for analyzing specific capacity of multiaquifer wells, *U.S. Geological Survey Water-Supply Paper* 1536A, 25 pp., 1960.

3. Birman, J. H., Geothermal exploration for ground water, *Geol. Soc. Amer. Bull.,* v. 80, pp. 617–630, 1969.

4. Blankennagel, R. K., Geophysical logging and hydraulic testing, Pahute Mesa, Nevada Test Site, *Ground Water,* v. 6, no. 4, pp. 24–31, 1968.

5. Bredehoeft, J. D., and I. S. Papadopulos, Rates of vertical ground water movement estimated from earth's thermal profile, *Water Resources Research,* v. 1, pp. 325–328, 1965.

6. Brown, D. L., Techniques for quality-of-water interpretations from calibrated geophysical logs, Atlantic Coastal area, *Ground Water,* v. 9, no. 4, pp. 25–38, 1971.

7. Brown, R. H., et al., *Ground-water studies,* Studies and Repts. in Hydrology 7, UNESCO, Paris, vars. pp., 1972.

8. Callahan, J. T., et al., Television—a new tool for the ground-water geologist, *Ground Water,* v. 1, no. 4, pp. 4–6, 1963.

9. Cartwright, K., Groundwater discharge in the Illinois Basin as suggested by temperature anomalies, *Water Resources Research,* v. 6, no. 3, pp. 912–918, 1970.

10. Cartwright, K., Tracing shallow groundwater systems by soil temperatures, *Water Resources Research,* v. 10, pp. 847–855, 1974.

11. Croft, M. G., A method for calculating permeability from electric logs, *U.S. Geological Survey Prof. Paper* 750-B, pp. B265–B269, 1971.

12. Crosby, J. W., and J. V. Anderson, Some applications of geophysical well logging to basalt hydrogeology, *Ground Water,* v. 9, no. 5, pp. 12–20, 1971.

13. Doll, H. G., The S. P. log: theoretical analysis and principles of interpretation, *Trans. Amer. Inst. Min. and Met. Engrs.,* v. 179, pp. 146–185, 1949.

14. Frimpter, M. H., Casing detector and self-potential logger, *Ground Water,* v. 7, no. 6, pp. 24–27, 1969.

15. Garber, M. S., and F. C. Koopman, Methods of measuring water levels in deep wells, *U.S. Geological Survey Techniques of Water-Resources Investigations,* Bk. 8, Chap. A1, 23 pp., 1968.

16. Gorder, Z. A., Television inspection of a gravel pack well, *Jour. Amer. Water Works Assoc.,* v. 55, pp. 31–34, 1963.

17. Guyod, H., *Electrical well logging fundamentals,* Houston, 164 pp., 1952.

18. Guyod, H., and J. A. Pranglin, *Analysis charts for the determination of true resistivity from electric logs,* Houston, 202 pp., 1959.

19. Ineson, J., and D. A. Gray, Electrical investigations of borehole fluids, *Jour. Hydrology,* v. 1, pp. 204–218, 1963.

20. Jakosky, J. J., *Exploration geophysics,* Trija Publ. Co., Los Angeles, 1195 pp., 1950.

21. Johnson, A. I., An outline of geophysical logging methods and their uses in hydrologic studies, *U.S. Geological Survey Water-Supply Paper 1892,* pp. 158–164, 1968.

22. Jones, P. H., and T. B. Buford, Electric logging applied to ground water exploration, *Geophysics,* v. 16, pp. 115–139, 1951.

23. Jones, P. H., and H. E. Skibitzke, Subsurface geophysical methods in ground-water hydrology, *in Advances in Geophysics* (H. E. Landsberg, ed.), v. 3, Academic Press, New York, pp. 241–300, 1956.

24. Kelley, D. R., *A summary of major geophysical logging methods,* Bull. M61, Pennsylvania Geological Survey, Harrisburg, 82 pp., 1969.

25. Keys, W. S., Well logging in ground-water hydrology, *Ground Water,* v. 6, no. 1, pp. 10–18, 1968.

26. Keys, W. S., and R. F. Brown, The use of well logging in recharge studies of the Ogallala Formation in West Texas, *U.S. Geological Survey Prof. Paper 750-B,* pp. B270–B277, 1971.

27. Keys, W. S., and L. M. MacCrary, Application of borehole geophysics to water-resources investigations, *U.S. Geological Survey Techniques of Water-Resources Invs.,* Bk. 2, Chap. E1, 126 pp., 1971.

28. Keys, W. S., and L. M. MacCrary, Location and characteristics of the interface between brine and fresh water from geophysical logs of bore-holes in the Upper Brazos River Basin, Texas, *U.S. Geological Survey Prof. Paper 809-B,* 23 pp., 1973.

29. Kirby, M. E., Improve your work with drilling-time logs, *Johnson National Drillers Jour.,* v. 26, no. 6, pp. 6–7, 14, 1954.

30. Lao, C., et al., *Application of electric well logging and other well logging methods in Hawaii,* Tech. Rept. 21, Water Resources Research Center, Univ. of Hawaii, Honolulu, 108 pp., 1969.

31. LeRoy, L. W., *Subsurface geologic methods,* 2nd ed., Colorado School of Mines, Golden, 1156 pp., 1951.

32. Lynch, E. J., *Formation evaluation,* Harper and Row, New York, 422 pp., 1962.

33. McCardell, W. M., et al., Origin of the electric potential observed in wells, *Trans. Amer. Inst. Min. and Met. Engrs.,* v. 198, pp. 41–50, 1953.

34. Meyer, W. R., Use of a neutron probe to determine the storage coefficient of an unconfined aquifer, *U.S. Geological Survey Prof. Paper 450E,* pp. E174–E176, 1963.

35. Morley, L. W. (ed.), *Mining and groundwater geophysics/1967,* Economic Geology Rept. 26, Geological Survey Canada, Ottawa, 722 pp., 1970.

36. Norris, S. E., The use of gamma logs in determining the character of unconsolidated sediments and well construction features, *Ground Water,* v. 10, no. 6, pp. 14–21, 1972.

37. Norris, S. E., and A. M. Spieker, Ground-water resources of the Dayton area, Ohio, *U.S. Geological Survey Water-Supply Paper* 1808, 167 pp., 1966.

38. Parizek, R. P., and S. H. Siddiqui, Determining the sustained yield of wells in carbonate and fractured aquifers, *Ground Water*, v. 8, no. 5, pp. 12–20, 1970.

39. Patten, E. P., Jr., and G. D. Bennett, Application of electrical and radioactive well logging to ground-water hydrology, *U.S. Geological Survey Water-Supply Paper* 1544-D, 60 pp., 1963.

40. Patten, E. P., Jr., and G. D. Bennett, Methods of flow measurement in well bores, *U.S. Geological Survey Water-Supply Paper* 1544-C, 28 pp., 1962.

41. Peterson, F. L., *Neutron well logging in Hawaii*, Tech. Rept. 75, Water Resources Research Center, Univ. of Hawaii, Honolulu, 42 pp., 1974.

42. Peterson, F. L., and C. Lao, Electric well logging of Hawaiian basaltic aquifers, *Ground Water*, v. 8, no. 2, pp. 11–18, 1970.

43. Pickell, J. J., and J. G. Heacock, Density logging, *Geophysics*, v. 25, pp. 891–904, 1960.

44. Pickett, G. R., The use of acoustic logs in the evaluation of sandstone reservoirs, *Geophysics*, v. 25, pp. 250–274, 1960.

45. Poland, J. F., and R. B. Morrison, An electrical resistivity-apparatus for testing well-waters, *Trans. Amer. Geophysical Union*, v. 21, pp. 35–46, 1940.

46. Pryor, W. A., *Quality of groundwater estimated from electric resistivity logs*, Circ. 215, Illinois State Geological Survey, 15 pp., 1956.

47. Schneider, R., An application of thermometry to the study of ground water, *U.S. Geological Survey Water-Supply Paper* 1544-B, 16 pp., 1962.

48. Sorey, M. L., Measurement of vertical groundwater velocity from temperature profiles in wells, *Water Resources Research*, v. 7, pp. 963–970, 1971.

49. Stewart, D. M., The rock and bong techniques of measuring water levels in wells, *Ground Water*, v. 8, no. 6, pp. 14–18, 1970.

50. Trainer, F. W., Temperature profiles in water wells as indicators of bedrock fractures, *U.S. Geological Survey Prof. Paper* 600-B, pp. B210–B214, 1968.

51. Turcan, A. N., Jr., *Calculation of water quality from electrical logs—theory and practice*, Water Resources Pamphlet 19, Louisiana Geological Survey, Baton Rouge, 23 pp., 1966.

52. Turcan, A. N., Jr., and A. G. Winslow, Quantitative mapping of salinity, volume, and yield of saline aquifers using borehole geophysical logs, *Water Resources Research*, v. 6, pp. 1478–1481, 1970.

53. Vonhof, J. A., Water quality determination from spontaneous-potential electric log curves, *Jour. Hydrology*, v. 4, pp. 341–347, 1966.

54. Working Group on Nuclear Techniques in Hydrology, *Nuclear well logging in hydrology*, Tech. Repts. Ser. 126, Intl. Atomic Energy Agency, Vienna, 90 pp., 1971.

55. Wyllie, M. R. J., *The fundamentals of well log interpretation*, 3rd ed., Academic Press, New York, 238 pp., 1963.

Artificial Recharge of Groundwater

CHAPTER 13 ··

In order to increase the natural supply of groundwater, people artificially recharge groundwater basins. Artificial recharge may be defined as augmenting the natural movement of surface water into underground formations by some method of construction, by spreading of water, or by artificially changing natural conditions. A variety of methods have been developed, including water spreading, recharging through pits and wells, and pumping to induce recharge from surface water bodies.[34,69,78,86] The choice of a particular method is governed by local topographic, geologic, and soil conditions, the quantity of water to be recharged, and the ultimate water use. In special circumstances land value, water quality, or even climate may be an important factor.

Concept of Artificial Recharge

Artificial recharge projects are designed to serve one or more of the following purposes:

1. Maintain or augment the natural groundwater as an economic resource.
2. Coordinate operation of surface and groundwater reservoirs.
3. Combat adverse conditions such as progressive lowering of groundwater levels, unfavorable salt balance, and saline water intrusion.
4. Provide subsurface storage for local or imported surface waters.

458

5. Reduce or stop significant land subsidence.

6. Provide a localized subsurface distribution system for established wells.

7. Provide treatment and storage for reclaimed wastewater for subsequent reuse.

8. Conserve or extract energy in the form of hot or cold water.

Thus, in most situations, artificial recharge projects not only serve as water-conservation mechanisms but also assist in overcoming problems associated with overdrafts.[19,21] The role of artificial recharge in groundwater management is described in Chapter 9; control of seawater intrusion is described in Chapter 14.

To place water underground for future use requires that adequate amounts of water be obtained for this purpose. In some localities storm runoff is collected in ditches, basins, or reservoirs for subsequent recharge. Elsewhere recharge water is imported into a region by pipeline or aqueduct from a distant surface water source. A third possibility involves utilization of treated wastewater.

Recharging began in Europe early in the nineteenth century and in the United States near the end of the century; since then recharge installations have steadily increased throughout the world.[13,22,35] Recharge basins form integral parts of many Swedish municipal water supply systems.[36] Artificial recharge is widely practiced in Germany to meet industrial and municipal water demands. In the Netherlands, water supply systems for Amsterdam, Leiden, and The Hague include basins for recharging surface water into coastal sand dunes.[10] Today, in California alone, some 276 artificial recharge projects operate in areas where groundwater has been extensively exploited.[62,75]

Recharge Methods

A variety of methods have been developed to recharge groundwater artificially. The most widely practiced methods can be described as types of *water spreading*—releasing water over the ground surface in order to increase the quantity of water infiltrating into the ground and then percolating to the water table.[51] Although field studies of spreading have shown that many factors govern the rate at which water will enter the soil, from a quantitative standpoint, area of recharge and length of time water is in contact with soil are most important. Spreading efficiency is measured in terms of the recharge rate, expressed as the velocity of downward water movement over the wetted area.

Spreading methods may be classified as basin, stream channel, ditch and furrow, flooding, and irrigation. These, together with tech-

**TABLE 13.1   Distribution of Artificial Recharge
Projects in California by Method of Recharge
(after Richter and Chun[62])**

Method	Percent of Recharge Projects	Percent of Recharged Water
Basin	54	58.4
Stream channel	15	29.5
Ditch and furrow	8	9.4
Pit	7	1.3
Well	12	1.0
Flooding	4	0.4
	100	100.0

niques employing pits and recharge wells, are described in the following sections. Table 13.1 lists the distribution of projects in California by method of recharge.

Basin Method.   Water may be recharged by releasing it into basins formed by construction of dikes or levees or by excavation.
Generally, basin sizes and shapes are adapted to land surface slope.
Silt-free water aids in preventing sealing of basins during submer-
gence. Most basins require periodic maintenance to improve infiltration rates by scarifying, disking, or scraping the bottom surfaces
when dry.[76] Where local storm runoff is being recharged, a single
basin will normally suffice, but where streamflow is being diverted
for recharge, a series of basins, often parallel to the natural stream
channel, becomes advantageous.[8] Water from the stream is led by a
ditch into the uppermost basin. As the first fills, it spills into the
second, and the process is repeated through the entire chain of basins. From the lowest basin, any excess water is returned to the
stream channel. Figure 13.1 illustrates a typical plan of a multiple-
basin recharge project. This method permits water contact over 75
to 90 percent of the gross area.

Multiple basins provide for continuity of operation when certain
basins are removed from service for drying and maintenance. Furthermore, where streamflow from storm runoff is being spread, a
series of basins has the advantage that upper basins can be reserved
for settling silt. Figure 13.2 shows an aerial photograph of an extensive series of recharge basins in Los Angeles, California.

Basins, because of their general feasibility, efficient use of space,
and ease of maintenance, are the most favored method of recharge

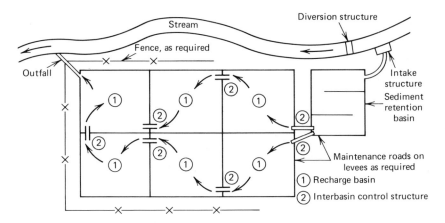

Fig. 13.1 Typical plan of a multiple-basin recharge project diverting water from a stream (after Amer. Soc. Civil Engrs., *Ground Water Management,* Man. and Repts. on Engrng. Practice 40, 1972).

(see Table 13.1). Long-time recharge rates vary widely; Table 13.2 summarizes representative rates obtained for basins in the United States. A study of artificial recharge projects in California indicated that natural ground slope can serve as a convenient guide for estimating long-time rates.[62] For alluvial soils in the slope range of 0.1 to

TABLE 13.2 Representative Spreading Basin Recharge Rates (after Todd[78])

Location	Rate, m/day
Santa Cruz River, Ariz.	0.3–1.2
Los Angeles County, Calif.	0.7–1.9
Madera, Calif.	0.3–1.2
San Gabriel River, Calif.	0.6–1.6
San Joaquin Valley, Calif.	0.1–0.5
Santa Ana River, Calif.	0.5–2.9
Santa Clara Valley, Calif.	0.4–2.2
Tulare County, Calif.	0.1
Ventura County, Calif.	0.4–0.5
Des Moines, Iowa	0.5
Newton, Mass.	1.3
East Orange, N.J.	0.1
Princeton, N.J.	<0.1
Long Island, N.Y.	0.2–0.9
Richland, Wash.	2.3

Fig. 13.2 Aerial view of spreading basins adjoining the San Gabriel River, Los Angeles, California, and temporary finger dikes within the river channel (courtesy Los Angeles County Flood Control District).

10 percent, the long-time infiltration rate W in meters per day is given by

$$W = 0.65 + 0.56\,i \qquad (13.1)$$

where i is the natural ground slope in percent. Individual rates were found to vary within a factor of 2 of this estimate.

    Stream-Channel Method. Water spreading in a natural stream channel involves operations that will increase the time and area over which water is recharged from a naturally losing channel. This involves both upstream management of streamflow and channel modifications to enhance infiltration. Upstream reservoirs enable erratic runoff to be regulated and ideally to limit streamflows to rates that do not exceed the absorptive capacity of downstream channels.

Fig. 13.3 Channel spreading with rock-and-wire check dams in Cucamonga Creek near Upland, California (courtesy D. C. Muckel).

Improvements of stream channels may include widening, leveling, scarifying, or ditching to increase infiltration. In addition, low check dams and dikes can be constructed across a stream where a wide bottom occurs; these act as weirs and distribute the water into shallow ponds occupying the entire streambed (see Fig. 13.3). These structures are normally temporary, consisting of river-bottom material, and sometimes protected by vegetation, wire, or rocks. Such works quickly collapse when high streamflows occur; if permanent structures are placed in a channel, it is important that they do not create a flood hazard.

In Fig. 13.2 can be seen L-shaped finger levees in the stream channel, each of which impounds water. These are simply constructed annually by bulldozer at the end of the high-streamflow season.

Channel spreading can also be conducted without a specific spreading works. In streams having storage reservoirs primarily for flood control, releases of clear water may be entirely recharged into downstream reaches. A majority of the spreading works in and near Los Angeles County are part of an integrated water conservation and flood protection plan.

Ditch-and-Furrow Method. In this method water is distributed to a series of ditches, or furrows, that are shallow, flat-bottomed, and closely spaced to obtain maximum water-contact area. One of three

Fig. 13.4 Spreading ditches in Tujunga Wash, Los Angeles, California (courtesy City of Los Angeles Department of Water and Power).

basic layouts is generally employed: (1) contour, where the ditch follows the ground contour and by means of sharp switchbacks meanders back and forth across the land; (2) tree-shaped, where the main canal successively branches into smaller canals and ditches; and (3) lateral, where a series of small ditches extend laterally from the main canal.[51]

Ditch widths range from 0.3 to 1.8 m. On very steep slopes checks are sometimes placed in ditches to minimize erosion and to increase the wetted area.

Gradients of major feeder ditches should be sufficient to carry suspended material through the system. Deposition of fine-grained material clogs soil surface openings. Although a variety of ditch plans have been devised, a particular plan should be tailored to the configuration of the local area. A collecting ditch is needed at the lower end of the site to convey excess water back into the main stream channel. The method is adaptable to irregular terrain but seldom provides water contact to more than about 10 percent of the gross area. Figure 13.4 shows typical spreading ditches on an alluvial plain.

Flooding Method. In relatively flat topography, water may be diverted to spread evenly over a large area. In practice, canals and earth distributing gullies are usually needed to release the water at intervals over the upper end of the flooding area. It is desirable to form a thin sheet of water over the land, which moves at a minimum velocity to avoid disturbing the soil. Tests indicate that highest infiltration rates occur on areas with undisturbed vegetation and soil covering. Compared with other spreading methods, flood spreading costs least for land preparation. In order to control the water at all times, embankments or ditches should surround the entire flooding area. To obtain maximum efficiency, a person should be on the grounds during flooding operations, because frequently movement of a few shovelsful of dirt will effectively increase the wetted area.

Irrigation Method. In irrigated areas water is sometimes deliberately spread by irrigating cropland with excess water during dormant, winter, or nonirrigating seasons. The method requires no additional cost for land preparation because the distribution system is already installed. Even keeping irrigation canals full will contribute to recharge by seepage from the canals. Where a large portion of the water supply is pumped, the method has the advantage of raising the water table and consequently reducing power costs. Consideration needs to be given to the effects of the leaching action of the percolating water both in carrying salts from the root zone to groundwater and in removing soil nutrients, thereby possibly reducing crop yields.

Pit Method. A pit excavated into a permeable formation serves as an ideal facility for groundwater recharge. Because the cost of excavation and removal of material is high, use of abandoned excavations, such as gravel pits, is most economic. In areas where shallow subsurface strata, such as hardpans and clay layers, restrict the downward passage of water, pits can effectively reach materials with higher infiltration rates.[40,48]

Besides the small capital cost of pits constructed for a purpose other than recharge, the steep sides provide a high silt tolerance. Silt usually settles to the bottom of the pit, leaving the walls relatively unclogged for continued infiltration of water.[8] Nonsilty water should be recharged whenever possible so as to minimize silt accumulation and periodic removal costs.[27] Attention to the geometry of a recharge pit is important in order to obtain the maximum infiltration rate.[24,67]

Recharge pits have been in operation at Peoria, Illinois, since 1951.[74] These penetrate the shallow sand and gravel aquifer serving well fields in the area and recharge chlorinated river water. One pit

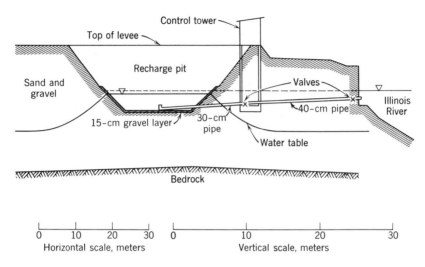

0   10   20   30   0   10   20   30

Horizontal scale, meters   Vertical scale, meters

Fig. 13.5 Cross section of a recharge pit at Peoria, Illinois (after Suter and Harmeson[71]).

is 9 m deep with side slopes of two horizontal to one vertical and 3 m below the river stage (see Fig. 13.5); another is 7 m deep with side slopes of three horizontal to one vertical and 2 m below the river stage. The bottom and side slopes are covered with a 15-cm layer of natural gravel ranging in size from 3.4 to 9.3 mm and with a permeability some 19 times greater than the underlying aquifer. The gravel is replaced whenever its silt accumulation reaches 0.97 kg/l of gravel; this occurs every three to four years. Chlorine is added at the rate of 3 to 5 mg/l. Because local groundwater is largely used for industrial cooling, recharge takes place only when the river temperature is below 18° C, which averages about six months per year. Recharge rates initially averaged about 23 m/day over the filtering surface. During 13 seasons of operation, rates gradually decreased to about 12 m/day. The reduction can be attributed to silt penetration into the upper portion of the aquifer.

In this connection laboratory studies by the Illinois State Water Survey[30] of the filtration efficiency of coarse media resulted in the equation

$$SS_o = 13.1 H^{-0.25} d^{0.5} Q_o^{0.33} SS_i^{1.33} \qquad (13.2)$$

where SS_o is suspended solids concentration (mg/l) transmitted through the filter layer, H is filter layer thickness (cm), d is mean diameter (mm) of particles forming the filter layer, Q_o is the rate of recharge (m/day), and SS_i is suspended solids concentration of the recharged water. This relation can serve as a guide for both the design and operation of a recharge pit.

Recharge Well Method. A *recharge well* may be defined as a well that admits water from the surface to freshwater aquifers.* Its flow is the reverse of a pumping well, but its construction may or may not be the same. Well recharging is practical where deep, confined aquifers must be recharged, or where economy of space, such as in urban areas, is an important consideration.[77]

If water is admitted into a well, a cone of recharge will be formed that is similar in shape but is the reverse of a cone of depression surrounding a pumping well. The equation for the curve can be derived in a similar manner to that for a pumping well (see Chapter 4). For a confined aquifer with water being recharged into a completely penetrating well at a rate Q_r, the approximate steady-state expression

$$Q_r = \frac{2\pi K b (h_w - h_0)}{\ln(r_0/r_w)} \tag{13.3}$$

is applicable. Symbols are identified in Fig. 13.6a. For a recharge well penetrating an unconfined aquifer (see Fig. 13.6b),

$$Q_r = \frac{\pi K (h_w^2 - h_0^2)}{\ln(r_0/r_w)} \tag{13.4}$$

By comparing the discharge equations for pumping and recharge wells, it might be anticipated that the recharge capacity would equal the pumping capacity of a well if the recharge cone has dimensions equivalent to the cone of depression. Field measurements, however, rarely support this reasoning; recharge rates seldom equal pumping rates. The difficulty lies in the fact that pumping and recharging differ by more than a simple change of flow direction.

As water is pumped from a well, fine material present in the aquifer is carried through the coarser particles surrounding the well and into the well. On the other hand, any silt carried by water into a recharge well is filtered out and tends to clog the aquifer surrounding the well.[58,59] Similarly, recharge water may carry large amounts of dissolved air, tending to reduce the permeability of the aquifer

---

*Recharge wells are also known as *disposal wells* and *drain wells.* They should be distinguished from *injection wells,* described in Chapter 8, which recharge brines and toxic industrial wastes to deep, saline-water aquifers.

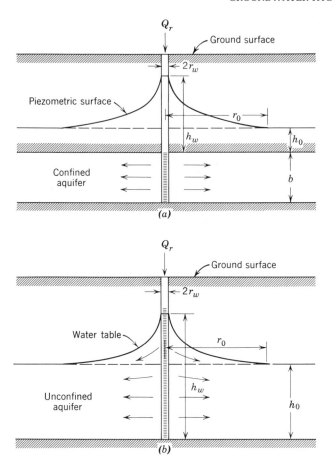

Fig. 13.6 Radial flow from recharge wells penetrating
(a) confined and (b) unconfined aquifers.

by air binding. Recharge water may contain bacteria, which can
form growths on the well screen and the surrounding formation,
thereby reducing the effective flow area. Chemical constituents of
the recharge water may differ sufficiently from the normal groundwater to cause undesired chemical reactions—for example, deflocculation caused by reaction of high-sodium water with fine soil
particles. These factors all act to reduce recharge rates; as a result,
well recharging has been limited to a few areas where experience,
particularly with water treatment and redevelopment of wells, has
shown the practicality of the method.[33,83]

TABLE 13.3   Average Well Recharge
Rates (after Todd[78])

Location	Rate, m^3/day
Fresno, Calif.	500–2200
Los Angeles, Calif.	2900
Manhattan Beach, Calif.	1000–2400
Orange Cove, Calif.	1700–2200
San Fernando Valley, Calif.	700
Tulare County, Calif.	300
Orlando, Fla.	500–51,000
Mud Lake, Idaho	500–2400
Jackson County, Mich.	200
Newark, N.J.	1500
Long Island, N.Y.	500–5400
El Paso, Texas	5600
High Plains, Texas	700–2700
Williamsburg, Va.	700

Measured inflow rates of recharge wells at several locations in the United States are tabulated in Table 13.3. In most cases these figures represent average rates based on continued operation. Recharge wells, like spreading areas, may show initial large intake rates followed by nearly constant or slowly decreasing values. Greatest intake rates are found in extremely porous formations such as limestones and lavas. Detailed studies of recharge wells have been conducted in Arkansas,[70] Minnesota,[61] and Oregon.[26]

It should be noted that supply wells can alternate as recharge wells, as shown in Fig. 13.7a. During World War II, distilleries in Louisville, Kentucky, recharged municipal water into their own pumping wells as an emergency measure to alleviate a serious groundwater overdraft. In a few localities recharge wells replace sewer systems to dispose of storm runoff. A noteworthy example is Orlando, Florida, where wells penetrate from 35 to 300 m into limestone, have casings to depths of 20 to 120 m, and range from 12 to 45 cm in diameter.[81] Although considerable rubbish is carried into the wells, the cavernous limestone formations seldom become clogged. The water table lies about 10 to 15 m below ground surface. Because these formations do not provide adequate filtration, a danger of pollution exists to any nearby water supply wells.

Recharge wells serve as convenient means for disposal of septic tank effluent, excess irrigation water, and surface runoff into the

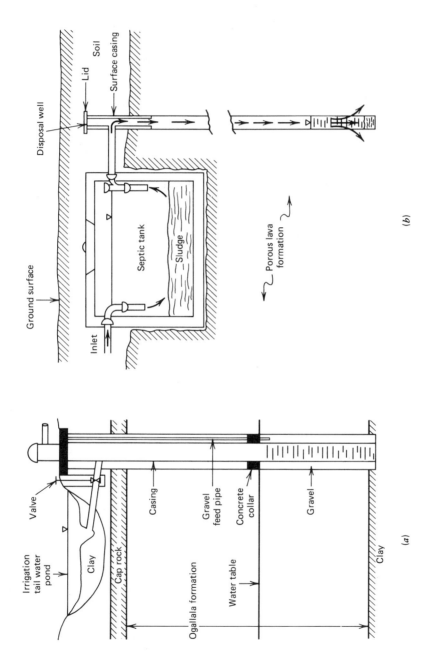

Fig. 13.7 Examples of recharge well designs. (a) Combined irrigation and recharge well in alluvium in the High Plains of Texas. (b) Recharge well for disposal of septic tank effluent into a lava formation in Central Oregon (after Hauser and Lotspeich[33] and Sceva[64]).

permeable volcanic terrains of the northwestern United States.[55,64] A typical domestic wastewater disposal system including a recharge well is shown in Fig. 13.7*b*. More than 2000 recharge wells are located in the 130,000 ha of agricultural land within the Snake River plain of southern Idaho.[28] These wells are typically 10 to 30 cm in diameter, 30 to 50 m deep, and are capable of accepting flows up to 20,000 m^3/day. The geology of the area consists of alternating layers of fractured basalt. A study of the effect of these disposal wells on water quality revealed that groundwater moved rapidly through fractures and channels in the basalt formations, that bacterial pollution persisted underground, and that turbidity was reduced by downward percolation.

An extensive series of recharge wells has been successfully operated since 1953 along the coasts of Los Angeles and Orange Counties, California. These wells create and maintain a pressure ridge of fresh water to control seawater intrusion (see Chapter 14). Experience gained on the project demonstrated that gravel-packed wells operate most efficiently; a typical dual-aquifer well is shown in Fig. 13.8. Favorable recharge rates have been maintained by chlorination and deaeration of the water supply and by a comprehensive well-maintenance program involving periodic pumping of the wells. It was also found that a concrete seal should be provided on the outside of the casing where it passes through the impermeable zone above the confined alluvial aquifer to prevent upward movement of water.

Finally, it should be mentioned that field and laboratory studies have demonstrated the feasibility of temporary storage of fresh water in saline water aquifers through wells first recharged and later pumped.[17,25,44] The efficiency of the procedure increases with each recharge-storage-withdrawal cycle. The technique has application in flat coastal areas underlain by saline water aquifers where no surface reservoir sites are available to provide freshwater supplies on a year-round basis.

Incidental Recharge.   Incidental, or unplanned, recharge occurs where water enters the ground as a result of a human activity whose primary objective is unrelated to artificial recharge of groundwater.[22] Included in this category is water from irrigation, cesspools, septic tanks, water mains, sewers, landfills, waste-disposal facilities, canals, and reservoirs. The quantity of incidental recharge normally far exceeds that deliberately accomplished by artificial recharge projects. Because several of these sources introduce polluted waters

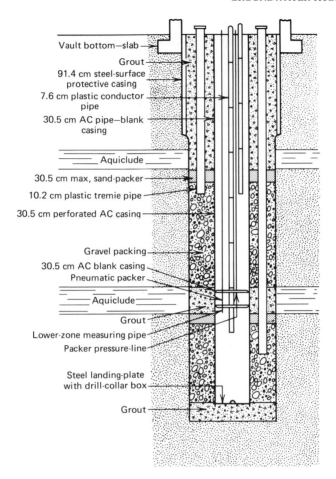

Vault bottom—slab
Grout
91.4 cm steel-surface protective casing
7.6 cm plastic conductor pipe
30.5 cm AC pipe—blank casing
Aquiclude
30.5 cm max, sand-packer
10.2 cm plastic tremie pipe
30.5 cm perforated AC casing
Gravel packing
30.5 cm AC blank casing
Pneumatic packer
Aquiclude
Grout
Lower-zone measuring pipe
Packer pressure-line
Steel landing-plate with drill-collar box
Grout

Fig. 13.8 Design of a dual-aquifer recharge well for control of sea water intrusion, Los Angeles County, California (after Department of Economic and Social Affairs[22]).

into the underground, degradation of the quality of groundwater can occur (see Chapter 8).

Research on Water Spreading

The economy of water spreading hinges on maintenance of a high infiltration rate. Typical rate curves, however, show a pronounced tendency to decrease with time. Determining the cause of this decrease and how to counteract it has led to extensive research programs. A variety of soil and water treatments and operational methods have been undertaken to study the problem.[7,63]

In Fig. 13.9 is a typical curve of recharge rate versus time. the initial decrease is attributed to dispersion and swelling of soil particles after wetting; the subsequent increase accompanies elimination of entrapped air by solution in passing water; the final gradual decrease results from microbial growths clogging the soil pores. Laboratory tests with sterile soil and water give nearly constant maximum recharge rates, thereby substantiating the effect of microbial growths.

Recharge rates generally decrease as the mean particle size of soil on a spreading area decreases. Efforts to maintain soil pores free for water passage have led to additions of organic matter and chemicals to the soil as well as to growing vegetation on the spreading area.[51.] Chemical soil conditioners, which tend to aggregate the soil, show promise in soils of certain textures. Alternating wet and dry periods on a basin generally furnishes a greater total recharge than does continuous spreading in spite of the fact that water is in contact with the soil for as little as one-half of the total time. Drying kills microbial growths, and this, combined with scarification of the soil surface, reopens the soil pores.[52]

Other factors contribute to recharge rates. Studies in small ponds have confirmed that infiltration rates are directly proportional to head of water. Where less pervious strata lie below the surface stratum, the recharge rate depends on the rate of subsurface lateral flow. Hence, spreading only in narrow, widely spaced strips recharges nearly as much water as spreading over an entire area.[24,67] Sunlight, affecting bacterial action in soil, water temperature, and algal growths in water, has not been fully evaluated as a recharge factor. Water containing silt or clay is known to clog soil pores, leading to rapid reductions in recharge rates. Wave action in large,

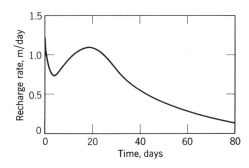

Fig. 13.9 Typical time variation of recharge rate for water spreading on undisturbed soil (after Muckel[51]).

Fig. 13.10 Aerial view of Leaky Acres Recharge Facility, Fresno, California. Here ten spreading basins with a total area of 47.4 ha serve as an important research center for artificial recharge (courtesy U.S. Agricultural Research Service).

shallow ponds can stir bottom sediments and seal pores that would otherwise remain open. Water quality can be an important factor; thus, recharging water of high sodium content tends to deflocculate colloidal soil particles and thereby hinder water passage. Because a high water table limits the downward flow of recharged water, this surface should be at least 3 to 6 m below spreading surfaces.

A detailed study of the environmental aspects of water spreading by Nightingale and Bianchi[53] at Fresno, California (see Fig. 13.10) documented impacts of recharge basins on soils, water quality, plant and animal communities, algae, insects, fish, birds, aquatic microfauna, and periphytic microorganisms.* They concluded that to maximize infiltration rates in basins:

1. Soil for levees should be brought from a borrow pit, and heavy construction equipment should be kept out of basins.

---

*It is interesting to note that 25 species of birds, 42 species of insects, and more than 51 species of algae were found in the recharge basins. Bottom-feeding fish, such as carp, were discovered to cause an increase in suspended material and hence contributed to soil surface sealing.

2. Recharge water should be introduced into a basin at its lowest point to prevent intrabasin erosion and clogging of the soil surface.

3. Trees and shrubs should be removed initially and the soil disked once.

4. Turbid water should be prevented from entering basins.

5. Aquatic vegetation should be minimized to prevent biological clogging of soils.

6. Recharge should be concentrated in summer when basin water temperature is highest.

Wastewater Recharge for Reuse

In recent years increasing attention has been focused on reuse of municipal wastewater. Almost all uses are nonpotable, such as for irrigation or industrial purposes, because of questionable health effects,[72] and one of the mostly widely favored is recharge to supplement groundwater resources.[16,20,66] Recharge of wastewater (usually after secondary treatment) improves its quality by removal of physical, biological, and some chemical constituents; provides storage until subsequent reuse; reduces seasonal temperature variations; and dilutes the recharged water with native groundwater. Land application practices involve irrigation, spreading, overland flow, and recharge wells.[15,81] Selection of a given system is governed by soil and subsurface conditions, climate, availability of land, and intended reuse of the wastewater. Each method is briefly described in the following sections (see also Chapter 8).

Irrigation Method. Effluent can be applied by sprinklers or surface irrigation techniques to irrigate cropland. Application rates are low, ranging from 0.05 to 0.2 m/week; also, only the portion not consumed by plants percolates downward to the water table. In humid regions, low evapotranspiration rates plus dilution with rainwater contributes good-quality water to the groundwater; in arid regions brackish water that degrades the groundwater can result.

Spreading Method. Recharging of effluent onto bare ground or native vegetation for infiltration and percolation functions as a tertiary treatment plant, producing reclaimed water for reuse. Application rates are high, ranging from 0.5 to 10 m/week, depending on local conditions. Thus, with a recharge rate of 2 m/week, for example, a basin area of 1.4 ha would suffice for 10,000 people.[15] Wastewater can be applied with surface irrigation or spray techniques, but spreading basins are often employed.[73] High-rate systems require deep permeable soils (sandy loams to loamy sands) and a water table that does not rise to ground surface. Flooding is conducted intermittently—for example, 2 to 14 days wet alternating with 5 to 20 days dry. Movement of effluent through the soil removes bacteria and viruses, almost all biochemical oxygen demand and suspended solids, up to 50 percent of nitrogen, and 60 to 95 percent of phosphorus. Because municipal use adds up to 300 mg/l of dissolved solids to water and this cannot be removed by recharging, wastewater can deteriorate the quality of groundwater unless adequate subsurface dilution is available.

Overland Flow Method. Where soils have low infiltration rates, such as clays and clay loams, wastewater is applied by irrigation or spray techniques to the upper end of sloping vegetated plots and allowed to flow in a shallow sheet to runoff collection ditches. Only a minor fraction of the applied water infiltrates; hence, this method contributes little to groundwater recharge.

Recharge Well Method. High-quality, tertiary-treated effluent can be placed underground through recharge wells. Best results are obtained with an effluent of BOD < 5 mg/l, suspended solids < 1 mg/l, phosphate < 1 mg/l, iron < 0.5 mg/l, and turbidity < 0.3 turbidity units. The high cost of recharging effluent of this quality into wells can only be economically justified where some special purpose such as control of land subsidence, control of seawater intrusion, or delayed development of a costly alternative water supply source can be served.[3,65]

The removal of pollutants from secondary effluent by recharging depends on the time and distance of travel underground and on the type and properties of the soils and subsurface formations. In general, for percolation through fine-textured alluvial deposits, bacteria and viruses are removed, nitrogen is reduced, soluble salts are not removed, and trace elements and heavy metals may be reduced.[72] To ensure that potable water results from recharged effluent, criteria governing the design of a recharge project should specify:

1. The degree of treatment required for wastewater before recharging.

2. A minimum vertical distance of percolation through an unsaturated soil zone above the water table.

3. A maximum amount of wastewater to be recharged and diluted by native groundwater.

4. A minimum residence time of reclaimed water underground before withdrawal for use.

5. A monitoring program of water quality for the groundwater influenced by recharging, together with limiting criteria for specified biological and chemical constituents.

One of the most fully documented studies of wastewater recharge has been conducted near Phoenix, Arizona, by the U.S. Agricultural Research Service.[14,15] Secondary effluent is recharged to obtain renovated water for irrigation. Wastewater is spread in six parallel basins, each 6 m wide and 210 m long and spaced 6 m apart (see Fig. 13.11). The soil is a sandy loam with the water table 3 m below ground surface. Basins are flooded for 14 days and rested for drying for 10 days in summer and 20 days in winter. Infiltration rates decrease from 0.75 m/day to 0.45 m/day during each flooding period; annual recharge totals 90 m. Wells adjacent to the basins pump the infiltrated effluent for use. Table 13.4 lists water quality data for the recharged effluent and for water pumped from a nearby well. This well water had moved through 2 m of unsaturated soil and 9 m of saturated soil during an underground detention time of 5 to 10 days. Although the pumped water does not meet drinking water standards, it is apparent that the recharging substantially reduced most constituents.

Two major Southern California wastewater reclamation plants involving groundwater recharge have been in successful operation for more than 15 years. One at Whittier Narrows in Los Angeles County

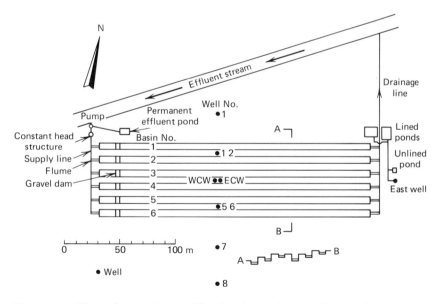

Fig. 13.11 Plan of experimental basins for recharge of treated wastewater near Phoenix, Arizona (after Bouwer[14]; reprinted from *Journal American Water Works Association*, Vol. 66, by permission of the Association; copyright © 1974 by American Water Works Association, 6666 West Quincy Avenue, Denver, Col. 80235).

**TABLE 13.4   Water Quality Data from Recharge
with Secondary Effluent at Phoenix, Arizona
(after Sopper and Kardos[71])**

Constituent	Recharged Effluent, mg/l	Sampled Well Water, mg/l
BOD	10–20	0–1
COD	30–60	10–20
TOC	10–25	1–7
Org.-N	2–6	0.3–0.7
NO_3-N	0–1	0.1–50
NO_2-N	0–3	0–1
NH_4-N	20–40	5–20
PO_4-P	7–12	4–8
F	3–5	2–2.5
B	0.7–0.9	0.7–0.9
Cu	0.1	0.02
Zn	0.2	0.1
Cd	0.008	0.007
Pb	0.08	0.07
Total salts	1000–1200	1000–1200
pH	7.7–8.1	6.9–7.2
Fecal coliforms per 100 ml	10^5–10^6	0–100

provides secondary treatment to 57,000 m³/day of wastewater and
then discharges it into spreading basins along the Rio Hondo and
San Gabriel River, where it percolates through permeable alluvium
to the groundwater.[47] Study of water quality aspects of the plant
revealed that drinking water standards were met after percolation
and dilution underground.* The other is at Santee in San Diego
County. Here treated effluent from an oxidation pond is pumped to
recharge basins having an area of 1.2 ha. From there the water flows
through natural sand and gravel strata for distances varying from
120 to 450 m to large collection ditches. The intercepted water is
chlorinated prior to entry into a series of recreational lakes or
to release in a swimming basin (after tertiary treatment). Detailed
studies of the project led to the conclusion that the oxidation pond
and percolation zone were efficient in removing bacteria and viruses.

---

*Recharge of treated wastewater has increased steadily in Los Angeles County; in
1973 an estimated 30 percent of all water spread came from this source.[72] For all of
California, 1.58×10^8 m³ of wastewater was reclaimed in 1976, while a potential of
31.5×10^8 m³ has been projected by 2000.

Recharge Mounds

When water percolates beneath a spreading basin, a mound in the water table is formed, as shown in Fig. 13.12. Clearly, the dimensions of this mound are governed by the basin size and shape, recharge rate and duration, and aquifer characteristics. Mound geometries have been computed by various investigators[4,11,29] based on complex mathematical analyses stemming from the generalized nonsteady groundwater flow equation (Eq. 3.76).* Most solutions are based on the usual assumptions of homogeneous and isotropic aquifers, vertical recharge at a uniform rate, the top of the mound does not contact the bed of the spreading basin, and the height of the mound is small in relation to the initial saturated thickness.[32]

The shape of a mound beneath a square recharge area can be expressed by dimensionless parameters as shown in Fig. 13.13. Here h is the mound height (see Fig. 13.12), S is the storage coefficient of the unconfined aquifer, W is the recharge rate, t is time since recharge began, L is the length of one side of the recharge area, T is transmissivity (KD) of the aquifer, and x is a coordinate distance from the center of the recharge area.

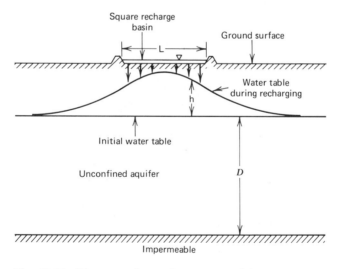

Fig. 13.12   Diagram of a recharge mound in a water table beneath a square spreading basin.

*Credit should be given to R. E. Glover of Denver, Colorado, for the original but unpublished solution to the mound geometry problem.

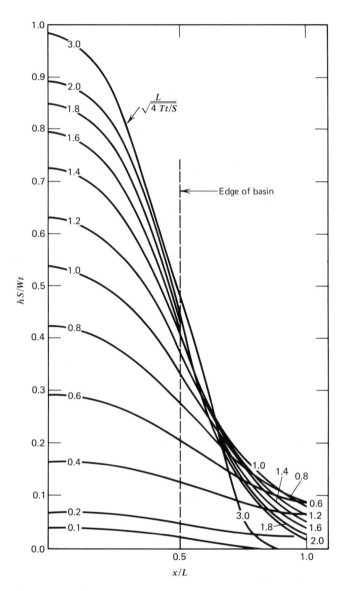

Fig. 13.13 Dimensionless graph defining the rise and horizontal spread with time of a water table mound beneath a square recharge area (after Bianchi and Muckel[8]).

Example. If water is spread in a square basin 100 m on a side at a uniform rate of 0.5 m/day, what will be the height of the ground-water mound at the edge of the basin after 15 days? Let $T = 800$ m^2/day and $S = 0.15$; then,

$$\frac{L}{\sqrt{4Tt/S}} = \frac{100}{\sqrt{4(800)(15)/0.15}} = 0.25$$

At the edge of the basin $x/L = 0.5$. Given these two dimensionless parameters, $hS/Wt = 0.070$ from Fig. 13.13. Thus,

$$h = \frac{0.070Wt}{S} = \frac{(0.070)(0.5)(15)}{0.15} = 3.50 \text{ m}$$

Similar solutions are available for circular and rectangular spreading basins and for basins above sloping water tables.[4,8] If recharge ceases at time t_o, dissipation of the mound can be calculated by superposing hypothetically on the flow system at $t = t_o$ a rate of uniform discharge equal to that of the percolation rate. The algebraic sum of the two mounds yields the mound shape at any time after the end of recharge. It follows that as time after t_o becomes large, h approaches zero.

Under uniform recharge conditions a mound will continue to grow until some control provides a limit. Two types of control can be recognized,[4] potential and lateral, as shown in Fig. 13.14. Potential control occurs when the mound builds up to the recharge surface; with a fixed maximum height the gradient and hence the recharge rate must decrease with time. Lateral control occurs when the mound intersects a constant surface water elevation such as a stream or lake; for a large horizontal extent of this boundary, the mound approaches an equilibrium with a constant recharge rate.

Artificial Recharge on Long Island, New York

Illustrative of the significant role that artificial recharge plays in groundwater is the situation on Long Island, New York. As of 1971 groundwater was the sole source of fresh water for more than 2.5 million people outside of the metropolitan New York portion, which occupies only the western end of the island.[54] The primary aquifer consists of unconsolidated coastal-plain deposits overlain by a thin upper sequence of glacial materials. Where groundwater is intensively developed, seawater intrusion has occurred in coastal areas.

Several forms of artificial recharge are practiced on Long Island. More than 2100 recharge basins dispose of about 231,000 m^3/day of storm runoff.[56,68] Many of these basins initially were abandoned

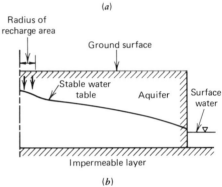

Fig. 13.14 Sketches of controls which limit the dimensions of a water table mound beneath a recharge area (*a*) Potential control. (*b*) Lateral control (after Todd[79]).

gravel pits, but now urban developers are required to provide basins both to accommodate increased runoff and to conserve water.* Practically all basins are unlined excavations in the upper glacial deposits; depths range from 3 to 6 m and areas from less than 0.1 to 12 ha. Figure 13.15 shows typical recharge basins in an urban housing development. Average infiltration rates for storm runoff ranged from 1.4 to 6.6 m/day on three monitored basins. Besides basins for storm runoff, an estimated 200 basins dispose of some 114,000 m³/day of industrial and commercial wastes (mostly cooling water) into the ground.

Incidental recharge from cesspools and septic tanks on Long Island was estimated at 450,000 m³/day. In some places where this has degraded groundwater quality, sewer systems have been constructed; the conversion to sewering reduces recharge, lowers the water table, and can extend seawater intrusion. Other incidental recharge stems from pressurized water supply pipe distribution systems. Assuming a 10 percent leakage factor, an estimated 378,000 m³/day is recharged from this water source.

---

*Ultimately, the total number of recharge basins on Long Island will approach 5000 based on zoning laws regarding recharge basins and a current maximum density of 1.5 basins/km² in urban areas.

Fig. 13.15 Aerial photograph of a residential portion of Syosset, New York, showing four basins for recharging surface runoff to groundwater (courtesy U.S. Geological Survey).

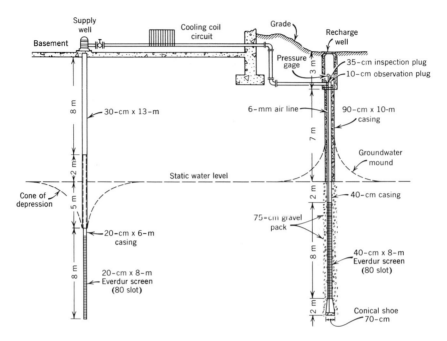

Fig. 13.16 Typical supply and recharge well installation for industrial cooling purposes on Long Island, New York (after Johnson[37]; reprinted from *Journal American Water Works Association,* Vol. 40, by permission of the Association; copyright © 1948 by American Water Works Association, 6666 West Quincy Avenue, Denver, Col. 80235).

More than 1000 recharge wells return about 189,000 m³/day to groundwater in compliance with governmental regulations specifying that groundwater pumped for air-conditioning and industrial cooling must be returned to the originating aquifer. In the past dug wells or pits were constructed for recharging, but these have gradually been replaced by drilled wells, gravel packed and extending to below the water table. A typical supply and recharge well installation is shown in Fig. 13.16. Clogging of recharge wells results from precipitation of iron compounds, biological growths, and air bubbles, necessitating periodic redevelopment by pumping and/or surging.

Artificial recharge on Long Island can be summarized as shown in Table 13.5 as of about 1971. It is apparent that this quantity of water significantly affects the subsurface hydrologic regime of the island, both quantitatively and qualitatively. And looking ahead, a series of experiments are underway at two Long Island locations, River-

TABLE 13.5   Artif :ial Recharge on
Long Island, New York (after Parker,
et al.,[54] and Seaburn and Aronson[68])

Water Source	Recharge Rate, m³/day
Recharge basins	
Storm runoff	231,000
Industrial wastewater	114,000
Cesspools and septic tanks	450,000
Leaking water mains	378,000
Recharge wells	189,000
Total	1,362,000

head[2] and Bay Park,[42,57,84] to determine the economic and technical feasibility of recharging treated municipal wastewater into wells for the purpose of controlling seawater intrusion.

Induced Recharge

Direct methods of artificial recharge described above involve the conveyance of surface water to some point where it enters the ground. Distinguished from these is the method of induced recharge, accomplished by withdrawing groundwater at a location adjacent to a river or lake so that lowering of the groundwater level will induce water to enter the ground from the surface source. The schematic cross sections of a river valley in Fig. 13.17 show flow patterns with and without induced infiltration from a stream. On the basis of this definition, wells located directly adjacent to and fed largely by surface water serve as means of artificial recharge. The hydraulics of wells located near streams is described in Chapter 4; the construction and operation of wells, collectors, and galleries is in Chapter 5.

Induced infiltration where supplied by a perennial stream ensures a continuing water supply even though overdraft conditions may exist in nearby areas supplied only by natural recharge. The method has proved effective in unconsolidated formations of permeable sand and gravel hydraulically connected between stream and aquifer. The amount of water induced into the aquifer depends on the rate of pumping, permeability, type of well, distance from surface stream, and natural groundwater movement. It is important that the velocity of the surface stream be sufficient to prevent silt deposition from sealing the streambed.

Studies of water quality have shown that induced recharge can furnish water free of organic matter and pathogenic bacteria.[38,41] Because surface water commonly is less mineralized than ground-

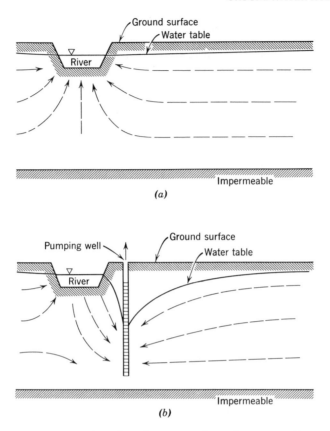

Fig. 13.17 Induced recharge resulting from a well pumping near a river. (*a*) Natural flow pattern. (*b*) Flow pattern with pumping well.

water, water obtained by induced infiltration, being a mixture of two water sources, possesses a higher quality than natural groundwater. Variations in water quality following installation of a collector well along the Mississippi River are shown in Fig. 13.18. Improvement of iron and chloride contents and total hardness are indicated. In a similar manner, temperatures of water from induced recharge will lie between those of the surface and underground sources. Figure 13.19 illustrates this effect for a group of collectors along the Ohio River. The mixture of the nearly constant groundwater temperature with river water has a temperature range less than that of the river and lags the seasonal river temperature oscillations by about 2½ months. Many cities bordering rivers in the Mississippi River basin have placed municipal well fields along riverbanks because of the assured supply of high-quality water.

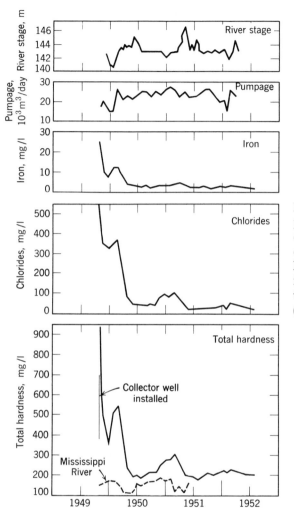

Fig. 13.18 River stage, pumpage, iron content, chloride content, and total hardness of water from a collector along the Mississippi River (after Klaer[41]).

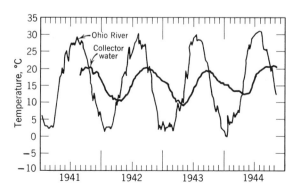

Fig. 13.19 Temperature of the Ohio River and water from a group of collector wells at Charlestown, Indiana (after Kazmann[38]).

Artificial Recharge for Energy Purposes

In recent years concern for energy resources has focused attention on the possibility of storing high-quality by-product heat in the form of hot water in aquifers for later utilization. The concept is best suited to large-scale applications, such as total-energy utilities that cogenerate electricity and heat. The demand for heat is seasonal, for space heating in winter and for absorption-cycle air-conditioning in summer. To match heat production with demand, storage for periods of about 3 months during spring and fall seasons is required.

Surplus hot water can be recharged through *heat storage wells* into confined aquifers.[49,50] The hot water will move radially outward from the well; because of density and viscosity differences it will tend to override the native groundwater and form an inverted truncated conic volume. Heat will be lost by conduction into the upper and lower layers confining the aquifer and by dispersion along the hot-cold interface. Residual heat left in the aquifer as the hot water is pumped to the surface for use increases the efficiency of subsequent cycles. Modeling of the flow system has indicated that more than three-fourths of the stored heat can be recovered after 90 days of storage. The combined advantages of energy conservation, low cost, and reduced environmental impact from waste heat disposal make heat storage wells attractive.[39] Research to demonstrate their economic and technical feasibility is currently underway; furthermore, employing the same principle, storage of cold water for later use in cooling is also under study.

Another aspect of artificial recharge and energy, also in the research stage, is the production of geothermal energy from hot dry rock. Two closely spaced wells more than 1000 m deep are drilled. Water recharged into one well is heated by the hot, fractured rock formations at the bottom of the well and pumped to the surface from the adjacent well. A demonstration plant to study this heat extraction technique has been completed in New Mexico.

References

1. Aronovici, V. S., et al., Basin recharge of the Ogallala aquifer, *Jour. Irrig. Drain. Div.*, Amer. Soc. Civil Engrs., v. 98, no. IR1, pp. 65–76, 1972.
2. Baffa, J. J., Injection well experience at Riverhead, N.Y., *Jour. Amer. Water Works Assoc.*, v. 62, pp. 41–46, 1970.
3. Baier, D. C., and G. W. Wesner, Reclaimed waste for groundwater recharge, *Water Resources Bull.*, v. 7, pp. 991–1001, 1971.
4. Baumann, P., Technical development in ground water recharge, *in Advances in Hydroscience*, (V. T. Chow, ed.) v. 2, pp. 209–279, Academic Press, New York, 1965.

5. Bear, J., and C. Braester, Flow from infiltration basins to drains and wells, *Jour. Hydraulics Div.*, Amer. Soc. Civil Engrs., v. 92, no. HY5, pp. 115–134, 1966.
6. Bear, J., and M. Jacobs, On the movement of water bodies injected into aquifers, *Jour. Hydrology*, v. 3, pp. 37–57, 1965.
7. Behnke, J. J., Clogging in surface spreading operations for artificial ground-water recharge, *Water Resources Research*, v. 5, pp. 870–876, 1969.
8. Bianchi, W. C., and D. C. Muckel, *Ground-water recharge hydrology*, ARS 41-161, Agric. Research Service, U.S. Dept. Agric., 62 pp., 1970.
9. Bianchi, W. C., et al., A case history to evaluate the performance of water-spreading projects, *Jour. Amer. Water Works Assoc.*, v. 70, pp. 176–180, 1978.
10. Biemond, C., Dune water flow and replenishment in the catchment area of the Amsterdam water supply, *Jour. Instn. Water Engrs.*, v. 11, pp. 195–213, 1957.
11. Bittinger, M. W., and F. J. Trelease, The development and dissipation of a ground-water mound, *Trans. Amer. Soc. Agric. Engrs.*, v. 8, pp. 103–104, 106, 1965.
12. Boggess, D. H., and D. R. Rima, Experiments in water spreading at Newark, Delaware, *U.S. Geological Survey Water-Supply Paper 1594-B*, 15 pp., 1962.
13. Bourguet, L., Inventaire internationale des aménagements d'alimentation artificielle—depouillement et synthèse des réponses, *Bull. Intl. Assoc. Sci. Hydrology*, v. 16, no. 3, pp. 51–102, 1971.
14. Bouwer, H., Renovating municipal wastewater by high-rate infiltration of groundwater recharge, *Jour. Amer. Water Works Assoc.*, v. 66, pp. 159–162, 1974.
15. Bouwer, H., et al., Land treatment of wastewater in today's society, *Civil Engrng.*, v. 48, no. 1, pp. 78–81, 1978.
16. Braunstein, J. (ed.), *Underground waste management and artificial recharge*, Amer. Assoc. Petrol. Geol., 2 vols., 931 pp., 1973.
17. Brown, D. L., and W. D. Silvey, Artificial recharge to a freshwater-sensitive brackish-water sand aquifer, Norfolk, Virginia, *U.S. Geological Survey Prof. Paper 939*, 53 pp., 1977.
18. Brown, F. F., and D. C. Signor, Groundwater recharge, *Water Resources Bull.*, v. 8, pp. 132–149, 1972.
19. Brown, R. F., and D. C. Signor, Artifical recharge—state of the art, *Ground Water*, v. 12, pp. 152–160, 1974.
20. California Dept. Water Resources, *Waste water reclamation*, Bull. 189, The Resources Agency, Sacramento, 43 pp., 1973.
21. Davis, G. H., et al., Use of ground-water reservoirs for storage of surface water in the San Joaquin Valley, California, *U.S. Geological Survey Water-Supply Paper 1618*, 125 pp., 1964.
22. Dept. Econ. and Social Affairs, *Ground-water storage and artificial recharge*, Natural Resources/Water Ser. 2, United Nations, New York, 270 pp., 1975.

23. Dvoracek, M. J., and S. H. Peterson, Artificial recharge in water resources management, *Jour. Irrig. Drain. Div.*, Amer. Soc. Civil Engrs., v. 97, no. IR2, pp. 219–232, 1971.

24. Dvoracek, M. J., and V. H. Scott, Ground-water flow characteristics influenced by recharge pit geometry, *Trans. Amer. Soc. Agric. Engrs.*, v. 6, pp. 262–265, 267, 1963.

25. Esmail, O. J., and O. K. Kimbler, Investigation of the technical feasibility of storing fresh water in saline aquifers, *Water Resources Research*, v. 3, pp. 683–695, 1967.

26. Foxworthy, B. L., Hydrologic conditions and artifical recharge through a well in the Salem Heights area of Salem, Oregon, *U.S. Geological Survey Water-Supply Paper* 1594-F, 56 pp., 1970.

27. Goss, D. W., et al., Fate of suspended sediment during basin recharge, *Water Resources Research*, v. 9, pp. 668–675, 1973.

28. Graham, W. G., et al., *Irrigation wastewater disposal well studies— Snake Plain aquifer*, Rept. EPA-600/3-77-071, U.S. Environmental Protection Agency, Ada, Okla., 51 pp., 1977.

29. Hantush, M. S., Growth and decay of groundwater-mounds in response to uniform percolation, *Water Resources Research*, v. 3, pp. 227–234, 1967.

30. Harmeson, R. H., et al., Coarse media filtration for artificial recharge, *Jour. Amer. Water Works Assoc.*, v. 60, pp. 1396–1403, 1968.

31. Harpaz, Y., Artificial ground-water recharge by means of wells in Israel, *Jour. Hydraulics Div.*, Amer. Soc. Civil Engrs., v. 97, no. HY12, pp. 1947–1964, 1971.

32. Haskell, E. E., Jr., and W. C. Bianchi, Development and dissipation of ground water mounds beneath square recharge basins, *Jour. Amer. Water Works Assoc.*, v. 57, pp. 349–353, 1965.

33. Hauser, V. L., and F. B. Lotspeich, Artificial groundwater recharge through wells, *Jour. Soil and Water Conservation*, v. 22, pp. 11–15, 1967.

34. Intl. Assoc. Sci. Hydrology, *Artificial recharge and management of aquifers*, Publ. 72, 523 pp., 1967.

35. International Assoc. Sci. Hydrology, *International survey on existing water recharge facilities*, Publ. 87, 762 pp., 1970.

36. Jansa, V., *Artificial replenishment of underground water*, Intl. Water Supply Assoc., Second Cong., Paris, 105 pp., 1952.

37. Johnson, A. H., Ground-water recharge on Long Island, *Jour. Amer. Water Works Assoc.*, v. 40, pp. 1159–1166, 1948.

38. Kazmann, R. G., River infiltration as a source of ground-water supply, *Trans. Amer. Soc. Civil Engrs.*, v. 113, pp. 404–424, 1948.

39. Kazmann, R. G., Underground hot water storage could cut national fuel needs 10%, *Civil Engrng.*, v. 48, no. 5, pp. 57–60, 1978.

40. Kelly, T. E., Artificial recharge at Valley City, North Dakota, 1932 to 1965, *Ground Water*, v. 5, no. 2, pp. 20–25, 1967.

41. Klaer, F. H., Jr., Providing large industrial water supplies by induced infiltration, *Min. Engrng.*, v. 5, pp. 620–624, 1953.

42. Koch, E., et al., Design and operation of the artificial-recharge plant at

Bay Park, New York, *U.S. Geological Survey Prof. Paper* 751-B, 14 pp., 1973.

43. Knapp, G. L., *Artificial recharge of groundwater—a bibliography,* WRSIC 73-202, Water Resources Sci. Info. Center, Washington, D.C., 309 pp., 1973.

44. Kumar, A., and O. K. Kimbler, Effect of dispersion, gravitational segregation, and formation stratification on the recovery of freshwater stored in saline aquifers, *Water Resources Research,* v. 6, pp. 1689–1700, 1970.

45. Laverty, F. B., et al., Reclaiming Hyperion effluent, *Jour. San. Engrng. Div.,* Amer. Soc. Civil Engrs., v. 87, no. SA6, pp. 1–40, 1961.

46. McCormick, R. L., Filter-pack installation and redevelopment techniques for shallow recharge shafts, *Ground Water,* v. 13, pp. 400–405, 1975.

47. McMichael, F. C., and J. E. McKee, *Wastewater reclamation at Whittier Narrows,* Publ. 33, California State Water Quality Control Board, Sacramento, 100 pp., 1966.

48. McWhorter, D. B., and J. A. Brookman, Pit recharge influenced by subsurface spreading, *Ground Water,* v. 10, no. 5, pp. 6–11, 1972.

49. Meyer, C. F., Status report on heat storage wells, *Water Resources Bull,* v. 12, pp. 237–252, 1976.

50. Meyer, C. F., and D. K. Todd, Conserving energy with heat storage wells, *Environmental Sci. and Technology,* v. 7, pp. 512–516, 1973.

51. Muckel, D. C., *Replenishment of ground water supplies by artificial means,* Tech. Bull. 1195, Agric. Research Service, U.S. Dept. Agric., 51 pp., 1959.

52. Nightingale, H. I., and W. C. Bianchi, Ground-water recharge for urban use: Leaky Acres Project, *Ground Water,* v. 11, no. 6, pp. 36–43, 1973.

53. Nightingale, H. I., and W. C. Bianchi, *Environmental aspects of water spreading for ground-water recharge,* Tech. Bull. 1568, Agric. Research Service, U.S. Dept. of Agric., 21 pp., 1977.

54. Parker, G. G., et al., Artificial recharge and its role in scientific water management, with emphasis on Long Island, New York, *Proc. Natl. Symposium on Ground-Water Hydrology,* Amer. Water Resources Assoc., pp. 193–213, 1967.

55. Price, D., et al., Artificial recharge in Oregon and Washington, 1962, *U.S. Geological Survey Water-Supply Paper* 1594-C, 65 pp., 1965.

56. Prill, R. C., and D. B. Aaronson, Flow characteristics of a subsurface-controlled recharge basin on Long Island, New York, *U.S. Geological Survey Jour. Research,* v. 1, pp. 735–744, 1973.

57. Ragone, S. E., Geochemical effects of recharging the Magothy aquifer, Bay Park, New York, with tertiary-treated sewage, *U.S. Geological Survey Prof. Paper* 751-D, 22 pp., 1977.

58. Rahman, M. A., et al., Effect of sediment concentration on well recharge in a fine sand aquifer, *Water Resources Research,* v. 5, pp. 641–646, 1969.

59. Rebhun, M., and J. Schwartz, Clogging and contamination processes in recharge wells, *Water Resources Research,* v. 4, pp. 1207–1217, 1968.

60. Reed, J. E., et al., Induced recharge of an artesian glacial-drift aquifer

at Kalamazoo, Mich., *U.S. Geological Survey Water-Supply Paper* 1594-D, 62 pp., 1966.

61. Reeder, H. O., et al., Artificial recharge through a well in fissured carbonate rock, West St. Paul, Minnesota, *U.S. Geological Survey Water Supply Paper* 2004, 80 pp., 1976.

62. Richter, R. C., and R. Y. D. Chun, Artificial recharge of ground water reservoirs in California, *Jour. Irrig. Drain. Div.*, Amer. Soc. Civil Engrs., v. 85, no. IR4, pp. 1–27, 1959.

63. Ripley, D. P., and Z. A. Saleem, Clogging in simulated glacial aquifers due to artificial recharge, *Water Resources Research*, v. 9, pp. 1047–1057, 1973.

64. Sceva, J. E., *Liquid waste disposal in the lava terranes of Central Oregon,* U.S. Federal Water Pollution Control Admin., Corvallis, Oregon, 2 vols., 162 pp., 1968.

65. Schicht, R. J., Feasibility of recharging treated sewage effluent into a deep sandstone aquifer, *Ground Water,* v. 9, no. 6, pp. 29–35, 1971.

66. Schmidt, C. J., et al., A survey of practices and regulations for reuse of water by groundwater recharge, *Jour. Amer. Water Works Assoc.*, v. 70, pp. 140–147, 1978.

67. Scott, V. H., and G. Aron, Aquifer recharge efficiency of wells and trenches, *Ground Water*, v. 5, no. 3, pp. 6–14, 1967.

68. Seaburn, G. E., and D. A. Aronson, Influence of recharge basins on the hydrology of Nassau and Suffolk Counties, Long Island, New York, *U.S. Geological Survey Water-Supply Paper* 2031, 66 pp., 1974.

69. Signor, D. C., et al., Annotated bibliography on artificial recharge of ground water, 1955–67, *U.S. Geological Survey Water-Supply Paper* 1990, 141 pp., 1970.

70. Sniegocki, R. T., et al., Testing procedures and results of studies of artificial recharge in the Grand Prairie Region, Arkansas, *U.S. Geological Survey Water-Supply Paper* 1615-G, 56 pp., 1965.

71. Sopper, W. E., and L. T. Kardos, *Conference on recycling treated municipal wastewater through forest and cropland,* Rept. EPA-660/2-74-003, U.S. Environmental Protection Agency, Washington, D.C., 463 pp., 1974.

72. State Water Resources Control Board, Dept. Water Resources, and Dept. Health, *A "state-of-the-art" review of health aspects of wastewater reclamation for groundwater recharge,* Water Information Center, Huntington, N.Y., 240 pp., 1978.

73. Stone, R., and W. F. Garber, Sewage reclamation by spreading basin infiltration, *Trans. Amer. Soc. Civil Engrs.*, v. 117, pp. 1189–1217, 1952.

74. Suter, M., and R. H. Harmeson, *Artificial ground-water recharge at Peoria, Illinois,* Bull. 48, Illinois State Water Survey, Urbana, 48 pp., 1960.

75. Task Group on Artificial Ground Water Recharge, Artificial ground water recharge, *Jour. Amer. Water Works Assoc.*, v. 55, pp. 705–709, 1963.

76. Task Group on Artificial Ground Water Recharge, Design and operation

of recharge basins, *Jour. Amer. Water Works Assoc.*, v. 55, pp. 697–704, 1963.

77. Task Group on Artificial Ground Water Recharge, Experience with injection wells for artificial ground water recharge, *Jour. Amer. Water Works Assoc.*, v. 57, pp. 629–639, 1965.

78. Todc', D. K., Annotated bibliography on artificial recharge of ground water through 1954, *U.S. Geological Survey Water-Supply Paper 1477*, 115 pp., 1959.

79. Todd, D. K., The distribution of ground water beneath artificial recharge areas, *Intl. Assoc. Sci. Hydrology Publ.* 57, pp. 254–262, 1961.

80. Todd, D. K., Economics of ground water recharge, *Jour. Hydraulics Div.*, Amer. Soc. Civil Engrs., v. 91, no. HY4, pp. 249–270, 1965.

81. Unklesbay, A. G., and H. H. Cooper, Jr., Artificial recharge of artesian limestone at Orlando, Florida, *Econ. Geol.*, v. 41, pp. 293–307, 1946.

82. U.S. Environmental Protection Agency, *Land application of wastewater*, Rept. EPA 903-9-75-017, U.S. Environmental Protection Agency, Philadelphia, Penn., 94 pp., 1975.

83. Valliant, J., Artificial recharge of surface water to the Ogallala Formation in the High Plains of Texas, *Ground Water*, v. 2, no. 2, pp. 42–45, 1964.

84. Vecchioli, J., Experimental injection of tertiary-treated sewage in a deep well at Bay Park, Long Island, N.Y.—a summary of early results, *Jour. New England Water Works Assoc.*, v. 86, pp. 87–103, 1972.

85. Warner, D. L., and J. H. Lehr, *An introduction to the technology of subsurface wastewater injection,* Rept. EPA-600/2-77-240, U.S. Environmental Protection Agency, Ada, Okla., 344 pp., 1977.

86. Water Research Assoc., *Proceedings, Artificial Groundwater Recharge Conf.*, 2 vols., 481 pp., Medmenham, England, 1971.

Saline Water Intrusion in Aquifers

CHAPTER 14 ·······································

Saline water is the most common pollutant in fresh groundwater. Intrusion of saline water occurs where saline water displaces or mixes with freshwater in an aquifer. The phenomenon can occur in deep aquifers with the upward advance of saline waters of geologic origin, in shallow aquifers from surface waste discharges, and in coastal aquifers from an invasion of seawater. The interrelations of two miscible fluids in porous media have been studied extensively both theoretically and under field conditions. Therefore, management techniques that enable development of fresh water and at the same time control of saline intrusion are available.

Occurrence of Saline Water Intrusion

Saltwater intrusion into fresh groundwater formations generally results inadvertently from activities of man.* Most large sources of fresh groundwater are in close proximity to the sea, to natural bodies of saline groundwater, or to salts from effluent wastes released by

---

*It should be noted that saline groundwater can represent a valuable resource, particularly in arid inland regions. Potential uses include industrial processes such as: cooling; irrigation, where the mineral content is moderate; and desalination for local domestic purposes.

human activities.[41] Typically, shallow freshwater overlies saline water because the flushing action during recent times removes salts from ancient marine deposits. But at greater depths groundwater movement is much less so that displacement of saline water is slower. Furthermore, at depths of several thousand meters, brines are normally encountered. The U.S. Geological Survey estimates that approximately two-thirds of the United States is underlain by aquifers known to produce water containing more than 1000 mg/l of salt.[22]

Saline water in aquifers may be derived from any of several sources:[52]

1. Encroachment of seawater in coastal areas.
2. Seawater that entered aquifers during past geologic time.
3. Salt in salt domes, thin beds, or disseminated in geologic formations.
4. Water concentrated by evaporation in tidal lagoons, playas, or other enclosed areas.
5. Return flows to streams from irrigated lands.
6. Human saline wastes.

The mechanisms responsible for saline water intrusion fall into three categories.[21] One involves the reduction or reversal of groundwater gradients, which permits denser saline water to displace fresh water. This situation commonly occurs in coastal aquifers in hydraulic continuity with the sea when pumping of wells disturbs the natural hydrodynamic balance. A second method stems from the destruction of natural barriers that separate fresh and saline waters. An example would be the construction of a coastal drainage canal that enables tidal water to advance inland and to percolate into a freshwater aquifer. The third mechanism occurs where there is subsurface disposal of waste saline water, such as into disposal wells, landfills, or other waste repositories (see Chapter 8).

The occurrence of saline water intrusion is extensive and represents a special category of groundwater pollution. A comprehensive listing of examples of intrusion has been prepared;[52] this reveals that the problem exists in localities of most parts of the United States. Seawater intrusion along coasts has received the most attention.[30,37] In the United States the coastal perimeter is dotted with intruded localities; states most severely affected include Connecticut,[7] New York, Florida, Texas, California,[11] and Hawaii.[34,58] Internationally, the problem has received attention in populated coastal areas in England,* Germany, the Netherlands,[5,21,55] Israel,[2] and Japan, among

---

*One of the earliest reports of intrusion was published in 1855 by Braithwaite,[6] who described the increasing salinity of water pumped from wells in London and Liverpool.

others. Many small oceanic islands are completely underlain with aquifers containing seawater; these pose special problems in meeting water supply demands.

Ghyben-Herzberg Relation Between Fresh and Saline Waters

More than 50 years ago two investigators,[20,28] working independently along the European coast, found that salt water occurred underground, not at sea level but at a depth below sea level of about 40 times the height of the fresh water above sea level. This distribution was attributed to a hydrostatic equilibrium existing between the two fluids of different densities. The equation derived to explain the phenomenon is generally referred to as the Ghyben-Herzberg relation after its originators.*

The hydrostatic balance between fresh and saline water can be illustrated by the U-tube shown in Fig. 14.1. Pressures on each side of the tube must be equal; therefore,

$$\rho_s g h_s = \rho_f g(z + h_f) \tag{14.1}$$

where ρ_s is the density of the saline water, ρ_f is the density of the fresh water, g is the acceleration of gravity, and z and h_f are as shown in Fig. 14.1 Solving for z yields

$$z = \frac{\rho_f}{\rho_s - \rho_f} h_f \tag{14.2}$$

which is the Ghyben-Herzberg relation. For typical seawater conditions, let $\rho_s = 1.025$ g/cm^3 and $\rho_f = 1.000$ g/cm^3, so that

$$z = 40 h_f \tag{14.3}$$

Translating the U-tube to a coastal situation, as shown in Fig. 14.2, h_f becomes the elevation of the water table above sea level and z is the depth to the fresh-saline interface below sea level. This is a hydrodynamic rather than a hydrostatic balance because fresh water is flowing toward the sea. From density considerations alone, without flow, a horizontal interface would develop with fresh water everywhere floating above saline water. It can be shown[29] that where the flow is nearly horizontal, the Ghyben-Herzberg relation gives

---

*Almost unnoticed in the hydrologic literature is the much earlier contribution of Joseph DuCommun, a French teacher at West Point Military Academy. He first stated clearly and correctly the fresh-salt water balance existing in coastal aquifers (DuCommun, J., On the cause of fresh water springs, fountains, etc., *Amer. Jour. Science and Arts*, v. 14, pp. 174–176, 1828).

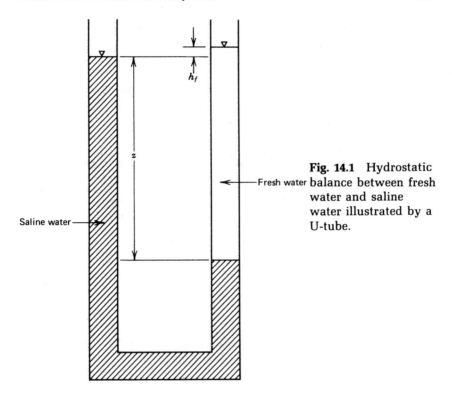

Fig. 14.1 Hydrostatic balance between fresh water and saline water illustrated by a U-tube.

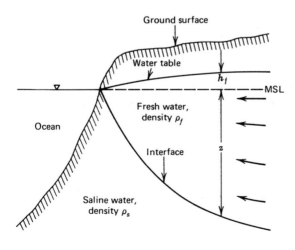

Fig. 14.2 Idealized sketch of occurrence of fresh and saline groundwater in an unconfined coastal aquifer.

satisfactory results. Only near the shoreline, where vertical flow components become pronounced (see Fig. 14.2), do significant errors in the position of the interface occur.

For confined aquifers the above derivation can also be applied by replacing the water table by the piezometric surface. It is important to note from the Ghyben-Herzberg relation that fresh-salt water equilibrium requires that the water table, or piezometric surface, (1) lie above sea level, and (2) slope downward toward the ocean. Without these conditions, seawater will advance directly inland.

Starting from the work of Hubbert,[29] the Ghyben-Herzberg relation has been generalized by Lusczynski and others[38,39] for situations where the underlying saline water is in motion with heads above or below sea level. The result for nonequilibrium conditions has the form

$$z = \frac{\rho_f}{\rho_s - \rho_f} h_f - \frac{\rho_f}{\rho_s - \rho_f} h_s \qquad (14.4)$$

where h_f is the altitude of the water level in a well filled with fresh water of density ρ_f and terminated at depth z, while h_s is the altitude of the water level in a well filled with saline water of density ρ_s and also terminated at depth z (see Fig. 14.3). When $h_s = 0$, the saline water is in equilibrium with the sea, and Eq. 14.4 reduces to Eq. 14.2.

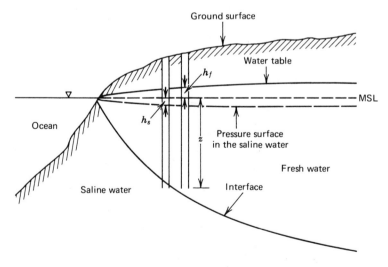

Fig. 14.3 Diagram illustrating differing heads for nonequilibrium conditions in fresh and saline water in an unconfined coastal aquifer.

Shape of the Fresh-Salt Water Interface

Recognizing the approximations inherent in the Ghyben-Herzberg relation, more exact solutions for the shape of the interface have been developed from potential flow theory.[15,18] The result by Glover[18] has the form

$$z^2 = \frac{2qx}{\Delta\rho K} + \left(\frac{q}{\Delta\rho K}\right)^2 \qquad (14.5)$$

where z and x are as shown in Fig. 14.4, $\Delta\rho = \rho_s - \rho_f$, K is hydraulic conductivity of the aquifer, and q is the freshwater flow per unit length of shoreline. The corresponding shape for the water table is given by

$$h_f = \left(\frac{2\Delta\rho qx}{K}\right)^{1/2} \qquad (14.6)$$

The width x_o of the submarine zone through which fresh water discharges into the sea can be obtained for $z = 0$, yielding

$$x_o = -\frac{q}{2\Delta\rho K} \qquad (14.7)$$

The depth of the interface beneath the shoreline z_o occurs where $x = 0$ (see Fig. 14.4) so that

$$z_o = \frac{q}{\Delta\rho K} \qquad (14.8)$$

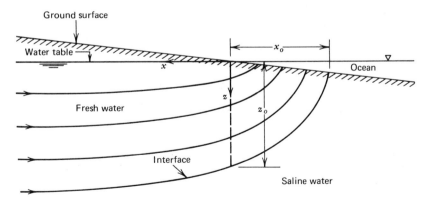

Fig. 14.4 Flow pattern of fresh water in an unconfined coastal aquifer.

Structure of the Fresh-Salt Water Interface

The sharp interfacial boundary described above between fresh and saline water does not occur under field conditions. Instead, a brackish transition zone of finite thickness separates the two fluids. This zone develops from dispersion by flow of the fresh water plus unsteady displacements of the interface by external influences such as tides, recharge, and pumping of wells.[61] In general, greatest thicknesses of transition zones are found in highly permeable coastal aquifers subject to heavy pumping. Observed thicknesses vary from less than 1 m to more than 100 m.* Figure 14.5 shows a transition zone in a highly permeable limestone aquifer in Miami, Florida. Note that the isochlors approach the base of the aquifer perpendicularly; this results because the flow parallels the base of the aquifer, thereby restricting vertical mixing.[18]

An important consequence of the transition zone and its seaward flow is the transport of saline water to the sea. This water originates from the underlying saline water; hence, from continuity considerations, there must exist a small landward flow in the saline water

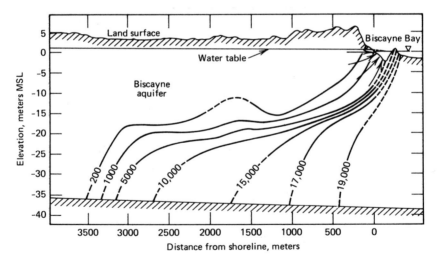

Fig. 14.5 Cross section through the transition zone of the Biscayne aquifer near Miami, Florida. Numbered lines are isochlors in mg/l (after Cooper, et al.[18]).

*As an extreme case, concentrated pumping in the Honolulu-Pearl Harbor area of Hawaii has created localized transition zones more than 300 m thick; these occupy essentially the entire vertical extent of the aquifer.[58]

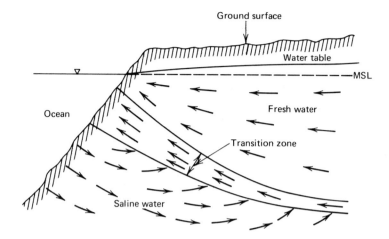

Fig. 14.6 Vertical cross section showing flow patterns of
fresh and saline water in an unconfined coastal aquifer
(after Todd[54]; reprinted from the *Journal American Water
Works Association,* Vol. 66, by permission of the
Association, copyright © 1974 by American Water Works
Association, 6666 West Quincy Avenue, Denver, Col. 80235).

region.[53] Figure 14.6 schematically illustrates the flow patterns in
the three subsurface zones. Field measurements at Miami[18] and
experimental studies[10] have confirmed the landward movement
of the saline water body. Where tidal action is the predominant
mixing mechanism, fluctuations of groundwater—and hence the
thickness of the transition zone—become greatest near the shoreline.

Within the transition zone the salinity of the groundwater increases progressively with depth from that of the fresh water to
that of the saline water. Typically the distribution of salinity with
depth varies as an error function, as shown in Fig. 14.7a.[18,57,58] It
then becomes advantageous to calculate the relative salinity S_R as a
percentage by

$$S_R = 100 \left(\frac{c - c_f}{c_s - c_f} \right) \tag{14.9}$$

where c is the salinity* at a particular depth within the transition
zone, and c_f and c_s are the salinities of the fresh and saline waters,
respectively.

---

*Salinity can be measured as total dissolved solids, chloride, or electrical con-
ductivity.

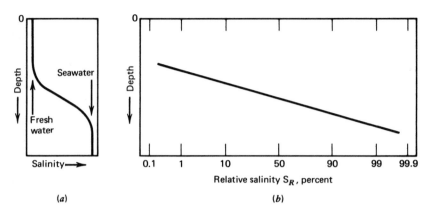

Fig. 14.7 Increase in salinity with depth through the transition zone. (a) Linear scale. (b) Probability scale.

Plotting values of S_R versus depth on probability paper will generally yield a straight line, as demonstrated in Fig. 14.7b. Because detailed data on the structure of a transition zone are difficult to obtain in the field, this graphic technique enables the zone to be estimated from any two point measurements of salinity. The 50 percent value of S_R, representing the midline of the transition zone, defines the position of the interface without mixing.

Where inhomogeneities occur in coastal aquifers, stratifications and irregularities in the distribution of fresh and saline waters occur.[26,43] Considerable research has been done on seawater intrusion in layered aquifers[17,48] and on unsteady movement of the transition zone.[3,25,33,44,47]

Upconing of Saline Water

When an aquifer contains an underlying layer of saline water and is pumped by a well penetrating only the upper freshwater portion of the aquifer, a local rise of the interface below the well occurs. This phenomenon, known as *upconing*, is illustrated by Fig. 14.8. Here the interface is horizontal at the start of pumping when $t = t_o$. With continued pumping the interface rises to successively higher levels until eventually it can reach the well. This generally necessitates the well having to be shut down because of the degrading influence of the saline water. When pumping is stopped, the denser saline water tends to settle downward and to return to its former position.

Upconing is a complex phenomenon and only in recent years has significant headway been made in research studies to enable criteria to be formulated for the design and operation of wells for skimming fresh water from above saline water.[1,4,59] From a water-supply standpoint it is important to determine the optimum location, depth, spacing, pumping rate, and pumping sequence that will ensure production of the largest quantity of fresh groundwater, while at the same time striving to minimize any underground mixing of the fresh saline water.

Most investigations of upconing have assumed an abrupt interface between the two fluids,[19,27] as shown in Fig. 14.8. This situation would pertain between immiscible fluids, but for miscible fluids such as fresh and saline groundwater a mixing, or transition, zone having a finite thickness occurs. Although an abrupt interface neglects the physical reality of a transition zone found in groundwater, the assumption has the advantage of simplicity. Furthermore, an inter-

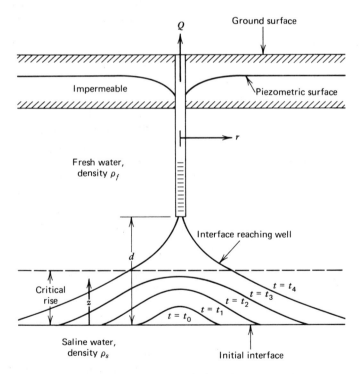

Fig. 14.8 Diagram of upconing of underlying saline water to a pumping well (after Schmorak and Mercado[79]).

face can be considered as an approximation to the position of the 50-percent relative salinity in a transition zone.

An approximate analytic solution for the upconing directly beneath a well, based on the Dupuit assumptions and the Ghyben-Herzberg relation, is given by[49]

$$z = \frac{Q}{2\pi dK(\Delta\rho/\rho_f)} \tag{14.10}$$

where $\Delta\rho = \rho_s - \rho_f$, K is the hydraulic conductivity, and all other quantities are defined by Fig. 14.8. This equation indicates an ultimate rise of the interface to a new equilibrium position that is directly proportional to the pumping rate Q.

Hydraulic model experiments have revealed that the relation in Eq. 14.10 holds only if the rise is limited.[31] If the upconing exceeds a certain critical rise (see Fig. 14.8), it accelerates upward to the well. The critical rise has been estimated to approximate $z/d = 0.3$ to 0.5. Thus, adopting an upper limit of $z/d = 0.5$, it follows that the maximum permissible pumping rate without salt entering the well is

$$Q_{max} \leq \pi d^2 K(\Delta\rho/\rho_f) \tag{14.11}$$

For anisotropic aquifers where the vertical permeability is less than the horizontal, a maximum well discharge larger than that for the isotropic case is possible.[14]

In the real-world situation, a transition zone with a finite thickness of brackish water occurs above the body of undiluted saline water. The water at the upper edge of the zone is essentially fresh water and moves accordingly. Upward movement of the almost-fresh water occurs readily along with the adjoining fresh water; consequently, even with a relatively low pumping rate, no limiting critical rise exists above which saline water will not rise. It follows that with any rate of continuous pumping, some saline water must sooner or later reach a well.

Examination of Eq. 14.10 supports the previous statements. In a transition zone the salinity changes gradually; hence, $\Delta\rho$ in an incremental width at the top of the zone approaches zero. As a result z must tend toward infinity, indicating that there can be no finite limit to z. Similarly, in Eq. 14.11 as $\Delta\rho$ approaches zero, so also must Q_{max} approach zero.

A comparison of the arrivals of salinity at a pumping well for an abrupt interface and for a transition zone is shown qualitatively in Fig. 14.9. With an abrupt interface, assuming $Q > Q_{max}$, the salinity appears later and increases more rapidly than with a transition zone. For $Q < Q_{max}$ there will be no salinity reaching the well in the abrupt

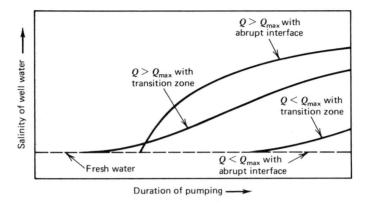

Fig. 14.9 Well water salinity curves for upconing of an abrupt interface and a transition zone (after Schmorak and Mercado[49]).

case; however, a gradual invasion of saline water will occur from a rising transition zone. The ultimate well-water salinity with upconing approaches an intermediate value between the extremes of the fresh and saline waters; empirical data indicate this lies in the range of 5 to 8 percent of the salt concentration in the saline water.[49]

It follows from the above analysis that upconing can be minimized by the proper design and operation of wells and galleries. For given aquifer conditions wells should be separated as far as possible vertically from the saline zone and pumped at a low uniform rate. Well and gallery designs for regions where a thin layer of fresh water overlies saline water are discussed in the section on oceanic islands. Field tests have shown that *scavenger wells*, which pump brackish or saline water from below the fresh water, can also successfully counteract upconing.[36]

The correspondence of pumping rate and salinity in Fig. 14.10 dramatically illustrates the effects of upconing caused by increased pumping to meet annual midsummer demands in Honolulu.

Fresh-Salt Water Relations on Oceanic Islands

Most small oceanic islands are relatively permeable, consisting of sand, lava, coral, or limestone, so that seawater is in contact with groundwater on all sides. Because fresh groundwater originates entirely from rainfall, only a limited quantity is available. A freshwater lens, shown schematically in Fig. 14.11, is formed by the radial movement of the fresh water toward the coast. This lens floats on

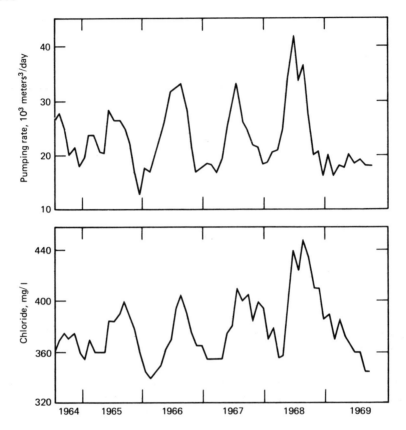

Fig. 14.10 Effect of annual variation in pumping rate on upconing of underlying saline water in Honolulu, Hawaii. Chloride contents were measured in an observation well at a depth of 128 m and located 430 m from the pumping well (after Todd and Meyer[56]).

the underlying salt water; its thickness decreases from the center toward the coast.[16,23]

From the Dupuit assumptions and the Ghyben-Herzberg relation, an approximate freshwater boundary can be determined. Assume a circular island of radius R, as shown in Fig. 14.11, receiving an effective recharge from rainfall at a rate W. The outward flow Q at radius r is

$$Q = 2\pi r K(z + h) \frac{dh}{dr} \qquad (14.12)$$

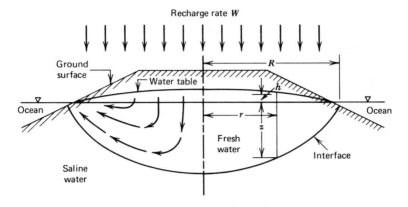

Fig. 14.11 Freshwater lens in an oceanic island under natural conditions.

where K is the hydraulic conductivity and h and z are defined in Fig. 14.11. Noting that $h = (\Delta\rho/\rho)z$ and that from continuity $Q = \pi r^2 W$, then

$$z\,dz = \frac{Wr\,dr}{2K\left[1 + \dfrac{\Delta\rho}{\rho}\right]\left[\dfrac{\Delta\rho}{\rho}\right]} \tag{14.13}$$

Integrating and applying the boundary condition that $h = 0$ when $r = R$,

$$z^2 = \frac{W(R^2 - r^2)}{2K\left[1 + \dfrac{\Delta\rho}{\rho}\right]\left[\dfrac{\Delta\rho}{\rho}\right]} \tag{14.14}$$

Thus, the depth to salt water at any location is a function of the rainfall recharge, the size of the island, and the hydraulic conductivity. For almost all island conditions, it can be shown that this approximate solution is indistinguishable from more exact solutions by potential theory.[18]

Tidal, atmospheric, and rainfall fluctuations (see Chapter 6), together with dispersion, create a transition zone along the interface in an oceanic island. The close proximity of this boundary zone to the water table can introduce saline water into a well by upconing. Therefore, care must be exercised in development of underground water supplies so that pumping causes a minimal disturbance to the fresh-salt water equilibrium. To avoid the danger of entrainment of saline water, island wells should be designed for minimum drawdown, just skimming fresh water from the top of the lens. If small-diameter wells are employed, they should be shallow, dispersed, and pumped at low uniform rates.

In areas where water tables are shallow, an infiltration gallery (see Chapter 5), consisting of a horizontal collecting tunnel at the water table, is advantageous.[46] Drawdowns of a few centimeters can in many instances furnish plentiful water supplies. Installations of infiltration galleries for local water supplies exist on Bermuda and the Bahamas in the Atlantic Ocean and on the Gilbert and Marianas islands in the Pacific Ocean.[40,42] Where water tables are deep, dug wells or shafts are sunk to the water table with horizontal tunnels (adits) extending outward to intercept the uppermost layer of fresh water. Examples of these can be found in Barbados, Guam,[24] and Hawaii.[60]

Illustrative of the occurrence of fresh water on an oceanic island is Barbados (Fig. 14.12). Here a thin highly permeable coral limestone layer serves as the aquifer. Recharge from rainfall percolates through solution channels along the bottom of the limestone until it reaches sea level. From there to the coast the fresh water floats in a layer (known locally as *sheet water*) above the underlying saline water. Water is extracted from large-diameter dug wells connected to horizontal adits at the water table.

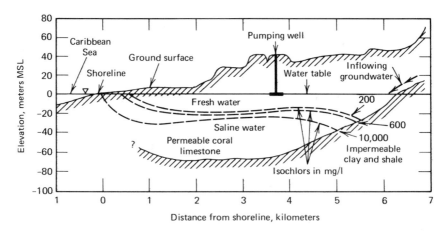

Fig. 14.12 Geologic profile showing interrelations of fresh and saline water in an unconfined aquifer at Bridgetown, Barbados (courtesy Waterworks Department, Government of Barbados).

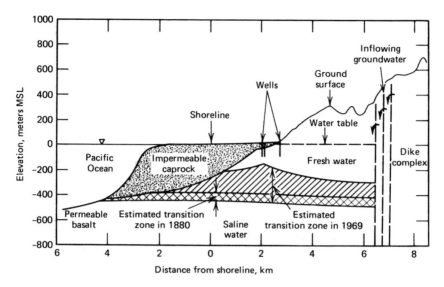

Fig. 14.13 Geologic profile showing interrelations of fresh and saline water at Honolulu, Hawaii. The thickening of the transition zone resulted from pumping begun after 1880 (after Todd and Meyer[56]).

Another example of the balance between fresh and saline groundwater on an oceanic island is Honolulu (Fig. 14.13). Permeable basalt forms the aquifer, but impermeable caprock (erosional material and marine deposits) acts as a groundwater dam. Before wells were first drilled in 1880, groundwater discharged as springs either at the inland or at submarine boundaries of the caprock. At that time the transition zone was narrow and nearly horizontal. Subsequent development of water by wells has lowered the water table and expanded the transition zone so that the volume of fresh water has been substantially reduced. Depths of production wells have decreased gradually from 450 m to 85 m as upconing has progressively increased.[56]

Seawater Intrusion in Karst Terranes

Coastal aquifers consisting of karstic limestone pose special problems of seawater intrusion. Irregular fissures and solution openings enable seawater to enter the aquifer in configurations that may differ appreciably from those for more homogeneous aquifers.[50,51] Unique features sometimes found in karst are intermittent brackish

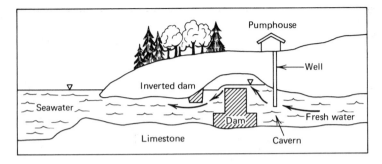

Fig. 14.14 Schematic diagram of an underground dam in a coastal limestone cavern to prevent mixing of fresh and saline water (courtesy *Business Week*, 1977).

springs, which can result from channels connecting with the sea or where saline heads under high tides exceed inland freshwater heads, causing seawater to be discharged inland.

A common feature in the karst regions surrounding the Mediterranean Sea are solutions channels, which discharge fresh water as submarine springs. But pumping of these channels to prevent wastage of fresh water often yields saline water within hours because seawater can freely enter the channel if the freshwater flow is reduced. To overcome this problem a two-part dam, illustrated in Fig. 14.14, can effectively prevent the entry of seawater into a pumping well located upstream. The technique has been successfully tested in the limestone caverns of the Port-Miou River near Marseilles, France.

Control of Saline Water Intrusion

Methods for controlling intrusion vary widely depending on the source of the saline water, the extent of intrusion, local geology, water use, and economic factors. Table 14.1 summarizes the generally recognized methods for controlling intrusion from various sources. Because as little as 2 percent of seawater in fresh water can render water unpotable, considerable attention has been focused on methods to control seawater intrusion. Alternative methods are discussed in the following paragraphs.[8,12,54] Measures for coping with upconing were described in earlier sections.

Modification of Pumping Pattern. Changing the locations of pumping wells, typically by dispersing them in inland areas, can aid in reestablishing a stronger seaward hydraulic gradient. Also,

TABLE 14.1   Methods for Controlling Saline Water Intrusion

Source or Cause of Intrusion	Control Methods
Seawater in coastal aquifer	Modification of pumping pattern
	Artificial recharge
	Extraction barrier
	Injection barrier
	Subsurface barrier
Upconing	Modification of pumping pattern
	Saline scavenger wells
Oil field brine	Elimination of surface disposal
	Injection wells
	Plugging of abandoned wells
Defective well casings	Plugging of faulty wells
Surface infiltration	Elimination of source
Saline water zones in freshwater aquifers	Relocation and redesign of wells

reduction in pumping of existing wells can produce the same beneficial effect.

Artificial Recharge. Groundwater levels can be raised and maintained by artificial recharge, using surface spreading for unconfined aquifers and recharge wells for confined aquifers. This necessitates development of a supplemental water source.

Extraction Barrier. An extraction barrier is created by maintaining a continuous pumping trough with a line of wells adjacent to the sea.[13] Seawater flows inland from the ocean to the trough, while fresh water within the basin flows seaward toward the trough, as shown in Fig. 14.15. The water pumped is brackish and normally is discharged into the sea.

Injection Barrier. This method maintains a pressure ridge along the coast by a line of recharge wells.[9,35] Injected fresh water flows both seaward and landward, as indicated in Fig. 14.16. High-quality imported water is required for recharge into wells. A combination of injection and extraction barriers is feasible; this reduces both recharge and extraction rates but requires a larger number of wells.[12]

Subsurface Barrier. Construction of an impermeable subsurface barrier parallel to the coast and through the vertical extent of the aquifer can effectively prevent the inflow of seawater into the

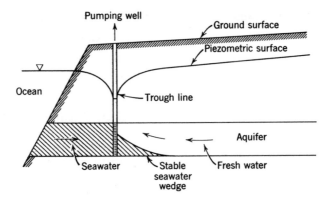

Fig. 14.15 Control of seawater intrusion by an extraction barrier forming a pumping trough paralleling the coast.

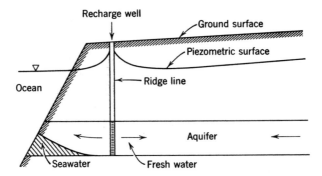

Fig. 14.16 Control of seawater intrusion by an injection barrier forming a pressure ridge paralleling the coast.

basin (see Fig. 14.17). Materials to construct a barrier might include sheet piling, puddled clay, emulsified asphalt, cement grout, bentonite, silica gel, calcium acrylate, or plastics. Chief problems are construction cost and resistance to earthquakes and chemical erosion.

Examples of Seawater Intrusion

Of the many localities along the coast of the United States experiencing seawater intrusion, those on Long Island, Miami, and Los Angeles are perhaps most significant. Brief descriptions of each of these situations follow.

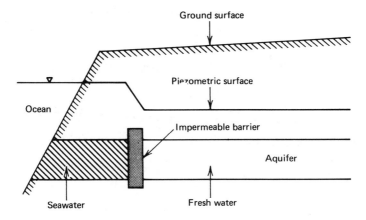

Fig. 14.17   Control of seawater intrusion by an
impermeable subsurface barrier paralleling the coast.

Long Island, New York.   Long Island is underlain by a wedge-shaped mass of unconsolidated sediments that extends more than 600 m in depth. Major intrusion has occurred in western Long Island as a result of development of groundwater and of a decrease in recharge by improved drainage and sewer systems. By the mid-1930s pumping had lowered water levels at the western end to as much as 10 m below sea level and caused extensive intrusion. Along the southern coast similar development and reduction of recharge are causing active intrusion at depth.[39,43]

Intrusion has been largely controlled by abandoning pumping and using an imported water supply and by requiring that water pumped from industrial wells be recharged back into the ground after use. Farther east in Nassau County, intrusion is not yet a serious problem; however, to prevent this tendency an extensive program to salvage storm runoff and recharge it through infiltration basins has been initiated (see Chapter 13). In addition, experimental work is underway to reclaim wastewater for recharge to form an injection barrier. Figure 14.18 shows a cross section of the distribution of saline water in southwestern Nassau County; the advance of the saline wedge inland varies from 3 to 60 m per year depending on local pumping conditions.

Miami, Florida.   Intrusion of seawater into the permeable limestone aquifer in the Miami area stemmed from an extensive system of drainage channels started in the 1930s. These were constructed to permit urbanization of low-lying areas; however, they lowered the

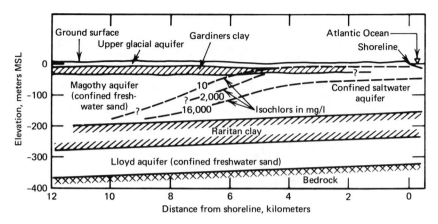

Fig. 14.18 Geologic profile showing interrelations of fresh and saline water at Far Rockaway, Nassau County, Long Island (after Perlmutter and Geraghty[43]).

water table excessively and allowed the inland movement of seawater both within the aquifer and within the channels. Several well fields in Miami and Ft. Lauderdale had to be abandoned. The marked increase in intrusion over a 55-year period is clearly demonstrated in Fig. 14.19.

Efforts to control the movement of seawater began in 1940 with construction of salinity-control dams in the drainage canals to halt the upstream advance of seawater and to maintain freshwater levels sufficiently high to replenish the aquifer.[32] By 1968 the saline front was dynamically stabilized 3 to 13 km inland by these measures. Legislation now prevents the construction of additional tidal canals.

Los Angeles, California. Intrusion appeared in the early 1930s along the west coast of Los Angeles County and the rate accelerated rapidly with development of the area in the 1940s. The first and largest injection barrier project has been constructed along a 11-km portion of this shoreline. Some 94 recharge wells form a pressure ridge in the confined aquifers so that seawater is effectively separated from the overpumped basin inland.[9,35]

Figure 14.20 shows a profile of the piezometric surface perpendicular to the barrier during the years that this portion of the barrier was established. It can be seen how the landward gradient in the injected Silverado Aquifer was transformed into a pressure ridge extending above sea level.

A total of 267 observation wells are monitored to study effects of

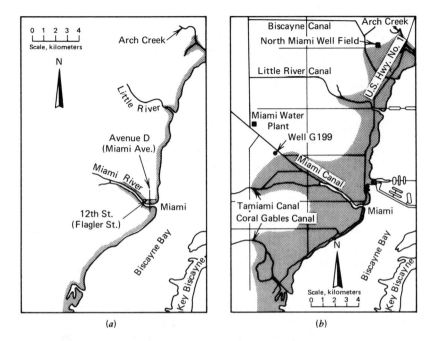

Fig. 14.19 Maps showing extent of seawater intrusion in and near Miami, Florida as of (a) 1904, and (b) 1959. Note the extent of intrusion associated with the drainage canals (after Kohout[32]; reprinted from the *Journal American Water Works Association,* Vol. 53, by permission of the Association, copyright © 1961 by American Water Works Association, 6666 West Quincy Avenue, Denver, Col. 80235).

the injection barrier. Filtered water is delivered to the distribution line with an average chlorine residual of 0.5 mg/l; under normal operating conditions 1.5 mg/l of chlorine is added before injection. Elevations of the piezometric head in the barrier are maintained 1 to 3 m above sea level in the major aquifer. The heads at midpoints between injection wells are measured weekly, and injection-well flow rates are adjusted as needed to maintain the barrier. In recent years the injection rate per well has averaged 1500 m³/day.

Recognition of Seawater in Groundwater

Analysis of groundwater samples collected in zones of seawater intrusion may show a chemical composition differing from a simple proportional mixing of seawater and groundwater. Modifications in composition of seawater entering an aquifer can occur by three

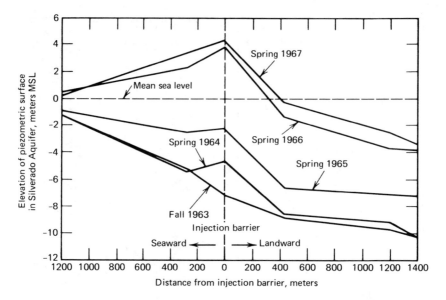

Fig. 14.20 Piezometric surface profiles perpendicular to the seawater intrusion injection barrier, Los Angeles County, after various time intervals. Note establishment of the pressure ridge following initiation of the barrier at the end of 1963 (courtesy Los Angeles County Flood Control District).

processes:[45] (1) base exchange between the water and the minerals of the aquifer, (2) sulfate reduction and substitution of carbonic or other weak acid radicals, and (3) solution and precipitation. Only the last process can change the total salt concentration; however, the first two processes, which require maintenance of ionic balance, can alter the percentage by weight of different salt components and thereby the total dissolved solids in milligrams per liter.

In order to avoid mistaken diagnoses of seawater intrusion as evidenced by temporary increases of total dissolved salts, Revelle[45] recommended the chloride-bicarbonate ratio as a criterion to evaluate intrusion.* Chloride is the dominant anion of ocean water, is unaffected by the above processes, and normally occurs in only small amounts in groundwater. On the other hand, bicarbonate is usually the most abundant anion in groundwater and occurs in only minor amounts in seawater. Although pollutants other than seawater can change the chloride-bicarbonate ratio, these would seldom be important in water collected from a well subject to intrusion.

*Actually the $Cl/(CO_3 + HCO_3)$ ratio is employed for practical purposes.

References

1. Ackermann, N. L., and Y. Y. Chang, Salt water interface during ground-water pumping, *Jour. Hydraulics Div.*, Amer. Soc. Civil Engrs., v. 97, no. HY2, pp. 223–232, 1971.
2. Bear, J., and G. Dagan, Intercepting freshwater above the interface in a coastal aquifer, *Intl. Assoc. Sci. Hydrology Publ.* 64, pp. 154–181, 1964.
3. Bear, J., and G. Dagan, Moving interface in coastal aquifers, *Jour. Hydraulics Div.*, Amer. Soc. Civil Engrs., v. 90, no. HY4, pp. 193–216, 1964.
4. Bennett, G. D., et al., Electric-analog studies of brine coning beneath freshwater wells in the Punjab region, West Pakistan, *U.S. Geological Survey Water-Supply Paper* 1608-J, 31 pp., 1968.
5. Biemond, C., Dune water flow and replenishment in the catchment area of the Amsterdam water supply, *Jour. Instn. Water Engrs.*, v. 11, pp. 195–213, 1957.
6. Braithwaite, F., On the infiltration of salt water into the springs of wells under London and Liverpool, *Proc. Inst. Civil Engrs.*, v. 14, pp. 507–523, 1855.
7. Brown, J. S., A study of coastal ground water with special reference to Connecticut, *U.S. Geological Survey Water Supply Paper* 537, 101 pp., 1925.
8. Bruington, A. E., Control of salt-water intrusion in a ground-water aquifer, *Ground Water*, v. 7, no. 3, pp. 9–14, 1969.
9. Bruington, A. E., and F. D. Seares, Operating a sea water barrier project, *Jour. Irrig. Drain. Div.*, Amer. Soc. of Civil Engrs., v. 91, no. IR1, pp. 117–140, 1965.
10. Cahill, J. M., Hydraulic sand-model study of the cyclic flow of salt water in a coastal aquifer, *U.S. Geological Survey Prof. Paper* 575-B, pp. 240–244, 1967.
11. California Dept. Water Resources, *Sea-water intrusion in California*, Bull. 63, Sacramento, 91 pp. plus apps., 1958.
12. California Dept. Water Resources, *Santa Ana Gap salinity barrier, Orange County*, Bull. 147-1, Sacramento, 178 pp., 1966.
13. California Dept. Water Resources, *Oxnard Basin experimental extraction-type barrier*, Bull. 147-6, Sacramento, 157 pp., 1970.
14. Chandler, R. A., and D. B. McWhorter, Upconing of the salt-water–freshwater interface beneath a pumping well, *Ground Water*, v. 13, pp. 354–359, 1975.
15. Charmonman, S., A solution of the pattern of fresh-water flow in an unconfined coastal aquifer, *Jour. Geophysical Research*, v. 70, pp. 2813–2819, 1965.
16. Chidley, T. R. E., and J. W. Lloyd, A mathematical model study of freshwater lenses, *Ground Water*, v. 15, pp. 215–222, 1977.
17. Collins, M. A., and L. W. Gelhar, Seawater intrusion in layered aquifers, *Water Resources Research*, v. 7, pp. 971–979, 1971.
18. Cooper, H. H., Jr., et al., Sea water in coastal aquifers, *U.S. Geological Survey Water-Supply Paper* 1613-C, 84 pp., 1964.

19. Dagan, G., and J. Bear, Solving the problem of local interface upconing in a coastal aquifer by the method of small perturbations, *Jour. Hydr. Research*, v. 6, pp. 15–44, 1968.

20. Drabbe, J., and Badon Ghyben, W., Nota in verband met de voorgenomen putboring nabij Amsterdam, *Tijdschrift van het Koninklijk Instituut van Ingenieurs*, The Hague, Netherlands, pp. 8–22, 1888–1889.

21. Ernest, L. F., Groundwater flow in the Netherlands delta area and its influence on the salt balance of the future Lake Zeeland, *Jour. Hydrology*, v. 8, pp. 137–172, 1969.

22. Feth, J. H., et al., Preliminary map of the conterminous United States showing depth to and quality of shallowest ground water containing more than 1000 ppm dissolved solids, *U.S. Geological Survey Hydrol. Inv. Atlas* HA-199, 31 pp., 1965.

23. Fetter, C. W., Jr., Position of the saline water interface beneath oceanic islands, *Water Resources Research*, v. 8, pp. 1307–1315, 1972.

24. Fil, J. F., Horizontal wells reduce salt-water intrusion in Guam's water supply, *Civil Engrng.*, v. 20, no. 7, pp. 32–33, 1950.

25. Hantush, M. S., Unsteady movement of fresh water in thick unconfined saline aquifers, *Bull. Intl. Assoc. Sci. Hydrology*, v. 13, no. 2, pp. 40–60, 1968.

26. Harris, W. H., Stratification of fresh and salt water on barrier islands as a result of differences in sediment permeability, *Water Resources Research*, v. 3, pp. 89–97, 1967.

27. Haubold, R. G. Approximation for steady interface beneath a well pumping fresh water overlying salt water, *Ground Water*, v. 13, pp. 254–259, 1975.

28. Herzberg, B., Die Wasserversorgung einiger Nordseebader, *Jour. Gasbeleuchtung und Wasserversorgung*, v. 44, pp. 815–819, 842–844, Munich, 1901.

29. Hubbert, M. K., The theory of ground-water motion, *Jour. Geol.*, v. 48, pp. 785–944, 1940.

30. Kashef, A. I., On the management of ground water in coastal aquifers, *Ground Water*, v. 9, no. 2, pp. 12–20, 1971.

31. Kawabata, H., Coning up of confined two-layers' liquids through porous media by pumping up, *Bull. Kyoto Gakugei Univ.*, Ser. B, no. 27, pp. 19–29, 1965.

32. Kohout, F. A., Case history of salt water encroachment caused by a storm sewer in Miami, *Jour. Amer. Water Works Assoc.*, v. 53, pp. 1406–1416, 1961.

33. Kohout, F. A., and H. Klein, Effect of pulse recharge on the zone of diffusion in the Biscayne aquifer, *Intl. Assoc. Sci. Hydrology Publ. 72*, pp. 252–270, 1967.

34. Lau, L. S., and J. F. Mink, A step in optimizing the development of the basal water lens of Southern Oahu, Hawaii, *Intl. Assoc. Sci. Hydrology Publ. 72*, pp. 500–508, 1967.

35. Laverty, F. B., and H. A. van der Goot, Development of a fresh-water

barrier in southern California for the prevention of sea water intrusion, *Jour. Amer. Water Works Assoc.*, v. 47, pp. 886–908, 1955.

36. Long, R. A., *Feasibility of a scavenger-well system as a solution to the problem of vertical salt-water encroachment*, Water Resources Pamphlet 15, Louisiana Geological Survey, Baton Rouge, 27 pp., 1965.

37. Louisiana Water Resources Research Inst., *Salt-water encroachment into aquifers*, Bull. 3, Louisiana State Univ., Baton Rouge, 192 pp., 1968.

38. Lusczynski, N. J., Head and flow of ground water of variable density, *Jour. Geophysical Research*, v. 66, pp. 4247–4256, 1961.

39. Lusczynski, N. J., and W. V. Swarzenski, Salt-water encroachment in Southern Nassau and Southeastern Queens Counties, Long Island, New York, *U.S. Geological Survey Water-Supply Paper* 1613-F, 76 pp., 1966.

40. Mather, J. D., Development of the groundwater resources of small limestone islands, *Quarterly Jour. Engrng. Geol.*, v. 8, pp. 141–150, 1975.

41. Newport, B. D., *Salt water intrusion in the United States*, Rept. EPA-600/8-77-011, U.S. Environmental Protection Agency, Ada, Okla., 30 pp., 1977.

42. Ohrt, F., Water development and salt-water intrusion on Pacific Islands, *Jour. Amer. Works Assoc.*, v. 39, pp. 979–988, 1947.

43. Perlmutter, N. M., and J. J. Geraghty, Geology and ground-water conditions in Southern Nassau and Southeastern Queens Counties, Long Island, N.Y., *U.S. Geological Survey Water-Supply Paper* 1613-A, 205 pp., 1963.

44. Pinder, G. F., and H. H. Cooper, Jr., A numerical technique for calculating the transient position of the saltwater front, *Water Resources Research*, v. 6, pp. 875–882, 1970.

45. Revelle, R., Criteria for recognition of sea water in ground-waters, *Trans. Amer. Geophysical Union*, v. 22, pp. 593–597, 1941.

46. Rochester, E. W., Jr., and G. J. Kriz, Potable water availability on long oceanic islands, *Jour. San. Engrng. Div.*, Amer. Soc. Civil Engrs., v. 96, no. SA5, pp. 1235–1248, 1970.

47. Rumer, R. R., and D. R. F. Harleman, Intruded salt-water wedge in porous media, *Jour. Hydraulics Div.*, Amer. Soc. Civil Engrs., v. 89, no. HY6, pp. 193–220, 1963.

48. Rumer, R. R., and J. C. Shiau, Salt water interface in a layered coastal aquifer, *Water Resources Research*, v. 4, pp. 1235–1247, 1968.

49. Schmorak, S., and A. Mercado, Upconing of fresh water-sea water interface below pumping wells, field study, *Water Resources Research*, v. 5, pp. 1290–1311, 1969.

50. Stringfield, V. T., and H. E. LeGrand, Relation of sea water to fresh water in carbonate rocks in coastal areas, with special reference to Florida, USA., and Cephalonia, Greece, *Jour. Hydrology*, v. 9, pp. 387–404, 1969.

51. Stringfield, V. T., and H. E. LeGrand, Effects of karst features on circulation of water in carbonate rocks in coastal areas, *Jour. Hydrology*, v. 14, pp. 139–157, 1971.

52. Task Committee on Salt Water Intrusion, Saltwater intrusion in the

United States, *Jour. Hydraulics Div.*, Amer. Soc. Civil Engrs., v. 95, no. HY5, pp. 1651–1669, 1969.

53. Todd, D. K., Salt water intrusion of coastal aquifers in the United States, *Intl. Assoc. Sci. Hydrology Publ.* 52, pp. 452–461, 1960.

54. Todd, D. K., Salt water intrusion and its control, *Jour. Amer. Water Works Assoc.*, v. 66, pp. 180–187, 1974.

55. Todd, D. K., and L. Huisman, Ground water flow in the Netherlands coastal dunes, *Jour. Hydraulics Div.*, Amer. Soc. Civil Engrs., v. 85, no. HY7, pp. 63–81, 1959.

56. Todd, D. K., and C. F. Meyer, Hydrology and geology of the Honolulu aquifer, *Jour. Hydraulics Div.*, Amer. Soc. Civil Engrs., v. 97, No. HY2, pp. 233–256, 1971.

57. Vacher, H. L., *Groundwater hydrology of Bermuda*, Public Works Dept., Govt. of Bermuda, 87 pp., 1974.

58. Visher, F. N., and J. F. Mink, Ground-water resources in Southern Oahu, Hawaii, *U.S. Geological Survey Water-Supply Paper* 1778, 133 pp., 1964.

59. Wang, F. C., Approximate theory for skimming well formulation in the Indus Plain of West Pakistan, *Jour. Geophysical Research*, v. 70, pp. 5055–5063, 1965.

60. Watson, L. V., Development of ground water in Hawaii, *Jour. Hydraulics Div.*, Amer. Soc. Civil Engrs., v. 90, no. HY6, pp. 185–202, 1964.

61. Wentworth, C. K., The process and progress of salt-water encroachment, *Intl. Assoc. Sci. Hydrology Publ.* 33, pp. 238–248, 1951.

METRIC UNITS AND ENGLISH EQUIVALENTS

Metric Units

Quantity	Unit	Symbol	Formula
Base units			
Length	meter	m	
Mass	kilogram	kg	
Time	second	s	
Electric current	ampere	A	
Thermodynamic temperature	kelvin	K	
Plane angle	radian	rad	
Solid angle	steradian	sr	
Derived units			
Acceleration	meter per second squared	—	m/s^2
Activity (of a radioactive source)	disintegration per second	—	(disintegration)/s
Angular acceleration	radian per second squared	—	rad/s^2
Angular velocity	radian per second	—	rad/s
Area	square meter	—	m^2

Metric Units

Quantity	Unit	Symbol	Formula
Density	kilogram per cubic meter	—	kg/m^3
Electric capacitance	farad	F	$A \cdot s/V$
Electric conductance	siemens	S	A/V
Electric field strength	volt per meter	—	V/m
Electric inductance	henry	H	W/A
Electric potential difference	volt	V	W/A
Electric resistance	ohm	Ω	V/A
Electromotive force	volt	V	W/A
Energy	joule	J	$N \cdot m$
Force	newton	N	$kg \cdot m/s^2$
Power	watt	W	J/s
Pressure	pascal	Pa	N/m^2
Quantity of electricity	coulomb	C	$A \cdot s$
Quantity of heat	joule	J	$N \cdot m$
Radiant intensity	watt per steradian	—	W/sr
Specific heat	joule per kilogram-kelvin	—	$J/kg \cdot K$
Stress	pascal	Pa	N/m^2
Thermal conductivity	watt per meter-kelvin	—	$W/m \cdot K$
Velocity	meter per second	—	m/s
Viscosity, dynamic	pascal-second	—	$Pa \cdot s$
Viscosity, kinematic	square meter per second	—	m^2/s
Voltage	volt	V	W/A
Volume	cubic meter	—	m^3
Work	joule	J	$N \cdot m$

Common Metric Equivalents

Length	Area	Volume
1 m = 10^3 mm	1 hectare (ha) = 10^4 m^2	1 liter (l) = 10^{-3} m^3
= 10^2 cm	= 10^{-2} km^2	
= 10^{-3} km		**Mass**
		1 kg = 10^3 g

Metric Prefixes

Prefix	Symbol	Multiplication Factor
tera	T	10^{12}
giga	G	10^9
mega	M	10^6
kilo	k	10^3
hecto	h	10^2
deka	da	10^1
deci	d	10^{-1}
centi	c	10^{-2}
milli	m	10^{-3}
micro	μ	10^{-6}
nano	n	10^{-9}
pico	p	10^{-12}
femto	f	10^{-15}
atto	a	10^{-18}

Metric-English Equivalents

Length
1 cm = 0.3937 in.
1 m = 3.281 ft
1 km = 0.6214 mi

Area
1 cm^2 = 0.1550 in.^{2}
1 m^2 = 10.76 ft^2
1 ha = 2.471 acre
1 km^2 = 0.3861 mi^2

Volume
1 cm^3 = 0.06102 in.^{3}
1 l = 0.2642 gal
  = 0.03531 ft^3
1 m^3 = 264.2 gal
  = 35.31 ft^3
  = 8.106 $\times$ 10^{-4} acre-ft

Mass
1 g = 2.205 $\times$ 10^{-3} lb (mass)
1 kg = 2.205 lb (mass)
  = 9.842 $\times$ 10^{-4} long ton

Flow Rate
1 l/s = 15.85 gpm
  = 0.02282 mgd
  = 0.03531 cfs
1 m^3/s = 1.585 $\times$ 10^4 gpm
  = 22.82 mgd
  = 35.31 cfs
1 m^3/day = 0.1834 gpm
  = 2.642 $\times$ 10^{-4} mgd
  = 4.087 $\times$ 10^{-4} cfs

Velocity
1 m/s = 3.281 ft/s
  = 2.237 mi/hr
1 km/hr = 0.9113 ft/s
  = 0.6214 mi/hr

Temperature
degree Celsius
  = kelvin − 273.15
  = (degree
    Fahrenheit − 32)/1.8

Pressure
1 Pa = 9.872×10^{-6} atmosphere
   = 1.000×10^{-5} bar
   = 0.01000 millibar
   = 10.00 dyne/cm^2
   = 3.346×10^{-4} ft H$_2$O (4°C)
   = 2.953×10^{-4} in. Hg (0°C)
   = 0.1020 kg (force)/m^2
   = 0.02089 lb (force)/ft^2

Heat
1 J/m^2 = 8.806×10^{-5} BTU/ft^2
      = 2.390×10^{-5} calorie/cm^2
1 J/kg = 4.299×10^{-4} BTU/lb (mass)
      = 2.388×10^{-4} calorie/g

Hydraulic Conductivity
1 m/day = 24.54 gpd/ft^2
1 cm/s = 2.121×10^4 gpd/ft^2
1 m/day = 1.198 darcy (for water
            at 20°C)
1 cm/s = 1035 darcy (for water at
            20°C)

Viscosity
1 Pa·s = 1.000×10^3 centistoke
      = 10.00 poise
      = 0.02089 lb (force)·s/ft^2
1 m^2/s = 1.000×10^6 centistoke
      = 10.76 ft^2/s

Force
1 N = 1.000×10^5 dyne
   = 0.1020 kg (force)
   = 0.2248 lb (force)

Power
1 W = 9.478×10^{-4} BTU/s
   = 0.2388 calorie/s
   = 0.7376 ft-lb (force)/s

Energy
1 J = 9.478×10^{-4} BTU
   = 0.2388 calorie
   = 0.7376 ft-lb (force)
   = 2.778×10^{-7} kw-hr

Water Quality
1 mg/l = 1 ppm
      = 0.0584 grain/gal
Equivalent weight of ion = atomic weight of ion/valence of ion
meq/l of ion = mg/l of ion/equivalent wt of ion
1 meq/l = 1 me/l
       = 1 epm
1 μS/cm = 1 μmho/cm
1 μS/cm = 0.65 mg/l   } Approximations for most natural
                               waters in the range of 100 to
       = 0.10 meq/l of cations} 5000 μS/cm at 25°C

Numerical Values for Physical Properties

Quantity	Metric	English
Gravitational acceleration, g (std., free fall)	9.807 m/s^2	32.2 ft/s^2
Density of water, ρ @ 50° F/10° C	1000 kg/m^3	1.94 slugs/ft^3
Specific weight of water, γ @ 50° F/10° C	9.807 × 10^{3}N/m^3	62.4 lb/ft^3
Dynamic viscosity of water, μ		
@ 50° F/10° C	1.30 × 10^{-3}Pa · s	2.73 × 10^{-5}lb · s/ft^2
@ 68° F/20° C	1.00 × 10^{-3}Pa · s	2.05 × 10^{-5}lb · s/ft^2
Kinematic viscosity of water, ν		
@ 50° F/10° C	1.30 × 10^{-6}m^2/s	1.41 × 10^{-5}ft^2/s
@ 68° F/20° C	1.00 × 10^{-6}m^2/s	1.06 × 10^{-5}ft^2/s
Atmospheric pressure, p (std.)	1.013 × 10^{5}Pa	14.70 psia

Index

METRIC-ENGLISH EQUIVALENTS

Length
1 cm = 0.3937 in.
1 m = 3.281 ft
1 km = 0.6214 mi

Area
1 cm^2 = 0.1550 in.^{2}
1 m^2 = 10.76 ft^2
1 ha = 2.471 acre
1 km^2 = 0.3861 mi^2

Volume
1 cm^3 = 0.06102 in.^{3}
1 l = 0.2642 gal
   = 0.03531 ft^3
1 m^3 = 264.2 gal
   = 35.31 ft^3
   = 8.106 $\times$ 10^{-4} acre-ft

Mass
1 g = 2.205 $\times$ 10^{-3} lb (mass)
1 kg = 2.205 lb (mass)
   = 9.842 $\times$ 10^{-4} long ton

Flow Rate
1 l/s = 15.85 gpm
   = 0.02282 mgd
   = 0.03531 cfs
1 m^3/s = 1.585 $\times$ 10^4 gpm
   = 22.82 mgd
   = 35.31 cfs
1 m^3/day = 0.1834 gpm
   = 2.642 $\times$ 10^{-4} mgd
   = 4.087 $\times$ 10^{-4} cfs

Velocity
1 m/s = 3.281 ft/s
   = 2.237 mi/hr
1 km/hr = 0.9113 ft/s
   = 0.6214 mi/hr

Temperature
degree Celsius
   = kelvin − 273.15
   = (degree
      Fahrenheit − 32)/1.8

Pressure
1 Pa = 9.872 $\times$ 10^{-6} atmosphere
   = 1.000 $\times$ 10^{-5} bar
   = 0.01000 millibar
   = 10.00 dyne/cm^2
   = 3.346 $\times$ 10^{-4} ft H$_2$O (4°C)
   = 2.953 $\times$ 10^{-4} in. Hg (O°C)
   = 0.1020 kg (force)/m^2
   = 0.02089 lb (force)/ft^2

Heat
1 J/m^2 = 8.806 $\times$ 10^{-5} BTU/ft^2
   = 2.390 $\times$ 10^{-5} calorie/cm^2
1 J/kg = 4.299 $\times$ 10^{-4} BTU/lb (mass)
   = 2.388 $\times$ 10^{-4} calorie/g

Hydraulic Conductivity
1 m/day = 24.54 gpd/ft^2
1 cm/s = 2.121 $\times$ 10^4 gpd/ft^2
1 m/day = 1.198 darcy (for water
      at 20°C)
1 cm/s = 1035 darcy (for water at
      20°C)

Viscosity
1 Pa·s = 1.000 $\times$ 10^3 centistoke
   = 10.00 poise
   = 0.02089 lb (force)·s/ft^2
1 m^2/s = 1.000 $\times$ 10^6 centistoke
   = 10.76 ft^2/s

Force
1 N = 1.000 $\times$ 10^5 dyne
   = 0.1020 kg (force)
   = 0.2248 lb (force)